Framing and Man
Organizations in th
Economy

This book examines the dominance and significance of lean organizing in the international economy. Scholars from each discipline see lean production as positive or negative; the book blends theory with practice by sorting out these different academic views and revealing how lean is implemented in different ways.

The first part synthesizes academic research from a range of disciplines—including, engineering, sociology, and management—to present the reader with an integrated understanding of the benefits and drawbacks of lean management. The second part links this theory to practice, with a set of case studies from companies like Apple, Google, Nike, Toyota, and Walmart that demonstrate how lean is implemented in a variety of settings. The book concludes with three models, explaining how Toyotism, Nikefication with offshoring, and Waltonism provide full or less complete models of lean production. It clearly presents the positive and negative aspects of lean and insights into the culture of lean organizations.

With its rich interdisciplinary approach, *Framing and Managing Lean Organizations in the New Economy* will benefit researchers and students across a range of classes from management, sociology, and public policy to engineering.

Darina Lepadatu is Professor of Sociology and International Conflict Management at Kennesaw State University in metro Atlanta, USA and the current President of the Georgia Sociological Association.

Thomas Janoski is Professor of Sociology at the University of Kentucky, and has taught at the University of California-Berkeley, where he received his PhD in Sociology, and Duke University, where he helped create the highly successful Program in Management and Marketing Studies.

Framing and Managing Lean Organizations in the New Economy

Darina Lepadatu and Thomas Janoski

NEW YORK AND LONDON

First published 2020
by Routledge
52 Vanderbilt Avenue, New York, NY 10017

and by Routledge
2 Park Square, Milton Park, Abingdon, Oxon, OX14 4RN

Routledge is an imprint of the Taylor & Francis Group, an informa business

Library of Congress Cataloging-in-Publication Data
Names: Lepadatu, Darina, author. | Janoski, Thomas, author.
Title: Framing and managing lean organizations in the new economy / Darina Lepadatu and Thomas Janoski.
Description: New York, NY : Routledge, 2020. | Includes bibliographical references and index.
Identifiers: LCCN 2019043715 (print) | LCCN 2019043716 (ebook) | ISBN 9781138499034 (hbk) | ISBN 9781138499102 (pbk) | ISBN 9781351015158 (ebk)
Subjects: LCSH: Lean manufacturing—Management. | Organizational effectiveness.
Classification: LCC HD58.9 .L467 2020 (print) | LCC HD58.9 (ebook) | DDC 658.4/013—dc23
LC record available at https://lccn.loc.gov/2019043715
LC ebook record available at https://lccn.loc.gov/2019043716

ISBN: 978-1-138-49903-4 (hbk)
ISBN: 978-1-138-49910-2 (pbk)
ISBN: 978-1-351-01515-8 (ebk)

Typeset in Bembo
by ApexCovantage, LLC

To

Elena and Constantin Neacsu, who were both teachers in Romania and instilled in me a love for learning, travel and intellectual curiosity for everything around me, and

Edy Neacsu, who stimulates me to search deeper connections between economy, society and politics.

DL

Edward Janoski who took me to the UAW toy sales in River Rouge just before Christmas when I was young, and

Laura Jannika, Donald Janoski, and the late Robert Janoski who have lived with the auto and steel industries.

TJ

Contents

Tables

Figures

Boxes

Abbreviations

AG	*Aktiensgesselschaft*, a joint stock company with limited liability (a corporation)
AHRD	Academy of HR Development
AMA	American Medical Association (US)
ANA	American Nursing Association (US)
AQP	Association for Quality and Participation (formerly known as the International Association for Quality Circles—IAQC)
ASQ	American Society for Quality, the current engineering association for quality
ASQC	American Society for Quality Control, renamed American Society for Quality
ASTM	American Society for Testing and Materials
BAA	*Bundesanstalt für Arbeit,* and then in the 2000s, *Bundesagentur für Arbeit*—Federal Employment Agency in Germany
BLI	Business leadership initiative at Ford Motor Company patterned on GE
CAD/CAM	Computer assisted design and computer assisted manufacturing
CalPERS	California Public Employees Retirement System, a very large interest group that Invests California pension funds for employees and workers; it can influence the stock market and CEO tenure
CEO	Chief Executive Officer, which is the most powerful manager in a corporation
CFO	Chief Financial Officer
COO	Chief Operating Officer
CPFR	Collaborative planning, forecasting, and replenishment program used at Walmart
DFMA	Design for Manufacture and Assembly are standardized designs of component parts across suppliers, which reduces the number of parts needed across different organizations
DMAIC	Define, Measure, Analyze, Improve, Control strategy in Six Sigma; it is somewhat similar to PDCA or PDSA in lean production

DMAIV	Define, Measure, Analyze, Improve, Verify strategy in Six Sigma
DNA	Deoxyribonucleic acid is a molecule that carries genetic instructions for the development, functioning, growth, and reproduction of all known organisms; it is used here as describing the foundations of an institution or process
DQP	Diversified Quality Production or the German approach to quality
DRUM	Dodge Revolutionary Union Movement
ee	Employees (used in tables to save space)
EU	European Union
FIFA	*Federation Internationale de Football Association*, or the controlling organization for international soccer or football
FLA	Fair Labor Association organized to protect foreign workers
FMC	Ford Motor Company
FPV	Ford Performance Vehicles of Australia
GE	General Electric Corporation
GERPISA	Research organization for the study of automobile production and wages based in Paris, France, or *Groupe d'études et de Recherche Permanent sur L'industrie et les Salariés de l'Automobile* (Group to Study and Research Industry and Salaries in the Automobile Industry)
GM	General Motors Corporation
GMAD	General Motors Assembly Division
HBUC	Historically Black Universities and Colleges
HR	HR department or discipline
IAQC	International Association of Quality Control (name changed to Association for Quality and Participation in 1987)
IG Metall	*Industriegewerkschaft Metall*, or the metal worker's union that handles automotive production. Pronounced "Ee-Gay-Me-Taal" with accent on last syllable
ILO	International Labor Organization headquartered in Geneva, Switzerland
ILPC	International Labour Process Conference
ILRA	International Labor Relations Association, associated with LERA and IRRA
IMVP	International Motor Vehicle Project at MIT
IPO	Initial public offering when a private company sells shares on the stock exchange.
IRRA	Industrial Relations Research Association, replaced by LERA (US)
ISO	International Organization for Standardization (ISO in French) located in Geneva, Switzerland; it is closely linked with Six Sigma
JIT	Just-in-time inventory system, a specific version of supply chain management

JMNESG	Japanese Multinational Enterprises Study Group, which studied 500 factories in 30 countries
JPM	Japanese production methods, very much like TPS or Toyotism
JUSE	Japanese Union of Scientists and Engineers
LERA	Labor and Employment Relations Association, replaces IRRA (US)
LIN	Lean Implementation Network at Ford Motor Company
L-L-L	"Lean, loyal, long-term" view of lean production that tends to emphasize the opposite of the "cutting" approach that neo-liberalism might envision
LMM	Lean Manufacturing Manager at Ford who reports to top management
MBO	Management by Objectives. The name for a process to evaluate employees; it was heavily criticized by Deming
MBWA	Management by walking around (*genchi genbutsu*, or 'go and see')
MCW	*The Machine that Changed the World* by Womack, Jones, and Roos (1990)
Mgt	Management (used in tables to save space)
M-Form	The multi-divisional form of organization used at GM with multiple divisions
MIT	Massachusetts Institute of Technology
MLB	Major League Baseball (US)
NBA	National Basketball Association (US)
NCAA	National Collegiate Athletics Association (US)
NHTSA	National Highway Transportation and Safety Administration (US)
NFL	National Football League (US)
NLRA	National Labor Relations Act that created the NLRB in 1933 and other items
NLRB	National Labor Relations Board (US)
NRP	Nissan Revival Plan under Carlos Ghosn as CEO
NTT	Nippon Telephone and Telegraph (Japanese phone company)
NMUK	Nissan Motors in the United Kingdom
NUMMI	New American United Motors Manufacturing, Inc., a project of GM and Toyota
NYSILR	New York School of Labor and Management Relations at Cornell University
OEM	Original equipment manufacturer (usually a big corporation that does assembly and marketing)
OKR	Objective Key Results. A process of employee and unit evaluation used by Google. It is superficially similar to MBO but praised as being much better
OPEC	Organization of Petroleum Exporting Countries
PDCA	Plan-Do-Check-Act (from Deming and Japan)
PDSA	Plan-Do-Study-Act (more connected to Shewhart)

PVMI	Program on Vehicle and Mobility Innovation at the Wharton School at the University of Pennsylvania
QCC	Quality Control Circle
QOS	Quality Operating System at Ford involving the UAW and management
QOSC	Quality Operating System Coordinators from the union and management
QR Codes	Quick Response codes that are two-dimensional improvements on bar codes
R&D	Research and Development department or spending
RFID	Radio frequency identification device
RMG	The Repetitive Manufacturing Group that sponsored research in lean production by the American auto industry in the late 1970s
SACOM	Students and Scholars Against Corporate Misbehavior, an organization against sweatshops
SCM	Supply chain management
SET	The second step of software development where the code is tested internally (see also SWE and TE)
SHRM	Society for HR Management
SIOP	Society for Industrial and Organizational Psychology
SPC	Statistical Process Control that measures conformance to quality standards
SWE	The first step in developing the code for new applications (SET) test the code (see also SET and TE)
SWOT	A strategic management strategy of identifying **S**trengths, **W**eaknesses, **O**pportunities and **T**hreats
SVOR	A strategic management strategy of identifying **S**trengths, **V**ulnerabilities, **O**pportunities and **R**isks
SUV	Sport and utility vehicle
TE	The third step of testing code from the perspective of the user (see also SWE and SET)
TMMK	Toyota Motor Manufacturing in Georgetown, KY
TPS	Toyota Production System, or Toyotism
UAW	United Autoworkers Union, American union that handles autoworkers except for Japanese transplants
USMC	The United States (US), Mexico (M) and Canada (C) trade agreement that replaced NAFTA
VoC	Varieties of Capitalism
VW	Volkswagen; largest automobile company in the world; based in Germany
WRAP	Worldwide Responsible Apparel Production

Japanese and German Terms

Bundesanstalt für Arbeit	German "Federal Institute for Labor or Work" until 2000
Bundesangenteur für Arbeit	German "Federal Agent for Labor or Work" after 2000
Chimuwaku	Japanese word for team is *Chimu*, and *Chimuwaku* means teamwork

The Three G's of *Kaizen*

Gemba	Japanese for "the actual place"—refers to management by walking around or getting to the actual place where work and quality is being (or not being) done
Genbutsu	Japanese for "go and see", which is related to *Gemba* and to the process of "managing by walking around" (MBWA)
Genjitsu	Japanese for "realness" or "the actual facts" that may be associated with *Gemba* ("the actual place") and *Genbutsu* ("go and see")
Heijunka	Japanese for "leveling" applied to leveling out demand or the production schedule to reduce overtime, waste, or under-production
Hitozukuri	Japanese for "making or forming people"
Hoshin Kanri	Japanese word for "policy action or deployment"—a method of strategic management for guaranteeing that the goals of the corporation drive the actions taking place on a day-to-day, month-to-month, or year-to-year basis
Jidoka	Japanese term for one of the pillars of *Kaizen*—means to focus on the causes of problems and to stop work immediately when things go wrong (the Ishikawa or fishbone diagram would be a *Jidoka* tool for discovering causes)

Kaizen — Japanese term for "continuous improvement"

Kanban — Japanese term that literally means "card", which were the cards attached to supplier's parts—more generally, it means the JIT inventory system

Karate — Japanese form of martial art with punches and kicks

Karojisatsu — Japanese term for suicide due to overwork

Karoshi — Japanese term for death due to overwork

Kata — Japanese term for a group of 10–30 movements that are memorized in each stage of white to black belt training; they become tacit knowledge in *karate*

Keiretsu — Japanese term for a group of corporations that do not do the same things, but cooperate together to keep each one of them afloat and profitable

Kotozukuri — Japanese work for "making things happen"—can be composed of both *Monozukuri*, which is "making things" or production; and *hitozukuri*, which is "making, forming, or socializing people"

Marugakae — Japanese term for being sponsored or financed by someone else who might be considered to be a patron

Madogiwaoku — Japanese word for "encompassing embrace"

Meister — German term that means "master" craftsperson, but also refers to a first or second line supervisor

Mitbestimmung — German word for codetermination, especially the codetermination law (literally, it means regulating together)

Mittelstand — German word for "middle" suppliers of parts for firms, usually privately and/or family-owned; some feel that they are the backbone of the German Model

Modell Deutschland — German term for the "German Model" or Diversified Quality Production

Monozukuri — Japanese for "production" or "making of things"—at Honda, it is used more in the sense of *kaizen* with continuous improvement; it can also mean a synthesis of technology, know-how, and enthusiastic spirit.

Nemawashi — Japanese word for achieving consensus through a process of getting everyone's approval for a proposed change; sometimes it is criticized for being slow

The Three Wastes

Muda — Japanese term for "waste" or futility—this tends to be the most general term. There are also 8–10 different indications of waste in a production facility

Mura — Japanese word for unevenness in production. It is often a scheduling problem concerning supplies or distribution of products; it could also refer to the stopping and starting of the assembly line

Muri — Japanese word for the overuse of equipment or people; it can also represent too many hours or putting parts on the line that are too heavy, etc.

Obeya — Japanese term for the "big room", but refers to working spaces that are open without offices or cubby holes

Oidashibeya — A windowless room for personnel the Japanese corporation would like to see quit; however, this "expulsion room" or "quitting room" houses these workers. They often do little useful work while they are there

Poka-yoke — Japanese phrase for removing possible or current mistakes in a process; "Yoke" means preventing something, and "Poka" means mistakes (used extensively by Shigeo Shingo)

Ringi — Japanese word for their problem-solving system that goes through many different employees to get cooperative agreement among organizational players

Sangen Shigui — Japanese term for problem-solving: "San" means three; "Gen" means reality that is in front of you, and "Shugi" means principles—it is the three-principled method for problem-solving in a real setting that involves *Genba, Genbutsu,* and *Genjitsu*

The Five S's

Seiri — Japanese word for sorting items in a specific location and removing items that do not belong there

Seiton — Japanese word for putting items in their proper place so they can be easily used

Seiso — Japanese term for cleaning, sweeping, or keeping a location clean (can also mean inspecting)

Seiketsu — Japanese term for standardizing processes

Shitsuke — Japanese term for self-discipline for workers

Takt time — German term for the time it takes to do one repetitive task on an assembly line (literally, the tick of a clock or a heartbeat)

Waigaya — Japanese word for "noise" or a type of meeting where everyone talks. It is oriented for fixing problems, but the word itself is Honda jargon, which means that discussion takes place in spontaneous and unstructured meetings

The Japanese Equivalent of the Five Whys

These Japanese words are not often used, but one often wonders what the five whys are in Japanese. There are six listed here since "why" and "how" are closely related.

Naze	Japanese term for "why"
Nani or *Nan*	Japanese term for "what"
Itsu	Japanese term for "when"
Do	Japanese term for "how"
Doko	Japanese term for "where"
Dare	Japanese term for "who"
Zaibatsu	Japanese term for the large industrial conglomerates that controlled the Japanese economy before World War II. The *Keiretsu* is a weaker form of this organization—*Chaebol* in Korea might be closer in meaning

Preface

Lean production is now embedded in manufacturing throughout the world and is spreading to the services of many other industries. It is so much a bedrock to many industries that it becomes almost unseen to many of the scholars, workers, and customers using it. It is the purpose of this book to look at the many disciplines that analyze lean production and some of the major corporations in the world that actually use or claim to use it. Some parts of lean production are almost universal, like JIT inventory in supply chain management, and others are a tough sell and must be reinvigorated, like QCCs or teamwork. Within sociology, our own discipline, lean production is little observed. With this book, we hope to bring lean production to the forefront of scholarly analysis and to make up for the lacunae in the field of sociology especially.

There are a number of misnomers about lean production. One is that it is mainly about cutting the number of employees; this can be linked to outsourcing and off-shoring. While there is some truth to outsourcing and temporary employment, we see lean production as a way to grow employment through innovation and new processes. In some ways, the very term "lean production" is somewhat unfortunate. We often refer to the three Ls of lean production—lean, loyal, and long-term—as being more accurate (Janoski and Lepadatu 2014). The lean of only cutting certainly does not reflect the loyal and long-term aspects of lean production. However, in the penultimate chapter, we will discuss how lean production has also been associated with neoliberalism in government and the spirit of deregulation and cutting government services. We emphatically claim that this is not the case. One might be tempted to use another term; however, lean production has gained so much currency that we do not want to blaze new trails with another perhaps unfamiliar term. The term Toyotism, like Fordism before it, is more accurate.

In the development of this book, we have many people to thank. William Canak of Middle Tennessee State University—the long-term president of the award-winning Tennessee Labor and Employment Relations Convention (TERRA)—enthusiastically included many of our topics and presentations at the TERRA, and also connected to us to LERA people at the national level. We also thank Kim LaFevor, Dean of the College of Business,

Athens State University; three former HR managers at FMC; and a high-level UAW representative at Ford at the international level. We would like to thank Michael Samers of the University of Kentucky, Matt Vidal of Kings College of London, and Steven Vallas of Northeastern University. We would also like to thank Michael Burawoy of the University of California-Berkeley. Although the second author was not in his seminar on the division of labor at Berkeley, he was very much influenced by it through discussions with fellow students over the years. We thank Tommaso Pardi for including us in the GERPISA conferences in Puebla, Mexico in 2017, Paris, France and Sao Paolo, Brazil in 2018. Parts of this project were presented by the second author at the University of Kentucky Sociology Department colloquium in the fall of 2011, and by both of us at ASA Convention sessions in New York City in 2013, Seattle in 2016, and Philadelphia in 2018.

We'd like to thank Patricia E. White of the National Science foundation and Jan Stets of the University of California, Riverside who served three years at NSF for their help on The Maturing of Lean Production grant (NSF-ARRA 0940807) that provided summer support and travel money for both authors. Their support on the rewrites of a number of NSF proposals was very helpful and supportive. We also thank the anonymous reviewers, who though sometimes quite critical, helped reshape our focus and broaden our approach to many different disciplines. The NSF grant planted a seed that flourished and grew into multiple book projects: *Dominant Divisions of Labor* (2014), the forthcoming *International Handbook of Lean Production*, and this book project.

We thank the authors from many different disciplines who worked with us on *The International Handbook of Lean Organization* (forthcoming). They include Katsuki Aoki, Richard Schonberger, John Paul MacDuffie, Chris Smith, Matt Vidal, Michael Ballé, Abbot McGinnis, William Cooper, David Parsley, Tommaso Pardi, Doctor John Toussaint, Timo Anttila, Tomi Oinas, Armi Mustosmäki, Kenneth Grady, Pär Ahlström, Jean Cunningham, Mary Poppendieck, Byoung-Hoon Lee. James P. Womack, Daniel T. Jones, Paul Stewart, Adam Mrozowicki, Valeria Pulignano, Michael Krzywdzinski, Lu Zhang, Reynold James, Jorge Carillo, Giuliano Marodin, and Elena Shulzhenko. We would especially like to thank A. J. C. Bose from India who influenced the project but had to withdraw due to a serious illness. We wish him good health.

Our thanks also go to the many managers, current workers, and former workers at many companies including: Toyota Motors Corporation at Georgetown, KY; Nissan Manufacturing at Smyrna, TN; Honda Motors at Ana, OH; Honda Motors at Marysville, OH; Honda Motors at Lincoln, AL; the FMC truck plant at Louisville, KY; the Chrysler Corporation plant in Toledo, OH; and the GM plants in the United States and Shanghai, China. We especially appreciated talking to managers and workers on the extensive tours we took of the long-time VW Assembly plant, and the new, state of the art Audi Assembly plant, both in Puebla, Mexico. Equally important were the tours of the VW Wolfsburg plant in Germany in connection to the ILPC that was held in Berlin. We also had previously taken tours of the BMW plant in Bavaria and Spartanburg, SC as well as the Ford Motor Truck plant at the Ford Rouge Complex in Dearborn, MI. These tours were particularly helpful in understanding and seeing firsthand how the production process operates. For instance, one cannot miss the Andon boards at Toyota and their absence at the Honda plant in Marysville. Further, the conferences often attached to these tours gave us a link to scholars in the area of lean production when our own discipline of sociology has so few people studying this field. We also would like to thank employees at Sun Microsystems in China and Oracle in Silicon Valley. We would like to thank the United Automobile Workers for their willingness to be interviewed and provide valuable information on lean production in the United States, Japan, and China. We also thank the managers and employees at Apple, Google, Nike, Walmart, Costco, Amazon, and many other facilities in the United States.

Closer to home, the first author, Darina Lepadatu, would like to thank Kennesaw State University in metro-Atlanta for the Enhanced Tenured Faculty Leave provided during the data collection and writing phase of the book. She is appreciative of the many professional families that have supported her intellectual journey from the elite Petru Rares College in Piatra Neamt, Romania, to University of Bucharest, University of Kentucky, and Kennesaw State University, as well as the professional networks of information technology (IT) specialists that have facilitated the Silicon Valley interviews.

At the University of Kentucky, the second author, Thomas Janoski, would like to thank Patrick Mooney, who, as chair of the Sociology Department, supported our project by administering the NSF grant and providing the coffee cups that we gave out as a small token of our thanks to those we interviewed. He would also like to thank Claire Renzetti, the current chair at the Sociology Department at the University of Kentucky, and Mark Kornblum, the Dean of Arts and Sciences, who provided a phased retirement plan that was like having a six-month sabbatical every year for five years. This provided much of the time and resources to finish the project and this book. And in retrospect, he would like to thank his fellow piston shooters at the Chrysler Engine Plant in Trenton, MI for the insights that he gained working on the assembly line in the late 1960s.

Finally, we would like to thank Ronald Burt of the University of Chicago who advised us in 2001 that "you have a major auto company in your backyard—why don't you play to your comparative advantage?"

Atlanta and Lexington

1 Introduction

The Old and New Divisions of Labor

Are James Womack, Daniel Jones, and Daniel Roos correct in *The Machine that Changed the World* (1990) that lean production is the new model of the division of labor in the world? The division of labor is about how work is divided among workers, organizations, and industries. It has been observed since the time of Adam Smith (1776) and Emilé Durkheim (1997/1893)—farmers begetting bakers of bread and millers of wheat—about the actual procedures by which jobs are divided and organized so that they fit together in larger scale organizations. Adam Smith described the division of tasks involved in making a straight pin (obtaining wire, straightening it, cutting it, drawing it out, making a point, and putting a head on it). It inherently involves the division of work within a firm with greater specialization and repetition of tasks (1976/1776:8–9).[1] In fact, the Germans and many industrial engineers use "*takt* time" to describe the amount of time that it takes to do a repetitive task (for instance, 90 to 120 seconds on an assembly line). But the division of labor also consolidates some tasks through teamwork and job rotation, so it need not always be perceived negatively. In a sense, the division of labor refers to how people organize others to make things and deliver services. In most cases, it simply refers to dividing a project into smaller tasks, but it can also refer to the division of products between organizations (one firm makes pistons, another tie-rods, and an assembler puts them together as a car). Finally, it refers to country-level divisions of labor where one country produces one product, and another country produces a different product.[2] In this book, we focus on how this new division of labor is modeled by academics and organized by major corporations.

Although we will deal with some political issues in the conclusion, our presentation will concentrate on the distribution of tasks among workers. Many scholars see the division of labor as a fountain of progress, especially in earlier times, and a tsunami of boredom and alienation in some recent views. Marx viewed it as the source of inequality, and his followers have seen it as a plague descended upon the working people of the world with Taylorism and Fordism. But our focus will, for the most part, be on the division of work tasks in and between organizations. But we leave the battles of nation-states and production systems for Chapter 5.

We have entered into a new age of the division of labor. It is new not only because it is global but also because production is organized with much more focus on flexibility and quality. These processes have emerged from the old Taylorism and Fordism, surpassed a variety of competing ideas, and become part of the DNA of global production. So far, lean production has had a muted reception in the general social science literature as it is often seen as a specialty of the business management and industrial engineering departments. We widen its application and show how it fits into both manufacturing and service industries in a variety of disciplines.

We have two theses in this book. First, in Part I, we review the various disciplinary approaches to lean production—management, industrial engineering, sociology, labor process theory, labor and management relations, and human resources (HR)—to show that even though industrial engineering concentrates the most on lean production, all of these approaches have a partial approach to the topic that stresses particular issues and overlooks others. Hence, a multi-disciplinary approach is very useful. We also discuss the unique contributions of the more European approaches of the German "diversified quality production" and the French "productive models" theory. And we conclude that the management approach of shareholder value theory is inherently antithetical to lean production. Second, in Part II, we review a number of firm or industry-named models—Toyotism, McDonaldization, Nikeification, Waltonism, and Siliconism—as they apply to 11 major corporations in today's economy. Out of the many different conceptions of the new division of labor, we show that the present division of labor is best conceptualized as lean production and Toyotism, with two lesser forms of lean production that we call Nikefication and Waltonism (Janoski and Lepadatu 2014; Besser 1996; Berggren 1992). These three models are the most important forms of lean production: (1) Toyotism, as the highest form of lean production, (2) Nikeification, as a middle form of lean that is bisected by lean and socio-technical theory as well as simple Fordism in its off-shore production, and (3) Waltonism, as the lowest form of lean relying solely on just-in-time (JIT) inventory. We also discuss how some firms try to implement lean methods but are only moderately successful.

The remainder of this introduction now discusses the old division of labor, the new division of labor in Toyotism by defining lean production, and then we introduce our chapters to come.

What was the Old Division of Labor? Taylorism, Fordism, and Sloanism

Two men dominated the old division of labor. The first, Frederick Winslow Taylor, suffered headaches and poor eyesight so despite having passed the entrance exams to the Harvard University in 1874. He then became an apprentice pattern-maker and machinist at a pump-making factory in Philadelphia. He went on to become what we would now call a management consultant to the Watertown Arsenal and Bethlehem Steel, and later a professor at the Amos Tuck School of Business at Dartmouth University. His unbounded enthusiasm for machinery and factories led to his development of the theory of scientific management. While the industrial revolution, with theorization by Adam Smith's pin-making principles and Emile Durkheim's work on the division of labor, preceded what he had done in the breakdown of work, Taylor went further than others in specifying the division of labor's components of time and motion studies and piecework (Kanigel 1997).

The second, Henry Ford, grew up on a farm and although he had a basic education, he did not consider going to a university. His early life was based on tinkering until he started working for Thomas Edison. This gave him a chance to develop his ideas about automobiles, which was a newly emerging technology. He developed his ideas toward production on the basis of factories that he visited and also the disassembly processes at the Chicago stockyards. There is no evidence that Ford directly used any of Taylor's ideas—probably because Taylor died in 1915, just before Ford's influence was beginning to climb. Conversely, Taylor visited Ford's early plant before the Rouge facility was built (Ford lived to 1947). Ford was a self-taught autodidact who relied more on observation than books and was even a bit antagonistic toward higher education, though he did establish the Ford Trade School. Examples of both assembly lines and rational production existed as far back as the Venetian boat works, but Ford does not reference them and probably only had a passing knowledge of them (Ford 1922, 1926).

Consequently, these very different men are forever entwined as the two forceful personalities who shaped the modernist century of manufactured production in America and the rest of the world.

Taylorism and Scientific Management

In the Taylorism or scientific management approach, Frederick Taylor felt that the industrial order had been changed. In the past, man had been first, but in the future, the system must be first (Taylor 1911:2). This system had four principles. First, Taylor's engineers examined the work process in terms of how it was done. Using "time and motion studies", they would study the job by replacing informal or rule-of-thumb methods with scientific study of actual job processes and methods based on the results of a scientific study of a job's tasks. To do this, the engineers conducted experiments with the work process and developed more efficient ways of doing things including the physical movements of the body, thought processes of the mind, and optimal rest periods. Finally, they would provide "detailed instruction and supervision of each worker in the performance of that worker's discrete task" (Montgomery 1997:250).[3]

Second, some of these engineers, who in later decades became HR or personnel specialists, would scientifically select, train, and develop each employee rather than passively leaving them to train themselves. They classified workers to see if they fit the rigors or tedium of the work processes. They did not want workers whose abilities were limited and thus could not do the job, but they also did not want abilities to be too high so that workers would be bored with the job and want to do something else. Third, it targeted the wage system to create a payment system that strictly reinforced the optimal way of producing the product. Piecework wages were created that tied the number of products produced in an hour (or day) to increased wage rates if the worker exceeded a certain minimum. And fourth, it targeted the management system by dividing work nearly equally between different types of managers and clerks. The "foreman's empire" of the late 1800s was an anathema to Taylor because the foreman was essentially a subcontractor who had total control of the workplace (Nelson 1996:35–56). Instead, Taylor divided management up into nine different types of bosses or clerks: the route clerk, the information card clerk, the time and cost clerk, the shop disciplinarian, the gang boss, the speed boss, the repair boss, the inspector, and the overforeman (HR would evolve from some of the overforeman's tasks). Of course, Taylor's professional engineers also designed the whole process. This was called functional foremanship and to a large degree foresaw the functional divisionalization of organizations—separate supply, HR, production, quality control, and many other departments—that became common in the 20th century.

The end result of Taylorism was the individualization of work, so each worker has strong incentives to produce more products in order to get higher wages, and management controlled those incentives. It was a pure stimulus-response system that did not see work as existing beyond the immediate tasks before the worker. Engineers did the thinking about work techniques and designed the piecework system, and workers just followed orders.

While the Tayloristic system of scientific management might make sense from the perspective of a rare single-minded individual worker who wanted to make the most money possible, it did not work for most workers. The system had three weaknesses. First of all, Taylorism removed thinking from the workers' purview and actually sought out dull workers.

> One of the first requirements for a man who is fit to handle pig iron as a regular operation is that he should be so stupid and so phlegmatic that he more nearly resembles in his mental make-up the ox than any other type.
>
> (Taylor 1911:59)

This was exemplified in Taylor's testimony to Congress where he described the brawny Swede Schmidt who loaded pig iron (see Box 1.1).[4]

Box 1.1 Frederick Winslow Taylor and Scientific Management

Frederick Winslow Taylor (1856–1915) was a self-taught mechanical engineer who sought to improve industrial efficiency. Although he was an excellent student, Taylor avoided college and went to work as an apprentice in a steel company. He was soon promoted and focused on developing more and more efficient ways of doing things. He eventually came up with a theory of separating mental and manual work, perfecting time-and-motion studies to do so, and redesigning factory processes. In doing so, he was an early management consultant. Taylor was one of the leaders of the movement toward efficiency, and his ideas were influential in the progressive era from 1890 to 1920. Taylor explained his techniques of efficiency in *The Principles of Scientific Management* (1911), which, 110 years later, the Academy of Management voted "the most influential management book of the 20th century". His pioneering work was applied to the shop floor of numerous factories, and this eventually helped to create much of what is now known as industrial engineering. Taylor made his reputation on his work in extreme efficiency, which he named "scientific management".

After the industrial conflicts and unrest that followed his methods, he testified before Congress about his system. He used the example of the Swede Schmidt, who he referred to in derogatory terms, whom he then trained to set a record in loading pig iron ingots on a train.

Taylor: "Schmidt, are you a high-priced man?"
Schmidt: "Vell, I don't know vat you mean . . . "
Taylor: "You see that car?"
Schmidt: "Yes".
Taylor: "Well, if you are a high-priced man, you will load that pig iron on that car tomorrow for $1.85. Now do wake up and answer my question. Tell me whether you are a high-priced man or not".
Schmidt: "Vell, did I got $1.85 for loading dot pig iron on dot car tomorrow?"
Taylor: "Yes, of course you do . . . "
Schmidt: "Vell, dot's all right".

In actuality, Schmidt was a German named Henry Noll, rather than a Swede (Kanigel 1997:540–60; Braverman 1974:106–7; Wrege and Perroni 1974). Apparently Taylor changed the details for effect or perhaps not to offend the questioner.

> Taylor made his fortune patenting steel-process improvements. Taylor was also an athlete who won the 1881 doubles championship with Clarence Clark at what is now the US Tennis Association Open, and 19 years later, he finished fourth in golf at the Olympics in Paris.

Schmidt (aka Noll) received 61% more pay for moving 362% more pig iron compared to the average worker. Schmidt's case has become legendary, although the evidence shows that no other worker in history could break his record. Taylor's assistants wrote that "other workers broke down after two or three days", showing that Taylor's methods were not so scientific after all (Kanigel 1997). Although Taylor's pig-iron experiments had proven to be seriously flawed (Wrege and Perroni 1974), Taylor made the point that a "first-class worker" selected based on a scientific method can double or triple his productivity if properly motivated. Taylor did not care about dumbing the work down.

Second, Taylor thought workers would make more money under this system, but managers saw something else in the methods and pressured engineers to change the piecework rates so that when workers achieved high levels of production, the amount of money they actually received was ratcheted downward. Workers thought this was duplicitous and unfair, and labor conflicts led to Taylor's testimony before Congress. Nonetheless, more and more firms gradually came to adopt many aspects of the Tayloristic processes of work design, even if they did not always implement piecework. Third, the discovery of the informal group in the Hawthorne experiments at Western Electric showed that the social influence of the group had even more control over worker's performance than manager's exhortations or engineer's piecework charts. Taylor's theory had absolutely no conception of groups or norms except as a negative factor to be expunged.

While Taylor's proposed principles of standardization and specialization had a tremendous impact on industrial productivity throughout the entire 20th century, the basic premises of his managerial philosophy have been highly criticized from the beginning. In 1915, Robert Hoxie, the special investigator for the US Commission on Industrial Relations of the House of Representatives, reported on Taylor's management practices to show that scientific management was undemocratic because it did not involve workers in the fundamental parts of the production process, such as the setting of task, the wage rate or the general conditions of employment (Hoxie 1966). But American owners and most managers could care less about workplace democracy. The other major complaints against scientific management allude to the fact that the obsession with efficiency overshadows the fundamental social aspect of work (Mintzberg 1989) while the increased specialization leads to workers' deskilling, degradation of work, and alienation (Braverman 1974). Nevertheless, Taylorism marched on.

Fordism and Mass Automobile Production

The concept of Fordism—intrinsically tied to Taylorism or scientific management—took the world by storm in from 1918 to 1968 (Nye 2013). Since it used an assembly line, it could not make the pace of work the central aspect of payment like Taylorism; nonetheless, it shared many concepts of work design. Fordism consisted of both technical and social parts with a total of ten aspects. The technical aspects of Fordism had five principles. First, parts and processes underwent intensive standardization, with replaceable parts a key aspect of this approach. Second, these standardized parts were a natural fit with the assembly line which demanded the repetitive installation of exactly the same parts. Third, the assembly line relied on massive economies of scale to promote productivity in terms of the quantity of cars produced. And this of course, also produced higher revenues due to widespread sales based on a lower price for the car. The results were record levels of profitability. The massive vertically integrated plant on the Rouge River, which was largely built after World War I, embodied this massive scale of production (Ford 1988/1926; Sorenson 1956; Levinson 2002). Fourth, the standardization of material parts and assembly line technology led to a further standardization of human workers. This involved the deskilling of labor and the end of craft production with its careful fabrication of small numbers automobiles. Fifth, the production process then created a mass market for homogenized products, whether automobiles or hamburgers, by molding tastes for the same things (Ritzer 2019). The results led to the spread of automobile dealerships throughout the country with accompanying repair shops and gas stations. Later, it led to cookie cutter houses and fast food restaurants. All of this, of course, presumed the creation of a large road system that was capped off by the interstate highway system created by President Dwight Eisenhower. Fast food restaurants and inexpensive motels were then built at every exit.

There are five more social and political aspects of Fordism. First, Henry Ford increased wages with his unheard-of five-dollar day. Ford argued that he wanted his workers to be able to buy his product (1922, 1926). Previously, automobiles were for the rich and totally unaffordable by autoworkers. Ford's plan lowered the price of automobiles and increased the wages of his workers. But the "five-dollar day" was exaggerated in terms of who it applied to (i.e., there were moral requirements administered by the Ford Sociology Department). Second, the Fordist system became associated with the unionization of the auto industry by the United Autoworkers (UAW), and this led to collective bargaining for yet higher wages and an extensive array of benefits. While Ford and others did not intend to create this part of the Fordist system, it nevertheless became a standard feature in the industrialized North and Midwest. Third, the negotiation of uniform wages based on profits and productivity had a spillover effect on the surrounding industries—unionized or not. As such, worker demand for products and the production of goods and services increased in these areas in the United States. Fourth, Fordism is generally associated with Keynesian macro-economic policies that increased spending through a welfare state that existed along with collective bargaining. Ford's idea that every worker should be able to buy a car parallels the consumer focus of Keynesian economics (Harvey 1991, 2011). And fifth (and perhaps most distant to the actual production model): mass educational institutions (i.e., the American high school) providing undifferentiated workers who can be molded by short-term or on-the-job training (OJT). These workers naturally fit into the semi-skilled demands of the assembly line in manufacturing industries.

Sloanism as a Modification of Fordism

While Fordism is viewed as the dominant model of the 20th century, the actual Ford production process should not be considered the only version of it. Emma Rothschild refers to "Sloanism" as the combination of Fordism and the multi-divisional structure (M-Form) that combined some aspects of control with marketing (1973). Founded in 1908, General Motors (GM) challenged and surpassed the FMC as Henry Ford clung to his mass production model for too long—"You can have any color you want, as long as it is black". GM used the same assembly line and production process as Ford but introduced an element of flexibility through its divisional structure. Starting with Buick, GM bought other companies, and then produced a greater variety of automobiles with diverse divisions such as the Winton (1897–1937), Oakland (1917–1931), Oldsmobile (1897–2004), Buick (1897–present), Chevrolet (1911–present), GMC trucks and SUVs (1901–present), Pontiac (1926–2010), and Cadillac (1902–present). As explained by Alfred Chandler (1962), Peter Drucker (1946), and Alfred Sloan himself (1986), the M-form structure used a unified financial system to control profitability but allowed the divisions to produce a somewhat unique car that could be sold in differentiated markets. This approach clearly put GM into the lead and the FMC fell to a distant second.[5] It was not until Henry Ford II took over the reins of FMC after World War II that Ford was able to rebound.

The advantages of the M-form organization faded when GM's financial managers created the GMAD. This approach centralized the production of body and engine parts and pushed the Sloanism model back to classic mass-produced Fordism (Moberg 1978). The result was that the divisionally produced cars—Chevrolet, Buick, Oldsmobile, and Cadillac—started to look the same despite their very different prices. The model was clearly broken when an expensive Cadillac Cimarron (1982–1988) looked very much like a much cheaper and bottom-of-the-line Chevrolet Cavalier. In a process called "rebadging" by upgrading seats, dashboards, and chrome, GMAD paraphrased Henry Ford's statement about "any color so long as it is black": "You can get any color car with a wide range of features that you want, but underneath, each chassis was basically the same". Thus, we should not look at Fordism as invincible, even during the high period of its rule (i.e., the 1940s and 1950s). GM created some flexibility through the M-form organization that was successful for about a quarter of a century,[6] But then GMAD by-passed Sloanism with a return to Fordism (i.e., Cadillacs were built on Chevy undercarriages and frames).

Fordism also went beyond "the labor process" in the factory to create a theory of much wider scope. It described a whole way of life in the 1970s that included critique and some consensus (i.e., on wages and the expanding welfare state) (Gartman 2004:169–95; Jameson 1991; Harvey 1991, 2011; Farber 2002). It clearly had a cultural component as it affected consumption. The "big box stores" replaced "mom and pop stores" in the neighborhood, and rationalization went on to promote an imposing discipline of efficiency in all areas of life focused on the awaited "weekend" that would be filled with autos, toys, groceries and even rational psychotherapies that would improve one's personality.[7] It was the American way and it even entered into state theories through regulation theory (Boyer and Saillard 2001; Jessop 1990). In some ways, Braverman's *Labor and Monopoly Capital* (1974) was like the Owl of Minerva who only looks back to the past. Sociologists had not recognized the full implications of the Fordist age until it had peaked and went into decline with the first oil crisis in 1973. The edifice of scientific management would soon be surpassed by lean production.

Nonetheless, what is noteworthy is that the technical theory of scientific management led to two themes. First, there was a great excitement and enthusiasm for the theory and a desire to apply it to many areas of work from clerks to professors and medical doctors. For example, doctors in California even as far back as the 1940s witnessed the beginnings of Fordism in health maintenance organization (HMOs) through the Kaiser-Permanente Medical System. And second, the theory of scientific management expanded through its extension into Fordism into the rest of society, including trade unions, government policy, politics, and educational institutions. Since the decline of the Fordist model, there have been a number of theories to replace it, but none have taken off with the enthusiasm or scope that Fordism did after the Ford Rouge plant was built.

What caused the decline? Although Fordism—and to some extent, Taylorism—is still strong in the third world and in China, this approach declined mainly because of the Japanese approach to production improved quality. In the American automobile industry and many other manufacturing concerns, the quality of production began to suffer as workers became more and more dissatisfied with the sheer boredom of the production process, even though Henry Ford himself said that "repetitive labor is a terrifying prospect" (1922, Chapter 8). The mind-numbing process of doing the same thing 50 times an hour for eight or more hours a day led to "human beings" questioning whether this was all there was to most of their work lives. And this repetitive vision of the dominating and relentless machine even invaded the unconscious lives of workers when they slept and dreamed. Eventually, demands for the humanization of work in the 1960s and 1970s emerged, though these pressures did not go far as new models of work came upon the scene.

For workers, many of whom had been struggling peasants in Europe or the hired hands from American farms, high wages would be enough for a few decades. But as Americans and others gained more and more education, spending one's life on an assembly line was a nightmare. At the same time, many companies were pressured in the 1960s and 1970s to employ more disadvantaged workers who had not had the access to decent wages in the past, which led to social conflicts. Some of the rebellious baby boomers instigated sabotage on the line, and the rise of Marxist groups like the Dodge Revolutionary Union Movement (DRUM) led to conflicts and lower quality. At the same time, the Japanese used a method of production that stressed thinking. Participating blue collar workers and their reputation for quality rocketed Japanese auto sales to take over from American companies by the 1980s. The same thing happened in other industries, from mechanical pencils to television sets.

Box 1.2 Henry Ford and Alfred P. Sloan: Fordism and Sloanism

Henry Ford (1863–1947) founded the FMC and was the first to apply the assembly line approach to mass production in the auto industry. Ford did not invent the automobile or the assembly line, but he applied it on a massive scale at the Ford Rouge Plant to produce automobiles for a mass market. Ford converted the automobile from an upper class vehicle to a practical car that most Americans could afford, especially his workers, with their five-dollar day, which was unheard of at the time. Ford's Model T fundamentally changed transportation and the American road system—and beyond that, the manufacturing and even services industries. As the founder of the FMC, he became one of the first billionaires. He is credited with the larger term of "Fordism", which is the application of mass production to give workers higher wages so they can buy more inexpensive consumer goods.

Ford had a global vision that put consumerism at the center as the key to peace. He was intensely committed to the systematic lowering of costs, which resulted in innovations such as the FRANCHISING system that put car dealerships in North America and much of the world. Ford left most of his vast wealth to his heirs and the Ford Foundation for Social Justice, and arranged for his family to permanently control the company through preferred stock. Ford was also known for being a pacificist before and at the beginning of World War II, and for promoting antisemitism in the newspaper *The Dearborn Independent*.

He visited Germany to advise Adolf Hitler on the production of the Volkswagen and appeared to be sympathetic to the Nazis. This tarnished his reputation among many.

Alfred Pritchard Sloan, Jr. (1875–1966) was the president, chairman of the board, and CEO of the GM Corporation for three decades. Sloan helped GM grow from the 1920s through the 1950s with the introduction of the annual model change, divisional branding, industrial engineering, an emphasis on styling, and most significantly, financial control. This brought GM great success while Ford clung to the one model in one color. Sloan wrote *My Years with General Motors* in the 1950s to show his vision of the professional manager and the carefully crafted corporate organization that he created. Some consider it to be a milestone in the field of modern industrial management. Sloan is remembered for being a rational, tough, and successful manager who led GM to become the world's largest corporation. He is also remembered by his critics as cold and perhaps greedy. However, others saw him as a leader with great talent, responsibility, and generosity. His Sloan Foundation donated millions of dollars to form the Sloan School of Business at MIT. Sloan ran GM during a time that included the Great Depression, fascism, and World War II, and he was criticized for the Axis powers using GM's German subsidiary Opel for war production.

The Extension of Standardization into White Collar Work

In 1974, Harry Braverman revived the discourse on the division of labor in his book *Labor and Monopoly Capital.* He reasserted the Marxian concept of the "labor process", and pushed scientific management from the machine shop into the growing service sector with examples of clerical and professional workers. At the same time, Robert Cole was doing a participant observation in Japan on the Japanese production methods at Toyota and its suppliers that would be the earliest empirical works on what would later be called lean production. This shows the risk in proclaiming the "dominant division of labor" when, in fact, that system is beginning to be supplanted by another form of production. One has to be careful when claiming that the world is being driven by a new form of organizing work.

After Braverman extended of the labor process to white collar work, the criticism of deskilling and the labor process increased. Michael Burawoy (1979), Richard Edwards (1979), Dan Clawson (1980), and many others published important books in work and labor that were quite critical of the Fordist division of labor. The harshest criticism of Taylorism and Fordism comes from the Japanese. Konosuke Matsushita, founder of Panasonic, looked at Taylorism and Fordism as symbols of a flawed form of Western thinking. The separation of tasks between management and labor was an anathema to him. Matsushita commented on the US commitment to Fordism:

> We will win, and you will lose. You cannot do anything about it because your failure is an internal disease. Your companies are based on Taylor's principles. Worse, your heads are Taylorized too.
>
> (Labovitz et al. 1993:43)

Hajime Karatsu expressed similar sentiments in *Tough Words for American Industry* (1986). The various criticisms against Taylorism paved the road to new schools of thought in labor relations that bring the human factor back in the division of labor and propose a balance of the technical and social factors, most developed in sociotechnical systems theory. For instance, Swedish and German models are often more friendly to workers with a form of worker empowerment in problem solving and suggestion systems (Berggren 1992; Turner 1993).

Again, Fordism, and to some extent Taylorism, did not disappear, but on high-value added products, it went into a slow but consistently downward decline. What replaced it is a complex mix of models, which we will now sift and sort.

What is the New Division of Labor? Lean Production and Its Extension

Lean production starts with the quality control movement within the statistics and mathematics field in the United States. George S. Radford worked within a group of scholars interested in quality. He published *The Control of Quality in Manufacturing* as a first important step toward improving quality in production. He argued that "quality" needed to be improved to have a profitable enterprise or a reliable military (1922:4). This required that products be uniform, and the means to do this was through inspection at the process and final completion levels. A separate inspection department would do this. Although inspections used some statistics, it was a first and initial step at this time (see a further discussion of Radford and Shewhart in Box 1.3).

Box 1.3 George S. Radford and Walter A. Shewhart: The Originators of Quality Control

George S. Radford was one of the first scientists analyzing the processes of inspection to improve the quality of manufactured products. He used statistics in *The Control of Quality in Manufacturing* (1922) as a first important step toward improving quality in production. He emphasized that "quality" was a major aspect of competitive advantage, and that it needed to be improved to have a profitable enterprise or a reliable military. Uniformity of products was a major goal that the industry needed to recognize. He especially emphasized war time manufacturing and the lives that were at stake when poor quality harmed soldiers and sailors (e.g., prematurely exploding shells or duds in the face of the enemy). He stressed the usefulness of inspection, which had been largely ignored by major firms, and advocated for a separate inspection department that would engage in both "process inspection" and "final inspection". He developed the concept of sampling and included error rates and normal curves. However, his use of statistics was not very developed at this time and much of his analysis focused on precise measurements of tools and products.

Walter Andrew Shewhart (1891–1967) was a physicist, engineer, and statistician known as the originator of statistical quality control. He attended the University of Illinois at Urbana-Champaign and then received his doctorate in physics from the University of California, Berkeley in 1917. Shewhart joined the Western Electric corporation to help Bell Telephone improve the reliability and quality of their communication systems. When he arrived at Western Electric, the field of industrial quality only consisted of inspection and throwing out defects. Shewhart applied his statistical methods to the production of central station switching systems, and created the first schematic control hart for industrial engineering. This set forth all of the principles of statistical process control. Shewhart pointed out the importance of reducing variation in a manufacturing process and that non-conformance degraded quality. Shewhart concluded that while every process displays variation, some processes display controlled variation that is embedded in the process, while other variation may be uncontrolled and random. Thus, Shewhart framed the problem in terms of clear causes of variation and random variation. The control chart would differentiate the two. Keeping both under control is necessary for managing production and predicting quality. Shewhart's charts were adopted by the ASTM CAD/CAM and used during World War II to improve armament production. His work was summarized in his book *Economic Control of Quality of Manufactured Product* (1931) and expanded upon in *Statistical Method from the Viewpoint of Quality Control* (1939).

The Western Electric company involved in making telephone banks and receivers was especially interested in quality. Walter Shewhart worked for Western Electric, and was the first to definitely put statistics (i.e., statistical conformance) at the center of the quality control process.[8] He developed the first statistical industrial engineering that effectively tracked defects and their extent as a regular part of the production process. This provided the essential principles of statistical process control (SPC) of industrial engineering. Shewhart framed the problem in terms of specific causes of certain types of defects and the chance causes of defects due to random variation in product supplies and machinery. His control chart mapped out the extent of defects in as much detail as stock market statistics changing by the second. He discovered that observed variation in manufacturing data did not always behave the same way as data in nature. Shewhart concluded that while every process displays variation, some processes display controlled variation that is natural to the process, while others display uncontrolled variation that is not present in the process causal system at all times. Shewhart's control charts became widely adopted, and were recommended by the ASTM in 1933 and were used by the War Department to improve the quality of munitions and other war production. He published his work in *Economic Control of Quality of Manufactured Product* (1931) and *Statistical Method from the Viewpoint of Quality Control* (1939). Shewhart's methods were essentially what Edwards Deming and Joseph Juran brought to Japan that created their quality control movement. We will discuss these developments in the next chapter.

In the next section, we discuss lean production as the new division of labor defining the term more precisely and show how it developed in manufacturing and then how it spread to services from hospitals to Silicon Valley.

Defining Lean Production

Jostein Pettersen does an elaborate essay on defining lean production (2009). He takes definitions from nine different and important works on lean production and compares them on whether they include 33 different topics often associated with lean production. Table 1.1 is a slightly modified table that he presents with temporary employees added and statistical quality control reinterpreted as QCCs. The authors are reordered listing the highest total scores first (number of "yes" answers), and the characteristics grouped according to JIT, teams, and the other items.

The results show that Jeffrey Liker (2004) and Pascal Dennis (2002) have the most comprehensive definitions with other authors listing fewer items (usually QC circles and temporary workers). Taichi Ohno and Shigeo Shingo seem to concentrate on JIT; Kaouru Ishikawa is the opposite concentrating on teamwork and QCCs. In the next four sections, we discuss a synthesis of these different components of lean production in five parts: (1) a long-term view, (2) JIT inventory, (3) continuous improvement, (4) teamwork and flexibility, and (5) loyalty to employees.

(1) Long-Term Goals are what Lean production focuses on in terms of investments in plant and equipment, cooperative suppliers, trained and

Table 1.1 The Definition of Lean Production in Nine Studies
(Authors with the highest number of Yes's listed first)

Characteristics	*Liker*	*Dennis*	*Schon-berger*	*Womack et al.*	*Ohno*	*Bicheno*	*Feld*	*Monden*	*Shingo*
1-Standardized Work	Yes	Yes	Yes	—	Yes	Yes	Yes	Yes	Yes
2-Time & Work Studies	Yes	Yes	—	Yes	Yes	Yes	Yes	Yes	—
3-*Kaizen*/Continuous Imp.	Yes	Yes	Yes	Yes	Yes	Yes	Yes	Yes	Yes
4-Whys/Root Cause	Yes	—	—	—	Yes	Yes	—	—	—
5-Value Stream Mapping	Yes	Yes	—	Yes	—	Yes	—	—	—
6-*Poka yoke*	Yes	Yes	Yes	—	Yes	Yes	Yes	Yes	Yes
7-Prod. Leveling/*Heijunka*	Yes	Yes	—	Yes	Yes	Yes	Yes	Yes	Yes
8-JIT Inventory	Yes	Yes	Yes	Yes	Yes	—	Yes	Yes	Yes
9-Inventory Reduction	Yes	Yes	Yes	Yes	Yes	—	—	Yes	Yes
10-Supplier Involvement	Yes	Yes	—	Yes	—	Yes	Yes	Yes	Yes
11-Pull system/*Kanban*	Yes	Yes	Yes	Yes	Yes	Yes	Yes	Yes	—
12-Lead Time Reduction	Yes	Yes	—	—	—	—	Yes	Yes	—
13-Teamwork	Yes	Yes	—	Yes	Yes	Yes	Yes	—	—
14-QC Circles	**—**	**Yes**	**Yes**	**Yes**	**—**	**Yes**	**Yes**	**—**	**—**
15-Cross-training/Job Rotate	—	—	Yes	—	—	Yes	Yes	—	Yes
16-Employee Involvement	Yes	Yes	Yes	Yes	—	—	—	Maybe	—
17-Multi-manning	—	—	—	Maybe	Yes	—	—	Yes	Yes
18-5 's/Housekeeping	Yes	Yes	Yes	Yes	Yes	Yes	Maybe	Yes	—
19-Waste Eliminate, 7 *mudas*	Yes	Yes	Yes	Yes	Yes	—	Yes	—	Yes
20-Andon Cord	Yes	—	Yes	Yes	Yes	Yes	Yes	Yes	Yes

(*Continued*)

Table 1.1 (Continued)

Characteristics	*Liker*	*Dennis*	*Schon-berger*	*Womack et al.*	*Ohno*	*Bicheno*	*Feld*	*Monden*	*Shingo*
21-Small Lot Production	Yes	—	Yes	—	Yes	Yes	Yes	Yes	Yes
22-*Takt* Production	Yes	Yes	Yes	—	—	Yes	Yes	Yes	Yes
23-TPM/Prevent. Maint.	Yes	Yes	Yes	—	Yes	Yes	Yes	—	—
24-Autonomation/*Jidoka*	Yes	Yes	Yes	—	Yes	—	—	Yes	—
25-Stat. Qual. Control/SQC	—	—	Yes	Yes	—	Yes	Yes	Yes	Yes
26-Temporary employees	**—**	**Yes**	**Yes**	**—**	**—**	**Yes**	**Yes**	**Yes**	**—**
27-100% inspection	Yes	Yes	Yes	—	Yes	—	Yes	—	Yes
28-Less Set-up Time	Yes	Yes	Yes	Yes	Yes	Yes	Yes	Yes	Yes
29-Layout Adjustment	—	Yes	Yes	—	—	—	—	Yes	Yes
30-Policy Deploy/*Hosin kanri*	Yes	Yes	—	Yes	—	Yes	—	—	—
31-Process Synchronizing	Yes	—	Yes	—	—	—	—	—	Yes
32-Cellular Manufacture	—	—	—	—	Yes	Yes	Yes	Yes	Yes
33-Visual Control	Yes	Yes	Yes	—	Yes	Yes	Yes	Yes	Yes
Total	26	25	23	19½	19	19	18½	23	18

Sources: Liker (2004); Dennis (2002); Schonberger (1982); Womack and Jones (2003); Ohno (1988a, b); Bicheno (2004); Feld (2000); Monden (2011); and Shingo (1984).

Note: Based on Jostein Pettersen (2009) with the following changes: authors and characteristics reordered; QC circles and temporary workers are added as row headings; and we computed the 'total' which is not in Pettersen. The total is based on the number of yes responses, and the scoring is based on: Yes = 1, Maybe = 1/2. No or not present = —.

loyal employees, revenues and market share, and the all-important customer. Lean production in the auto industry stopped the short-term focus on quantitative profits and lack of quality as expressed in terms of "lemons". The idea is not to sell a cheap product that does not invite customer loyalty. So short-term goals like stock-price, dividends, and profits (including managerial salaries) take second place to long-term market share. In long-term goals, Liker mentions "Do the right thing for the company, the employees, the customer, and society as a whole". (2002:72). Produce a product in an industry that you can be proud of that will benefit society. Many corporations profess long-term goals of being dominant in their industry, but in the process they produce great amounts of waste in terms of cheap products or high wages for CEOs. Truly lean companies do not produce excessive salaries and dividends, short-term pumping up profits for high stock prices (e.g., for quarterly earnings reports), or other tactics like divestment, milking, and planned obsolescence.[9]

(2) **JIT Inventory** is part of the more general topic of supply-chain management. However, JIT is more specific in terms of wanting the parts exactly when they are need in the production process with minimal lead times. This is perhaps the most implemented part of lean production, since few if any corporations at this point in time do not try to minimize inventory and speed up deliveries (*Economist* 2019). JIT has moved from *Kanban* or physical card systems to highly computerized bar coding, and this has greatly aided the efficiency of this process. However, another part of this is developing long-term and cooperative relationships with suppliers while at the same time improving their quality and lowering their prices. Many corporations put suppliers in competitive market relationships, dropping them easily when there is a better offer.

(3) **Continuous Improvement** is a vital center to lean production. It has a constant commitment to *kaizen*. This is intimately involved in the process of statistical quality control with Ishikawa (or causal) diagrams, Pareto charts, graphs, bar charts, and normal curves of errors are brought together with QCCs (see Figure 1.1 for an Ishikawa diagram).

For teams to be analyzing statistics and new ideas created the direct opposite of Tayloristic divisions of mental and manual labor. In effect, it brought workers manual labor into a new arena of mental labor intended to create higher quality products. Continuous improvement also concerns the design of products and the production line itself. New products involve employees throughout the firm which, for blue collar employees or associates, means having their participation, especially concerning how hard it may be to assemble the product or produce the service.

(4) **Strong Teamwork** in QCCs is an especially important part of lean production concerning the conclusions of this book. As a result, we look more closely at teamwork. In Table 1.2, there are four types of

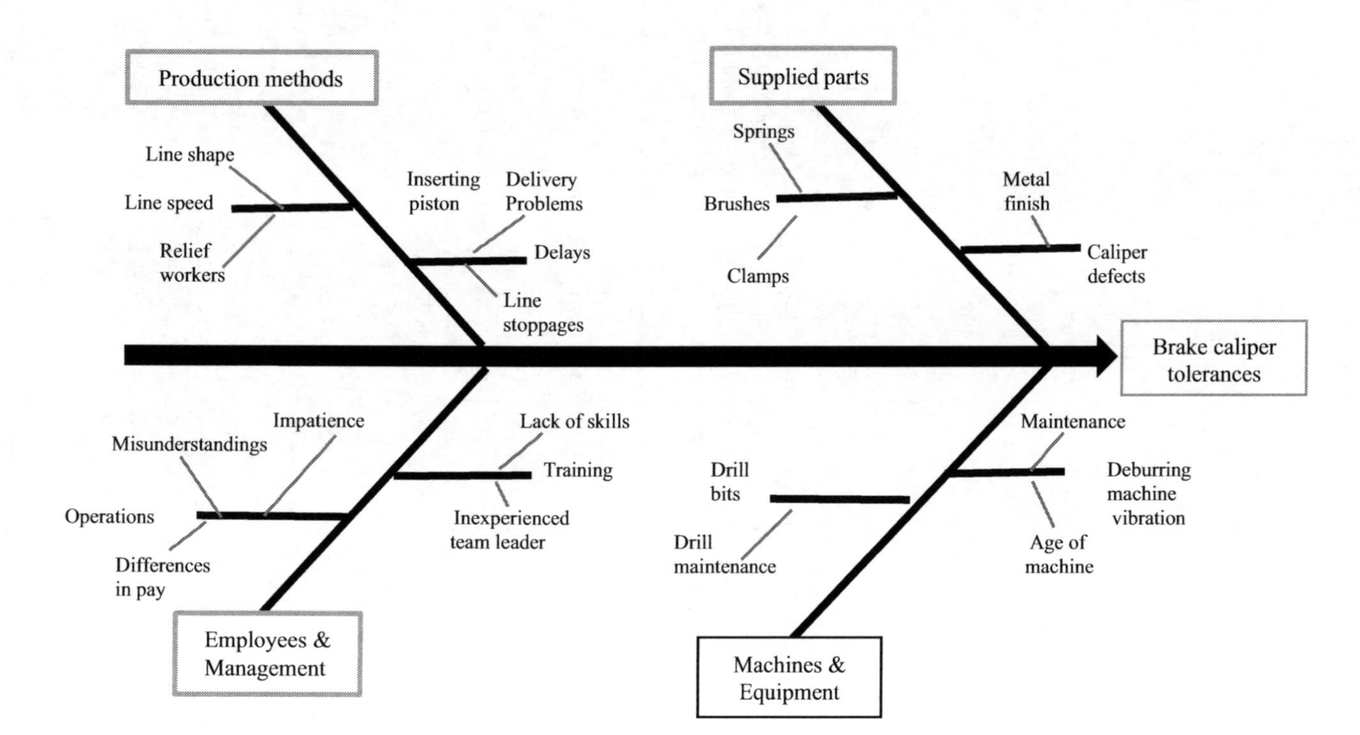

Figure 1.1 A Fishbone or Ishikawa Diagram for Causal Analysis

Table 1.2 Four Different Types of Teamwork
(Teams may vary from firm to firm, but these are general tendencies of each type of team)

Characteristics	*Managerial Teams*	*Quality Control Circles*	*Semi-autonomous teams*	*Network Teams*
1-Freedom	No	Maybe, limited to production	Yes	No
2-Voluntary membership	No	Yes	Yes	No
3-Wide Focus	No	No, focus on production	Yes	No
4-Self-selected team leader	No, manager	Yes, team selected	Yes	No, manager
5-Wide flexibility	Limited	Yes	Strong	Strong but episodic
6-Long length of time in existence	Yes	Yes	Moderate	Short

teams. Managerial teams have the least amount of freedom within the organization, and we see this as a more or less "functional team" in that workers more or less cooperate with each other but do not do serious analysis of their own behavior or future changes. Network teams are thrown together to solve a short-term problem, and are even less of a team in many ways than managerial teams. In this case, workers are plucked from their daily teams to serve on cross-functional teams to solve a particular problem. But for the most part, we focus on the problem-solving team on the shop or office floor. Later, we will discuss why teamwork is often the hardest element of lean production to maintain.

Coming from socio-technical theory to be discussed in Chapter 3, semi-autonomous teams have the most freedom, creativity, and self-direction and they are especially used in innovative firms creating new products like Google and Apple.

Quality control teams or circles come in the middle with a strong sense of teamwork, but a focus that is relatively restricted to the actual jobs that each team member is pursuing. Teamwork in QCCs is clearly focused on *kaizen* or continuous improvement of the product. Networked teams are more concerned with the relationship between suppliers and the main OEMand concerns how supply-chain management is concerned.

The Japanese contributed QCCs as an addition to quality control processes. It gives workers more control over their workplace and results in improvements due to the workers' unique and often tacit knowledge of the production process. What makes QCCs or teams distinctive and useful is that they specifically engage in statistical process control concerning their own work. Using an example of problems with brake calipers, they engage in problem-solving activity to locate the causes of problems in their work, often using the fishbone or Ishikawa charts (see Cole 1979 for tire and bumper examples).

Once they locate causes, they use bar charts to construct Pareto charts that list the most common causes of problems, which they then solve and chart the improvement against the previous situation (see A and B in Figure 1.2).

From this, they may plot the incidence of defects against particular causes using correlation coefficients and regression equations (see C in Figure 1.2). Further, they can plot the normal curves of defects against the target quality rates for each worker before and after an improvement (see Figure 1.3).

This type of activity and teamwork is not just functional teams with people cooperating together. Instead, it involves specific problem-solving techniques, which if carried out in a continuous fashion, make *kaizen* or continuous improvement a competitive advantage for the lean firm. So, one of our major points is that QCCs and semi-autonomous teams are "analytical teams" doing serious problem solving or *kaizen*, while managerial and some networked teams are more functional in nature, involving cooperation but only sporadic analysis.

(5) Loyalty to Employees and Suppliers is sometimes difficult to understand since much of Japan has dropped and the United States never fully embraced life-time employment. In Japan in the 1960s and even now, employees who have outlived their usefulness are often kept on. Ghosn especially reported on high level executives still being paid and having a staff, but having little or no authority or purpose within the company (Ghosn and Riès 2005). This is part of Japanese culture and not lean production as its principles have been more universalized.[10] Nonetheless, lean production has a strong sense of loyalty to employees in terms of selection and training. Toyota and Honda in Japan were often described as being overstaffed, but in many ways, these employees would be redirected toward new or prospective products and processes. In this sense, "lean" as simply cutting employees is rather foreign to Toyotism.

A further aspect of loyalty that is hard to understand is the use of temporary employees. However, temporary employees are used to protect the employment of permanent employees so that they do not have to be laid off or dismissed. In this sense, temporary employees are a loyalty buffer to protect permanent employees from the ups and downs of the market. This is rational for the company to protect these workers because the company puts forth considerable investment in their training for teamwork and different jobs within the company. Further investment is into the long hours and intensity with which they are expected to work.

For presentation and simplicity sake, we will use these five principles (well, knowing that Jostein Petersen, Jeffrey Liker, and Pascal Dennis present

A. Bar Charts: Operating Time Used in Finding Errors According to Location or Source

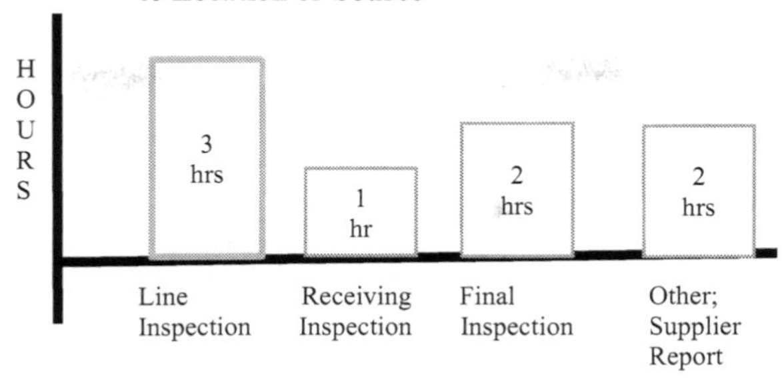

B. Pareto Chart: Before and After Improvement

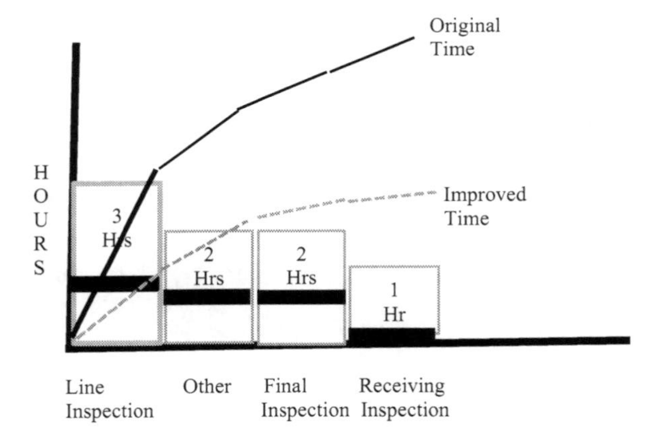

C. Plotting Errors and Using Correlations and Regression Equations

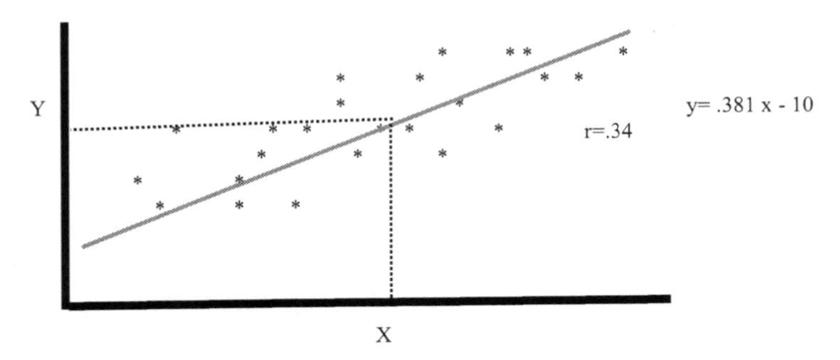

Y=Caliper tolerances, X=Deburring machine vibration

Figure 1.2 Charts and Graphs used in Statistical Process Control

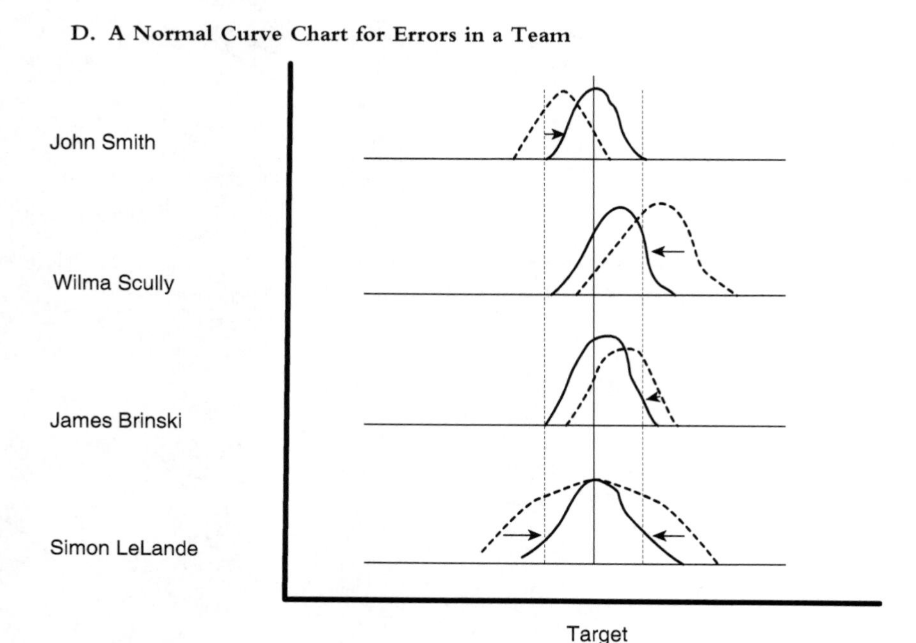

Figure 1.2 Continued

as many as 20 to 30) to examine five disciplines and the 11 corporations examined in this book. But we will also add to these principles as they might be used in different disciplines (e.g., for industrial engineering when the physical layout of the plant is important).

Data and Methods

This book was developed out of two previous books on lean production. The first one was an ethnography of a major Japanese auto-transplant—*Diversity at Kaizen Motors: Gender, Race, Age and Insecurity in a Japanese Auto Transplant* (2011). It was based on 115 interviews of past and present (at the time) workers.[11] The second book was a short theoretical piece called *Dominant Divisions of Labor: Models of Production that have Transformed the World of Work* (2014). This book of 71 pages of text outlined an expanded view of lean production in our first book. It was part of the "Pivot" series at Palgrave-MacMillan Press, which focused on new and emerging ideas. In it, we developed the ideas of Toyotism, Nikeification, and Waltonism, and since then, we have further developed these ideas in this book adding DQP to the mix. In the previous short book, we covered three corporations—Toyota, Nike, and Walmart—but in this expanded book, we go further, covering 11 corporations—Toyota, Honda, Ford, Nissan, Nike, McDonald's, Apple, Google, Walmart, Costco, and Amazon. So, the first book was an empirical ethnography, and the second book, while informed by the first one, was theoretical, and the present book builds on both.

The current book cannot be called an ethnography of 11 different companies because that would be well-nigh impossible. Instead, it is a synthesis of previous research interviews, theoretical proposals, and many new interviews, many of which were two-hour interviews, and others are a number of questions during and after seminars and presentations. While these latter field interviews may have pestered the employees and managers we talked to, they were surprisingly honest about many of the problems facing their plants. For example, at one seminar with visiting corporate representatives from many companies at Toyota, two associates were surprisingly critical of the company, much to the distress of the managers in charge. The managers questioned them immediately, and they said, "but you told us to be honest".

In terms of the number of interviews that we did, the total is about 261, with 170 being formal interviews and the rest being field interviews and participant observations in various public settings. Our interviews at the auto companies grew to 115 at the first auto company, and then 20 interviews were added at the other three auto companies (Honda, Nissan, and Ford). We also interviewed five workers at the Saturn Corporation in Tennessee; five at other GM plants in Ohio and two in China; three at Chrysler Corporation in Detroit, MI; three at Scania Motors in Sao Paulo, Brazil; four at Hyundai International Corporation in Alabama; three at Kia Motors Company in Georgia; two at Ducatti Motorcycles in Bologna in Italy; and six workers and managers at the main Toyota plant in Toyota City Japan. At VW, we talked to 15 workers in three locations: Chattanooga in the United States, Puebla in Mexico, and Wolfsburg in Germany. And we talked to five more workers at the brand new Audi plant in Puebla, Mexico. Two of the interviews were with members of the Supervisory Board (i.e., the German equivalent of the US Board of Directors) for the Wolfsburg plant, and a member of both the VW Global and Chattanooga Supervisory Boards. In sum, this work represents 261 formal and field interviews with additional observations inside numerous companies.

Our interviews at the merchandising companies—Walmart, Costco, and Amazon—were 10 workers per company, including one-on-one interviews, public seminars, tours, and participant observations as customers with both managers and workers. At Apple and Nike, we talked to five workers at each. Also, a number of authors have written detailed enthnographies on Walmart. Apple makes its employees sign confidentiality agreements and we found the Apple interviews to largely be optimistic and complementary of the company. We also had participant observations at the Apple store observing team meetings at opening hours and problem-solving on the floor. While they did tell us some information about hours and pay, which are largely public information, Apple employees never expressed anything critical about the company or unreleased products. We interviewed current and former workers and found that some could be quite critical—especially at Walmart and Amazon. One Amazon programmer said Amazon was the worst company he worked of the three high-tech companies he had experienced. We also interviewed a number of workers and managers at Sun Microsystems, Oracle, and Lexmark.

We only interviewed one person at Google, and we initially hesitated to include this corporation. However, we found that Google has been extraordinarily open in its software and also its inner workings. Unlike the very secretive Apple Corporation, Google allows their employees and managers to talk about how they work at Google, especially concerning the writing and de-bugging of computer code (e.g., CEO Eric Schmidt and VP James Whittaker). Because writing quality code at Google is so central to their business (and Apple's), we decided to include Google as somewhat of an "anti-Apple" corporation. Thus, our methods are synthetic, relying on our earlier ethnography, a theoretical book on the new division of labor, and additional formal and field interviews at a larger number of companies, which we camouflage to protect their identities. Only in this book are all the pieces of the puzzle to lean production put together in the larger global economic context.[12]

The focus of interviews changed over the 15 years of this project, but in every phase we were focused on lean production. At first, the interviews focused on diversity of employees on gender, race, age, and employment security. Secondly, with the NSF grant that we received, the focus changed to stress on the job in automobile companies, but quite frankly, workers most often did not claim any special stress except in the seminar mentioned above. Third, in expanding our focus to more auto and non-automotive companies, we focused field interviews more clearly on how much each company adhered to the tenets of lean production, especially teamwork.

In sum, our interviews are a combination of ethnographic and field interviews. They are with 75% workers or associates, and 25% with managers. It gave us a good picture of what was going on in each one of these companies. But again, only one plant had a complete ethnography.

To make comparisons between five disciplinary approaches and the 11 corporations, we use John Stuart Mill's methods of agreement and difference.[13] This is elaborated by Charles Ragin in *Redesigning Social Inquiry* (2008). The comparisons of disciplines and firms are based on binary "yes" or "no" responses, with a "maybe" being scored as half a point. These methods are most often used in comparing countries, but they can be equally useful in looking at disciplinary approaches and corporate use of lean production methods. Much of the data for our comparative methods comes from the interviews themselves, but additional data come from corporate financial reports and third-party quality assessments.

The Plan of the Book and its Next Chapters

We make our arguments that lean production is the new division of labor in two parts. In Part I, we discuss the various frameworks of lean disciplines in academia and the corporate world. Chapter 2 focuses on management and industrial engineering. While one might think that management is the chief discipline promoting lean production, it is actually industrial engineering that is at the forefront. Chapter 3 discuss the social science disciplines, though one cannot say that lean production is central to these areas. Sociology has focused mainly on automotive plants, and was especially interested in lean from 1970 to the early 1990s, but since then, interest has declined. Marxian theories of the labor process are highly critical of lean production. Socio-technical theory is a competing theory to lean production teams, and social psychologists are strongly involved in this area. This chapter ends with the diffusion of lean production, which affects a number of areas.

In Chapter 4, we look at two more specialized areas of broader disciplines that intensely interact with lean production. First, labor and management relations (sometimes called "industrial relations" and more recently "labor and employment relations") has approached this area concerning unions. But with more non-unionized Japanese transplants and other lean production facilities, labor and management relations have to encounter lean production with attention toward the protection of workers. Second, HR (HR) departments in management schools and corporate HR departments have witnessed a large expansion of the tasks and importance of HR as a field. They play a key role in dealing with lean production because they often have to conceptualize it for new employees and then are often called upon to represent or retrain workers when problems arise.

Chapter 5 looks at some competing theories to lean production in a wider comparative and globalized approach. We examine the hybrid of the German institutional system of works councils, codetermination, and advanced apprenticeship training as it encounters lean production. What is incorporated from Japan and elsewhere into the German system is called Diversified Quality Production (DQP). Second, we examine the French approach of "Productive Models" that has come out of the Regulation School and is akin to Marxist analysis. This is not a national system, but a classification of the productive and financial aspects of different corporate models. And third, we look at state-led capitalism in the Chinese case, which may seem far afield, but since so much manufacturing for Apple is done in China under state-led capitalism, its relationship to lean production needs to be examined.

In Part II, we examine the variations in the application of lean production in various corporations and industries. Chapter 6 is where we look at the differences between Toyota and Honda, which are the foremost practitioners of lean production. The "productive models" school sees these two corporations as representing two entirely different models. We discuss how they are similar and how they are occasionally different. But in either case, they are the primary models for lean production and Toyotism.

In Chapter 7, we examine what we call "semi-lean production", which constitutes corporations that implement parts but not all of the lean production model. Both Ford and Nissan in the auto industry fit this pattern, and McDonald's corporation in the fast food industry does even less so. We discuss why each one of these corporations does not fully implement lean production, especially QCCs and effective teamwork. The answers are more diverse than expected. In this chapter, we discuss "McDonaldization", which is a competing theory to lean production, and we argue that this is pure Fordism with only sporadic encounters with lean.

In Chapter 8, we look at some of the most successful and profitable corporations in the new economy. Nike Corporation embodies the first major company to be involved with off-shoring their manufacturing to foreign countries. And because they were the first corporation to completely off-shore, we call this model "Nikeification". The corporation is split between a high-paid home-base that does design and technology and a low-paid foreign-based labor force that actually makes the shoes. Apple Corporation is similar and we discuss their use of Foxconn as a supplier and actual producer of Apple manufactured products, while software is generally done in the United States. We also discuss Google, which is less a producer of manufactured product and more a creator of ingenious software. A critical part of this section is how lean production affects the writing and debugging of software, which is essential to quality. And we end this chapter by discussing the problems associated with off-shoring.

In Chapter 9, we shift gears away from the auto and electronics industries to corporations that, for the most part, do not make manufactured products. These are merchandising firms that buy products in order to resell them for a higher price (i.e., markup) with a more convenient distribution for customers. Walmart is the leader in pursuing an efficient supply chain management system, but it is also rather rough on its suppliers, pays low wages, and manipulates the hours of its employees. Nonetheless, it has had explosive growth. Costco is quite similar to Walmart in its supply chain and JIT inventory system, but it treats its employees quite differently—they offer higher pay and are often listed as one of the best employers to work for in the United States. Amazon is quite different because it grew from online selling of books to selling just about anything that can be delivered (they have not gotten into selling land and homes yet). The Amazon model is tough on employees in their vast warehouses, but surprisingly hands-off with its suppliers. These are also partial but quite distinct models of lean production.

Finally, in Part III, we synthesize in Chapter 10 the various models of lean production and discuss how they perform. This also involves the extension of lean production from the factory floor to lean production or Toyotism as an overall model of political economy. Just as Fordism came to be a wider theory of political economy with unionism and the Keynesian welfare state (both unintended by Henry Ford), lean production has similar tendencies with the current era of neo-liberalism. And in Chapter 11, the conclusion, we summarize our findings, discuss the consequences and limitations of lean production, and speculate on its future.

Notes

1. References to the division of labor are many in history, including Plato, Xenophon, Ibn Khaldan, William Petty, Karl Marx, Max Weber, and so on. Historical examples of the assembly line go back as far as the Venetian shipworks with the *Arsenale Nuovo*, and Eli Whitney's rifle assembly factory (1798 to 1825).
2. World systems theory covers many of these global issues, including war. Our focus will be on the organization of work rather than things like the battle for the core. However, we acknowledge how world systems theory is intimately connected to organizing work. We will only go as far as to discuss commodity chains and off-shoring (see Chapter 6, especially on Nikeification and Waltonism).
3. There are others who wrote on versions of scientific management at this time and even earlier: Charles Babbage, Henry Fayol (Fayolism), Frank and Lillian Gilbreth, Henry R. Towne and so forth. For brevity, we will only discuss Taylor.
4. The average worker could barely load 14 tons of pig iron for the pay of $1.15 a day. Schmidt moved as high as 52 tons for a sum of $1.70 a day.
5. Sloan is credited for saying that the business of GM is making money rather than automobiles. This has some resonance with the shareholder value approach to be discussed in the next chapter.
6. Robert Freeland (1996) disputes the efficiency of the M-Form organization based on its separation of tactical and strategic planning.
7. See David Moberg's dissertation—*Rattling the Golden Chains* (1978)—about the Lordstown GM plant in Ohio. This has taken on a new meaning in 2019 with the announcement that GM Lordstown would close.
8. The Bell Labs at Western Electric has its origins with the French government awarding Alexander Graham Bell 50,000 francs for the invention of the telephone. He developed Volta Laboratory in Washington, DC, which was then taken over in 1896 by the Western Electric Corporation that supplied AT&T with their telephone products. Under Frank B. Jewett as the president of Bell Labs, engineers and scientists developed further in a number of different locations from New York to Chicago. Many scientific discoveries came from these facilities, including the Hawthorne Studies at Western Electric written up by Elton Mayo (Gertner 2013).
9. "Milking a plant" is a strategy of not investing in maintenance or repairs, and then working a jerry-rigged plant for all the short-term profits one can squeeze out. Then, the plant is often sold as-is.
10. The Japanese employees who are no longer useful are asked to retire or leave, but if they do not, they are treated in two ways. One is *madogiwaoku,* which refers to the people who basically get out of the way and are "the people who sit by the windows" and simply put in their hours. The other is *oidashibeya* or the "chasing out room" which is putting people into small windowless rooms where they will hopefully leave, but if not, they surf the web or read semi-related work materials. Both techniques refer to the *obeya*, which is the large room in which Japanese project management takes place. The first process places you on the fringes of the *obeya*, and the second bans you from the room (Tabuchi 2013; Kopp 2014).

11. See Appendix 1 and 2 in Lepadatu and Janoski (2011) for interview details for the first study.
12. As a result, many of the terms and some of the phrases used in the earlier two books can be found in this book, but they are clearly expanded upon and developed in much more detail on five disciplines, 11 corporations, and many more interviews than our previous books.
13. We mention VW a number of times concerning interviews and diversified quality production in Germany; however, we are not counting VW as one of the corporations that we formally analyze in this book.

Part I

Theories of Lean Production

2 The Management and Industrial Engineering Approaches to Lean Production

The management and industrial engineering perspectives on lean production offer an overview of the improvements in terms of innovation, quality, JIT inventory, supplier relations, and how to implement lean models. The main development of lean production has been a mix of management and industrial engineering principles. In some ways, it is difficult to keep them apart, but we will point to the two disciplines in developing lean production in the next few sections of this chapter and then point to areas where they can differ considerably.

In the 1950s and 1960s, American industrial engineers worked in Japan to help their war-torn industries redevelop and adopt principles of quality based in statistical control. The most prominent were W. Edwards Deming and Joseph Juran, who published *The Quality Control Handbook* in 1951. Much of their work was influenced by a Japanese engineer working at Toyota and a Japanese management professor.

Box 2.1 W. Edwards Deming and Joseph Juran: American Advisors to Japan

William Edwards Deming (1900–1993) was an American engineer from Iowa and management consultant who was trained as an engineer at the Universities of Wyoming and Colorado; he received his PhD in mathematics and physics at Yale University. He worked at the Western Electric Hawthorne Works near Chicago, and later at the US Department of Agriculture. He helped develop many sampling techniques used by the US Department of the Census and the Bureau of Labor Statistics. Deming is best known for his work in Japan after WWII with the leaders of Japanese industry. General Douglas MacArthur used him as a census consultant to the Japanese government, but then he was asked to give a seminar on statistical process control to the Radio Corps. In August 1950 at the Hakone Convention

Center in Tokyo, Deming delivered a speech on what he called "Statistical Product Quality Administration". The Japanese Union of Scientists and Engineers (JUSE) then asked him to give them more training. Many in Japan say that Deming made important contributions to the post-war economic miracle in Japan. Japan then became the second-largest economy in the world based on their high quality products, which Deming helped bring about. The prestigious national quality control award in Japan is named "The Deming Prize", which the United States later copied with "The Baldrige Award". Deming subsequently advised a number of US companies on quality, including the FMC. In his book, *Out of the Crisis* (1986), Deming issued a challenge to American business to adopt quality control and many Japanese methods, which were elaborated with his 14 points.

Joseph Juran (1904–2008) was born in Romania and immigrated to the United States in 1912 when he was eight years old. He received a Bachelor of Science degree in electrical engineering from the University of Minnesota in 1924, and Juran then joined the Western Electric plant in Hawthorne. In 1925, industrial engineering started using control charts and statistical sampling at the Hawthorne Works. Juran joined the Inspection Statistical Department and was soon promoted to department chief in 1928 and division chief a year later. In 1937, he became chief industrial engineer at the Western Electric/AT&T headquarters in New York. During World War II, he worked in the Lend-Lease Administration and Foreign Economic Administration. In 1941, he used Vilfredo Pareto's "Pareto charts", listing the most frequent quality problems first and then the rest. Near the end of the war, he became a consultant and joined the industrial engineering department at New York University, where he taught courses in quality control. He published the first *Quality Control Handbook* in 1951, and many new editions and co-editors followed. The Japanese Union of Scientists and Engineers (JUSE) invited him to come to Japan in 1954 to help them develop higher quality products. He worked with executives from 10 manufacturing companies, and also lectured at four universities. In 1966, he learned about industrial engineering, which he then emphasized to Western manufacturers.

Taiichi Ohno at Toyota is largely responsible for a well-developed *Kanban* or JIT inventory system, which was later written up in 1978 in Japanese and 1988 in English. Kaoru Ishikawa developed the parallel idea of the QCCs that made the Japanese methods a step beyond the American approach of statistical quality control by giving these skills to worker teams. His *Guide to Quality Control* came out in 1968 in Japanese and was translated in 1985 (see also 1986). The Director of Manufacturing at GE, Armand Feigenbaum, wrote *Total Quality Control* in 1961, and Philip Crosby, a business leader with an unrelated education, wrote *Quality is Free* in 1977 (see also Crosby 1984, 1989, 1990 and 1996). But these methods were mainly contained within the manufacturing management domain.

Box 2.2 Japan's Originators of JIT Inventory and QCCs: Taiichi Ohno and Kaoru Ishikawa

Taiichi Ohno (1912–1990) was an industrial engineer and manager at Toyota Motors Corporation in Japan, who was actually born in China. He is often called the father of lean production of the Toyota Production System (TPS). He conceptualized the seven wastes, or *muda,* as part of this system. He wrote *Toyota Production System: Beyond Large-Scale Production*. He joined Toyoda Spinning in 1932 during the Great Depression, based on his relation to the leader Kiichiro Toyoda, the son of Toyota's founding father as an industrial engineer. He transferred to the Toyota Motor Company in 1943, where he worked as a shop-floor supervisor in the engine manufacturing shop of the plant and gradually rose to become an executive. Ohno's greatest contribution is in the development of the JIT inventory system, which reportedly came from observing the Piggly Wiggly Supermarket inventory system, especially of fresh fruits and vegetables. Ohno is widely recognized as the originator of lean production or TPS, and is in the Logistics Hall of Fame. However, he fell off the top executive track possibly because he talked and later wrote about TPS. He was sent to consult with Toyota Gosei, a supplier, but his fame spread throughout the West.

Kaoru Ishikawa (1915–1989) was an organizational theorist in engineering at the University of Tokyo. He worked as a technical officer in the Navy from 1939–1941, and then went to the Nissan Liquid Fuel Company. In 1947, he started as an associate professor of management at the University of Tokyo, and he later became the president of the Musashi Institute of Technology. He joined JUSE in 1949 in a quality control research group. He learned and then expanded on the management concepts of W. Edwards Deming and Joseph Juran in the Japanese system. He introduced QCCs in 1962 through JUSE. This led to training and concerns for quality throughout the organization (TQM) though he was not the originator of that term. Nippon Telephone and Telegraph (NTT) used quality circles and their successes led to their expanded use. Ishikawa would write five books on quality control with two on QCCs—*What is Total Quality Control: The Japanese Way* (1968 in Japanese, 1985 in English) and *How to Operate QC Circle Activities* (1980), though QC or Quality

Circle Headquarters is listed the author). He will probably be most remembered for developing the "fishbone diagram", or Ishikawa diagram, in 1968, which is commonly used to diagnose the root causes of production problems. Joseph Juran delivered his eulogy in 1989.

The impact of these two factors of cost and quality were intensified by the oil crises of 1973–1974 and 1980–1981. While Japanese automobiles were gaining market share in the United States in the 1970s, by the 1980s, the higher quality of their cars became painfully evident especially as American manufacturers began to cut costs, sometimes at the detriment of quality. By the 1980s, a flurry of influential books and articles came out extolling Japanese methods of production. From management, Robert Hall from the University of Indiana (1981, 1982, 1994, 1993) was "ahead of almost everyone else in North American in trying to understand Japanese systems" (Schonberger 1982, xi). William Ouchi wrote about meeting the Japanese challenge (1980). *Business Week*, the *Wall Street Journal,* and the rest of the business press was pressing management for answers to the comparative disadvantage for US automobiles.

In industrial engineering, there was an even larger response. Deming's book, *Out of the Crisis,* appeared in 1982, followed by new and expanded editions of Juran's *Quality Control Handbook* (reaching its fifth edition in 1999). Richard Schonberg's *Japanese Manufacturing Techniques* came out in 1982, followed by his *World Class Manufacturing* in 1986. It was clear that American automobile and electronics manufacturers had a comparative disadvantage in the marketplace and management and industrial engineering were pointing toward the solution to the problem with Japanese techniques. When the US government imposed quotas in 1981, the Japanese started to build plants in the United States, with Honda being the first in 1984, followed by Toyota in 1988. The success of the Japanese transplants in the United States laid waste to the idea that Japanese production methods were culturally specific and that they would not work in the United States.

The term "lean production" was first used in the industrial production and technology departments of MIT. John Krafcik coined the term in 1988 in the *Sloan Management Review*. But Japanese production methods and "lean production" became popularized with the multi-disciplinary team of James Womack (PhD in political science with an MA in transportation systems), Daniel Jones (BA in Economics), and Daniel Roos (PhD in civil engineering). Their evidence-based and best-selling book, *The Machine that Changed the World* (1990), not only explained the method but it stated that these principles could be applied to many other industries beyond automobile or electronics manufacturing (see also Womack and Jones 2004, 2005). Many firms tried to copy these Japanese methods and scholars tried to re-theorize their approaches to management to fit these new concepts (Keys and Miller 1984; Keys et al. 1994). After this, many books and articles have come out in management and especially industrial engineering journals, and the *Harvard Business Review* played a prominent role in popularizing lean production for a business audience.

In the 2000s, lean production spread into the services industries (Strang and Macy 2001; Graban 2009; Grunden 2008). This required some different approaches in that people providing services are different from machines producing products. For example, lean production in the health and medical industry has been taking off with considerable gusto. In some ways, it faces its own problems concerning the differences between professions from MD's to nurses, to lab technicians to nursing assistants and orderlies. We cannot follow all of these directions of diffusion, but note that lean has largely fulfilled some of the predictions of Womack, Jones, and Roos (1990) in the spread of its techniques; however, they are modified by culture and industry.

In the next sections, we first examine the two leading but differing approaches to lean production that fit the management and industrial engineering strains of thought. Second, we look at two major parts of lean in terms of networking in JIT systems and teams in QCCs. Third, we examine the management approach to introducing lean production in the normal flow of business and then in the development of new products. And finally, we examine the tension that exists between shareholder value theories in management that emphasize short-term results, and industrial engineering approaches that concentrate on process in the long-term.

The Differing Approaches Toward Lean

Lean production and Six Sigma are two similar but slightly different ways of approaching quality control. Lean production is more of a management term with firm control of processes and designations within an organization. Six Sigma is more professionally oriented, being outside the firm and ensconced in the ASQ and in the ISO. The ASQ and some other organizations provide professional certifications in Six Sigma, and the ISO accredits organizations that operate according to international standards (Murphy and Yates 2009).

Lean Production: Although the term is not generally used by the Japanese who invented it, "lean production" is a term that stuck throughout the world. Lee Krafcik (1988) first used the term lean production when he published an article on the topic while a graduate student at MIT (he later became the CEO of Hyundai-America and then Waymo) to describe the Japanese methods of production that had been successfully applied in manufacturing and service firms worldwide. But the term was popularized by Womack, Jones, and Roos in *The Machine that Changed the World* (1990) that was based on the IMPV at MIT and funded with more than a million dollars for a five-year study. This book made managers and the general public recognize the benefits of lean production. Christian Berggren (1992) introduced the term "Toyotism" to specifically target the system used at Toyota called the TPS. Lean production had also been described as a system of production that focuses on minimization of waste, continuous improvement (*kaizen*), and obsession with quality (Womack et al. 1990). Interestingly enough, the *kaizen* concept had become popular in Japan after the quality control training series offered by American management experts to help rebuild the Japanese industry after World War II. The Emperor of Japan awarded the *2nd Order Medal of the Sacred Treasure* to Edward Deming in 1960 for his efforts to spread the *kaizen* philosophy in Japan. Deming is celebrated as having had more impact on Japanese management than any other individual not of Japanese descent, and the Deming Prizes are awarded annually for achievement in quality in Japan (Petty Consulting 1991).

But two Japanese men went beyond Deming's original statistical control ideas and formed the core of the lean production breakthrough. At Toyota, Taiichi Ohno is credited with JIT inventory and breaking down the varieties of waste at work (*muda*). Although the idea was in Henry Ford's writings, Ohno found it not at a Ford plant but at the Piggly Wiggly grocery stores. He was then the first to apply it to manufacturing (Ohno 1988a, b). And Kaoru Ishikawa at the University of Tokyo discovered the fishbone diagram method and QCCs (Kondo 1994), which then laid the basis for strong teams in lean production. So, in formalizing lean production for manufacturing, teamwork and JIT inventory were largely indigenous Japanese ideas.

Continuous improvement is the mantra of the Toyotism. Even after the Toyota Motors company had become the world's largest automaker in the world in 2007, its President Katsuaki Watanabe was quoted saying that the Toyota DNA is to wreck your brain until you find a solution to problems:

> We've never tried to become number one in terms of volumes or revenues. Being the number one is about being the best in the world in terms of quality on a sustained basis. As long as we keep improving our quality, size will automatically follow.
>
> (Stewart and Raman 2007)

Lean production from a management and industrial engineering perspective can be summarized as follows:

(1) **Long-term philosophy:** Company decisions should be based on a long-term philosophy because managers want leaders and exceptional workers who thoroughly understand their work and company philosophy;
(2) **Standardization:** Tasks should be standardized on an assembly-line making them amenable to visual control using thoroughly tested technologies and processes;
(3) **JIT inventory:** JIT inventory systems create a production process with continuous flow, which will bring problems to the surface especially through a pull rather than push system;

(4) **Relations with Suppliers:** The company should create a trusted network of suppliers is integrated into the planning, design and production process including JIT;
(5) **Teamwork:** The firm should establish team cultures that produce quality the first time but stop the production process to fix problems using consensus to make slower but more implementable decisions;
(6) **Buffering Permanent Employees:** permanent employees are buffered by temporary employees who fill in for sick or injured team members and are let go during times of economic recession (Liker 2004; Liker and Ogden 2011; Lepadatu and Janoski 2011; Besser 1996).[1]
(7) **Design of the Workspace:** Use a pull system based on orders that operates on the basis of continuous flow (also affects JIT). Often this uses modular cells that concentrate common work rather than linear lines. Design also allows visual control involving management by walking around and the use of Andon boards and other stations in the plant. The design of the work processes helps just-in-time inventory, and teamwork. [2]

While the word *lean* connects to points 3 on JIT and 6 on buffering, the points about long-term philosophy, job rotation and flexibility, and quality control teams do not denote anything particularly connected to the word "lean". As a result, lean is not the best description of Toyotism processes. Perhaps "lean, long-term and loyal" (LLL) would be more appropriate, but since lean has such strong hold on the literature, we use it.

Lean production is different from Fordism in two additional ways. First, job rotation, cross-training, multiple skills and teamwork show the lean model as being antithetical to the rigid division of labor of Fordism (Jaffee 2001). W. Edwards Deming sees lean production as being totally different from Fordism since one of its main tenets is to "drive out fear", which allows the criticism of ineffectiveness without being afraid of losing your job. Second, lean production in the United States is accompanied by a weakening of labor unions and the development of labor flexibility. Whereas lean production was invented in Japan in the context of job security and life-time employment, in the United States the system seems to be sustained through the long-term employment of its core labor force of associates but also an expansion of precarious labor through temporary workers (Lepadatu and Janoski 2011, 2018; Bernier 2009).

The tenets of lean production can be presented in more detail. Jeffrey Liker, like Edwards Deming, presents 14 points of lean production (see Table 2.1).

First, base company decisions on long-term philosophy and not on short-term goals like price or getting the cheapest items. They use this long-term philosophy to develop thoroughly Toyotized leaders and exceptional workers who thoroughly understand the work, philosophy and teaching methods of their system (Liker 2004: items 1, 9 and 10). Long-term philosophy also applies to stock prices and investment, which avoid short-term

Table 2.1 Jeffrey Liker's 14 points of lean production (from *The Toyota Way* 2004)

(1) A long-term philosophy, even at the expense of short-term financial goals
(2) Continuous process flow to bring problems to the surface
(3) The Pull systems to avoid over production
(4) Level out the workload (*heijunka*)
(5) Culture of fixing fix problems to get quality right the first time
(6) Standardizing tasks for continuous improvement and employee empowerment
(7) Visual control so no problems are hidden
(8) Use thoroughly tested technology that serves your people and processes
(9) Develop leaders from the inside because they understand the work, live the philosophy, and teach it to others
(10) Develop exceptional teamwork (*chimuwaku*)
(11) Create and respect an extended network of partners and suppliers by integrating, helping and challenging them
(12) Manage by walking around to thoroughly understand the situation (*genchi genbutsu*)
(13) Make decisions slowly by consensus after considering all options. Then implement decisions rapidly (*Ringi* system)
(14) Become a learning organization through relentless reflection and continuous improvement (*kaizen*)

reporting pressures, and hence, dividends are low. Second, standardize tasks on an assembly-line and make them amenable to visual control using thoroughly tested technology for their people and processes. But at the same time reduce the number of job descriptions so that workers can rotate jobs and do many different tasks (Liker 2004: Items 6 and 8 with job rotation coming from elsewhere). Third, use JIT inventory to create a production process that has continuous flow, which will bring problems to the surface. This creates a pull system oriented toward customers to avoid over production and to level out the work (i.e., you do not produce a product until you are sure that you have an order). It also leads to flexibility and customization of products (Liker 2004: Items 2–4). Fourth, create a respected network of suppliers and partners and integrate them into the planning, design and production decision-making process including the JIT system (Liker 2004: Items 2–3, and 11). Fifth, make a team culture that produces quality the first time and stopping the production process to fix problems. Make decisions slowly by consensus, and implement them rapidly. Manage by walking around (MBWA) and going to see for yourself. (Besser 1996; Liker 2004: Items 5, 10, and 12–13). Sixth, buffer your permanent employees with temporary employees who can easily be let go when they are no longer needed. This requires a non-union or a company dominated union environment. While the word lean connects to points 3 on JIT and 6 on buffering, the points about long-term philosophy, job rotation and flexibility, and quality control teams do not denote anything particularly connected to lean. Again, we reiterate that "lean" is not the best description of the Toyotism process, but since it has such currency, we use it.

In terms of problems, lean production exerts a great deal of pressure and stress on workers, especially with mandatory overtime and exacting requirements concerning quality and cycle or *takt* times (i.e., the seconds that it takes to do a repetitive task). The work, despite job rotation, can be quite repetitive and carpal tunnel injuries are common, along with occasional back injuries and accidents. But while there have been some safety issues at Toyota (e.g., a worker died in 2011 and the brake problems in 2009), we believe that George Ritzer is dead wrong with his McDonaldization thesis in saying that lean production has declined in any way. Nonetheless, it is clear that lean methods can lead to higher stress.

On the positive side, workers in Japanese transplants appreciate the teamwork approach, the equalization of perks—combined cafeterias, no privilege parking spaces, and a low ratio of worker to management pay—and the job security (for permanent but not temporary workers) that this type of lean production offers to them. The recent book by Toyota worker Tim Turner and his fellow employees (2012) is a rather surprising account of the pride of Toyota workers around the country telling their own stories to the world.[3] The strongest down-side appears to come from workers who suffer injuries, who are often pressured to get back to work, and the temporary workers who do not have the promised job security that other permanent workers prize so much. Also, at one company presentation with a worker panel, a number of workers criticized the working hours. But given these stresses and conflicts, sociologists at this point need to recognize there are both positive and negative aspects of this new form of the division of labor.[4]

But all lean production is not Toyotism. In Schonberger's study of 1,500 firms, JIT processes can vary considerably, and Janoski and Lepadatu (2009) find that there is a wide variety of lean production in automobile plants. Not all companies use teams, and teamwork is usually the first item of lean production that drops out of the model. In particular, Nissan and Ford have many of the aspects of lean production outlined above, but they do not use particularly strong teams. At this point, lean production starts to look like flexible accumulation, but we cannot see it as neo-Taylorism because of the other elements of the system (see footnote below). Further, when looking at Nike, Apple, and the many computer and phone manufacturers, the off-shore production facilities in China and elsewhere do not use much in the way of teams. They use sophisticated JIT inventory procedures, standardization, and continuous flow, but not teams. On the other hand, the creative design teams of Silicon Valley at Microsoft, Intel, and Apple use cross-functional teams that are stronger teamwork programs than Toyotism. These teams are closer to the "semi-autonomous teams" of socio-technical theory, but they also exist within a lean production environment. One thing to keep in mind is that the implementation of industrial models of the division of labor is always attenuated by culture. The implementation of Taylorism and Fordism only partially took hold in Germany (Nolan 1994), Japan, and even the Soviet Union (i.e., Stakhanovitism based on heroes of production).[5] Each country has its twists on each production method, even though they praised the methods as a way to increase productivity. So, the same sorts of variances should be expected with lean production.

In a similar way, there is also a huge variation in how lean production has been adopted throughout the globe (Kochan et al. 1997b). Whereas higher involvement work practices are possible in some countries, others have strong union cultures that protect the workers from the lean pressure. Mediterranean lean production, for instance, is more hierarchical, granting team leaders a stronger role in the system, whereas the German lean model has to push for a stronger team culture (Turner 1993). Aoki (2008) shows that even for Japanese transplants in China, transferring lean production methods, can be a challenge. What goes on in Foxconn factories that make Apple products without direct Japanese management is much more Fordist than anything else? But overall, we are living in a mixed Fordist and lean production era, where firms do not choose anymore between the mass production or the lean production model, but rather implement parts of each system (Kochan et al. 1997b, c). From the lean production consultant perspective, this would be the fault of a company not being able to achieve a "true lean" production system.

An overall conclusion about the lean production system requires a combination of critical and praiseworthy points. On the one hand, criticism comes from lean production exerting a great deal of pressure and stress on workers, especially with mandatory overtime and exacting requirements concerning quality and cycle times. The work, despite job rotation, can be quite repetitive and carpal tunnel injuries are common, along with occasional back injuries and accidents. On the other hand, workers in Japanese transplants appreciate the teamwork approach, the equalization of perks, and the job security that this type of lean production offers to them (but this does not apply to temporary workers). In our account of diversity in a lean production at a Japanese auto transplant, there is stress under lean production, but the long hours and job intensification create a team culture that is not present in the Fordist environment (Lepadatu and Janoski 2011). The strongest downside appears to come from injured workers who are often pressured to get back to work, and the temporary workers who have no job security. But given these stresses and conflicts, scholars need to recognize both the positive and negative aspects of this new and dominant form of the division of labor.

Six Sigma and ISO

Six Sigma is strongly related to lean production and comes from the Japanese emphasis on quality; however, it takes a more hierarchical form based on extensive expertise. Six sigma means that the desired level of quality should be less than six standard deviations from the mean, which is a quite stringent quality control measure. This means that 99.99% of all production instances are to be free of defects. Six Sigma was introduced by engineer Bill

Smith while working at Motorola in 1980, with Motorola registering it with trademarks in 1991 and 1993. The CEO of GE, Jack Welch, made it a central principle of his approach to production in 1995. Six Sigma strategies improve the quality by processes of identifying and removing the causes of defects and minimizing the statistical variation of industrial engineering production. Six Sigma uses statistical analyses of empirical data using industrial engineering concepts. This creates a special infrastructure of certified experts in these methods. Their Six Sigma projects were carried out in steps with specific value targets—reduce process cycle time, reduce costs, increase customer satisfaction, and increase profits.

In Six Sigma, the precision of a manufacturing process can be described by a *sigma* rating indicating its yield or the percentage of defect-free products it creates. Six Sigma consists of three main points. First, continuous efforts to achieve stable and predictable process results (e.g., by reducing statistical variation) are of vital importance to business success. Second, manufacturing and business processes have characteristics that can be defined, measured, analyzed, improved, and controlled. Third, achieving sustained quality improvement requires commitment from the entire organization, particularly from top-level management. Three features of Six Sigma make it different from other quality control programs:

(1) A clear focus on programs that emphasize measurable financial returns for individual projects.
(2) A strong emphasis on management leadership (perhaps charismatic) and support.
(3) A clear commitment to making decisions on concrete data and statistical methods, rather than assumptions or trial-and-error approaches.

Six Sigma is used in statistical quality control, which evaluates quality. Organizations decide on the specific sigma level of error that is appropriate for their organization and industry. Management usually prioritizes areas of improvement. By the end of the 1990s, about two-thirds of the Fortune 500 corporations began Six Sigma initiatives to improve quality. In 2011, the ISO in Geneva defined the Six Sigma process in ISO 13053 (ISO 2011; Murphy and Yates 2009; Martin 2009) and other "standards" are often created by companies and universities that have certification programs for Six Sigma.

The Six Sigma processes follow two methodologies largely developed by W. Edwards Deming's PDCA cycle: (1) improving existing production processes, and (2) creating new products or production processes.[6] The first focuses on improving a production process that already exists and has five phases:

(1) Define the system, the customer requirements, and project goals
(2) Measure key aspects of the current process and collect data and calculate probabilities of success

(3) Analyze the data to identify cause-and-effect relationships; after that, they ensure that all factors have been considered. Seek out the root cause of defects
(4) Improve the current process based upon analysis techniques and experiments
(5) Control the process to detect and correct variation; they use statistical control systems (e.g., statistical process control), production boards, and visual workplaces to monitor the process

This is probably the most common application of Six Sigma.

The second process is based on product development or new production design, which are critical to both the quality of the product and its success in the market, especially in getting ahead of the competition.

(1) Design goals consistent with customer demands and the enterprise strategy
(2) Measure the characteristics that are critical to quality, product capabilities, production process capability, and especially risk
(3) Analyze the designs to improve products or come up with alternative designs
(4) Select an alternative that is best suited based on the previous step
(5) Verify the design, set up pilot studies, implement at trial production process, and if successful, pass it on the production

The main difference from the already-established procedure is that Steps 4 and 5 are different because there is no existing process with which to compare.[7]

In implementing Six Sigma, the emphasis is on professionalizing quality control and it has been more connected to the ASQ. Before Six Sigma, quality management in practice was only on the shop floor with a separate quality department. Formal Six Sigma programs tend to adopt a kind of marital arts ranking system to create hierarchy levels.

(1) Green Belts are the employees who take up Six Sigma implementation along with their other job responsibilities, operating under the guidance of Black Belts[8]
(2) Black Belts work under Master Black Belts in applying Six Sigma methodology to quality control projects; they focus on executing Six Sigma projects and leading special tasks
(3) Master Black Belts are in-house coaches who assist champions, black belts, and green belts on statistical tasks and project implementation
(4) Champions are responsible for implementing projects throughout the organization; the top executives use them as mentors to Black Belts
(5) Executive Leadership includes CEOs and top management; they set up the vision for Six Sigma projects and empower champions, master black belts, and black belts to develop new ideas for new improvements and reduce resistance

Certification programs exist to verify competence in quality control. Skill levels are frequently indicated by green to black belts. Many organizations in the 1990s started offering Six Sigma certifications to their employees. While there is no standard certification body, the ASQ and others offer courses (e.g., according to the 2018 website, ASQ Lean Sigma Green Belt training is $4,499 for classroom training). The ASQ requires candidates for Black Belt to pass a written exam and demonstrate that they have completed two quality projects or one project plus three years of quality experience.

Lean management and Six Sigma are very similar in terms of methodology and implementation, and Japanese production processes have influenced approaches. However, they are different on three counts. First, lean management is more focused on eliminating waste, ensuring efficiency, and developing effective teams on the production floor. Six Sigma is much more oriented toward statistics in reducing defects and variability. Second, the main difference is that lean production focuses more on quality control techniques that are diffused throughout the organization through teamwork processes. Six sigma may have teams, but whether they reach the shop floor level is often in question. More recently, "Lean Six Sigma" has combined many lean production and Six Sigma ideas, and this represents lean manufacturing with flow and waste issues, and Six Sigma, with variation and design approaches. Companies such as GE, Ford, Accenture, Verizon, and IBM have used Lean Six Sigma to focus transformation efforts on efficiency and growth. And third, lean production is much more organizationally controlled by corporations and firms, while Six Sigma is a professionally oriented system with some of its organization outside the firm (i.e., ISO and ASQ). From our perspective, lean Six Sigma is different from lean production because of its much weaker teamwork approach on the shop floor.[9]

Management attempted to make the leap to lean through two main mechanisms—JIT inventory (or supply chain management) and teamwork (McKinsey 2014). These two somewhat specialized areas are critical to lean production and will be reviewed below.

Management and Industrial Engineering on JIT

First, JIT inventory is largely the province of industrial engineering, though management has something to say on it occasionally. Taiichi Ohno (1988a, b), while at Toyota, developed JIT and focused almost exclusively on JIT inventory in the 1960s and 1970s. Richard Schonberger (2007, 2011, 2012, 2017) describes the five stages that Japanese production management (JPM) through JIT inventory has gone through.

Stage 1 of initial quality management (1950 to 1970, and reaching the United States in the late 1970s): The idea of *Kanban* or the card system of inventory control took place mainly within Toyota, with the card system placed on parts giving feedback on the need for additional parts. There was a difference between the single and dual *Kanban*, with dual having one card going to suppliers and the other card to the OEM manufacturer. These ideas reached the United States with Sugimori et al. (1977), as Ohno's book did not reach the United States until 1981. The process was created and improved upon at Ohno's direction (Schonberger 2007:406–8). The tools used also included statistical quality control methods with stopping the line, which slowed inventory usage.

Stage 2 of more developed JIT (1980 to 1991): In the 1980s, JIT became a well-known topic with Ohno's (1988a) book and Shingo's (1985, 1988, 2017) books being translated. The Repetitive Manufacturing Group (RMG), sponsored by the American auto industry, held conferences in 1979 and started the process of beginning JIT in American auto companies. A major effect was a meeting at the Kawasaki Plant in Nebraska. But the development of JIT occurred within the *Keiretsu* structures of networked firms in Japan; within the United States, the process of networked firms had to be built up and changed from market-based and weaker relationships between suppliers and OEM firms. [10]

Stage 3 of Western modifications (1991 to 2000): In 1990, JIT became well-known in the United States with *The Machine that Changed the World*. By this time, the major media networks were praising the Japanese production methods and severely questioning the American processes of Fordist and low-quality production. While implementing standard JIT procedures, three major changes occurred. First, bar coding entered the supply chains of American and Japanese production processes. This became a very fast way of keeping track of inventory. These bar codes were quite mono-functional at first, but they became more and more useful and complicated as time went on. This also impacted the cost accounting field, as instantaneous knowledge of parts usage revolutionized the monetary assessment of manufacturing processes (Cooper and Kaplan 1988). As in the past, the first implementation of these procedures came in grocery supermarkets in 1974; by 1984, 33% of firms used bar codes, and in 1994, QR codes were created for Toyota and then used by others.[11] By 2004, 80–90% of firms used bar codes. Second, DFMA was developed by Geoffrey Boothroyd and Peter Dewhurst (1990). This standardized designs of component parts across suppliers, and tremendously reduced the number of parts. Third, at the same time, computer-assisted design and computer-assisted manufacture (CAD/CAM) pervaded the design process, which did not take off until personal computers became widespread in the 1990s. All of this presents JIT with a vast array of new technologies.

Stage 4: Japan's "lost decade" and restructuring: This also occurs in the 1990s, but it is about the decline and stagnation of Japanese industry with forced layoffs in Japan. Sony slashed 17,000 jobs, Mitsubishi cut 14,600 jobs, and Nissan 21,000. (Schonberger 2007:413). Ford bought into Mazda, Renault into Nissan, GM had its NUMMI joint venture with Toyota, and there were other joint ventures and some mergers. US industry accepted outsourcing, but beyond this, off-shoring. Japan was reluctant to off-shore production to China and other low-cost countries, but in the 1990s, one further element was the development of modular design as a part of design for manufacturing and assembly (DFMA). This shrunk assembly line length, created U-shaped modules, and made plant sizes smaller. Japan was slow to adopt these processes, so their plant size began to be much larger than US plants (MacDuffie 2013; Sturgeon 2002; Hyer and Brown 1999; Hyer and Wemmerlöv 2001). All in all, this stage reduced the luster of the Japanese aura.

Stage 5: The state of the art of JIT. JIT became a combination of Western and some Japanese best practices with a kernel of Japanese JIT methods as its core. Measurement of JIT practices now focused on inventory turnover (the yearly value of cost of goods sold divided by the average value of inventory). In other words, value of inventory no longer in the firm due to its being sold, to the value of inventory currently in the plant. This became a lean inventory measure which is expected to go up as the cost of sold goods swamp current inventory on hand (or the reverse with inventories being swamped by sales and the ratio goes down).

Meanwhile, the network analysis of suppliers developed with a number of major propositions (Hearnshaw and Wilson 2013).

(1) The most efficient supply chain systems have shortest paths between companies
(2) The most useful connections between companies in a supply chain have hubs with high centrality
(3) Efficient supply chain systems have high clustering with dense networks
(4) The leanest supply chains tend to increase their lean characteristics
(5) However, the connectivity of firms in a supply chain reaches a network limit
(6) Efficient supply chains exhibit information that is shared through overlapping boundaries
(7) The distribution of important relations in a network follows a power law[12]

This sophisticated type of network analysis was unknown in the earlier stages of supply chain analysis, and to some extent, reflects a mature phase of JIT with more academic analysis in venues such as *The Journal of Supply Chain Management*, which began in 1965 and grew with JIT (Schonberger 1986, 2007, 2008, 2011; Hearnshaw and Wilson 2013, Taylor and Taylor 2008, Bicheno and Holweg 2016; Taiichi Ohno 1988a; Rich et al. 2006; Monden 2011).

The implementation of JIT inventory has been a good thing as firms have become more profitable and quick on their feet. However, JIT does have weaknesses. It is vulnerable to strikes, slowdowns, and major disasters. The large flooding in Thailand put great pressure on Japanese and US automobile manufacturers as parts became unavailable. The threat of trade difficulties with Brexit in the UK and tariff wars with China also create havoc in JIT systems. In more normal circumstances, delays can cause workers to be rifling through semi-trailers for critical parts to get the line going again. Thus, JIT systems need a relatively placid environment in order to work well.

Management and Industrial Engineering on Teamwork

Second, teamwork is a major component of lean production. QCCs were the first instance of Japanese production methods. In the 1960s, a broad movement developed in Japan around worker participation in management and making jobs of higher skill levels. Much of this developed in a social movement orientation with conferences where line-workers who had never spoken in public before would present on their experiences. In the United States, *The Harvard Business Review* and Robert E. Cole spread information on QCCs. Thomas Ling-Ping Tang with colleagues produced the most rigorous studies of QCCs (Tang et al. 1987, 1989, 1991, 1993, 1996; Tang and Butler 1997).

The contributions to teamwork came mainly from social psychologists in psychology departments who often did not refer to lean production at all. They have created a vast literature on teamwork. J. Richard Hackman produced a large amount of research on teams (2002, 2003; Hackman and Wageman 2007, 2005; Hackman and Edmondson 2008; Hackman and Katz 2010). Ruth Wageman at Harvard has also produced quite a bit on teamwork (Wageman 2001; Wageman et al. 2008, and Wageman et al. 2005). There are many others, especially with the Society for Industrial and Occupational Psychology (Lawler and Mohrman 1985). These are valuable contributions but they rarely refer to lean production or QCCs (Tang is the major exception). Also, the socio-technical literature discussed in the next chapter focuses on teamwork but this targets self-managing teams rather than QCCs.

Social psychologists also teach "organizational behavior" in management departments and they are the dominant producers of books and articles on teamwork. Most of the textbooks for college classes come from this crossover of social psychology and management (West 2012; Franz 2012; Dyer et al. 2015; Lafasto and Carl Larson 2001). Some of these studies of teamwork come under the title of group dynamics (Forsyth 2019, Levi 2017). Again, they rarely mention lean production, but they sometimes refer to Japanese concepts like *ringi* decision-making—multilevel and repeated comment in decision processes—that occurs within *nemawashi*—consensus building (Levi 2017; Kopp 2010, 2012).

In sum, social psychology and organizational behavior are the center of studies on teamwork (McGrath and Tschan 2004), and only a few major researchers directly connect teamwork with lean production. But then again, perhaps a third of these social psychologists are in management departments while the rest are in psychology or related social science departments. Teamwork is mentioned in industrial engineering research; however, engineers do not do a great deal of research on teamwork.

The Management Approach to Organizational Change and Strategic Management

The management literature has less specific material on lean production than industrial engineering; however, it does have an extensive literature on organizational change. This is important because implementing lean production usually requires two important aspects of organizational change: (1) to change work practices, leadership, and conceptions of production and quality, and (2) to maintain a comparative advantage by developing new and better products.

There is a plethora of organizational change literature in management that can be divvied into deep and surface change (Stephan et al. 2016:1256–7). For lean production purposes, deep change is quite essential and it can be seen as consisting of four parts. First, deep-level positive change consists of intrinsic or internal motivation that leads to effective capabilities and empowering opportunity structures and social structures. These changes evolve slowly but have pervasive and deep effects. By contrast, surface-level change comes from extrinsic or external motivation that has a weaker impact on capabilities that restructures decision-making environments that produce temporary improvements in a fast but fleeting way. They tend to be top-down changes imposed by managers, the board, stockholders, and even the government. Second, specific changes in work practices require various sorts of change mechanisms. These mechanisms focus on motivation, capability, and opportunity and may go through stages from ramping up to continual reinforcement. Third, there is the larger cooperative approach to organizational innovation. Prius, Taurus, and the soul of the new machine (Kidder 1981). These works provide a detailed look at how innovations were created with great enthusiasm and excitement. However, Schonberger's frustration-driven process of improvement (2018) discusses the dissatisfaction of current processes as being at the root of change.[13]

In terms of lean production, all three types of changes are important. While they might not refer to surface changes, lean production often relies on small and incremental changes that result is small improvements in products, processes, or eliminating waste. These are important. However, for an organization trying to implement lean production for the first time or trying to revive such an attempt, it must be recognized that lean production as a whole is not a small change. It involves the whole organization through TQM and often times, the implementation of teamwork and *kaizen* on a day-to-day basis is quite a challenge. The third type of change involving a wholly new product or process is a major change, and it is again difficult to get the whole. While it may seem that an elite group should do this, in lean production, this process again requires group participation (Morgan and Liker 2008; Schuh et al. 2017; Rossi et al. 2017; Shook 2008).

Lean production is a big, challenging change for most organizations. The management literature on change presents a number of resources to implement these changes, but they are more general. Julie Battilana and Tiziana Caseiaro (2012, 2013) develop a contingency theory of organizational change based on social networks. First, they find that structural closure in an organizational network makes change difficult, and that a more open organization with less structural closure allows "structural holes" that link up previously unconnected employees in an organization to start and adopt changes. Second, they argue that strong ties between employees can be used for a change agent to influence a fellow employee who is a fence-sitter to cooperate in change. However, if the fellow employee is a resistor, converting them to new projects will only work if the change is minimal. And of course, these minimal changes are easy to reverse. Eirin Lodgaard and colleagues (2016) focus on how top management, middle managers, and workers view barriers to implementing lean production. Top managers pointed to lean tools and practices, but workers focused on management challenges, while the middle managers cited a wide range of barriers (2016:598–9). Early auto implementation pointed to first-line supervisors who had been used to a command style and had great difficulty becoming a coach. Each one of these approaches to change are helpful in understanding how organizations adopt and maintain an active lean production orientation in their organization.

In general, while their work does not mention lean production, it does describe a more open work place with considerable participation between allies, opportunists, and resistors in a contingent environment. It clearly does not advocate organizational networks with closure where one unit "throws their work over the wall" to the next unit (i.e., the pejorative description of the non-lean organization in product development). As major examples of organizational change through innovation, consider the Toyota Prius project. (Toyota Global Newsroom, n.d. a, b, c.; Morgan and Liker 2006; Sobek 1997) and the Taurus project at Ford (Taub 1991). These changes required major efforts by the whole organization to create a new vision of a product and its production. Other examples are often used from Silicon Valley such as *The Soul of a New Machine* and the "internet of things" with the enthusiasm of a social movement in creating a new computer or inserting them into other machines (Fleming et al. 2012; Samaniego and Deters 2016; Kirk 2015).[14]

Related to contingency change is "strategic management". This is a central area in management that looks at the use of an organization's resources and abilities in terms of finance, production, marketing, and HR to achieve its existing and new goals and objectives in a context where competitors will change their behaviors based on your own (see Table 2.2).

More specifically, strategic management involves setting objectives for competitive advantage inside the organization and outside of it in the competitive environment. It involves the evaluation of strategies given that competing organizations may change in a dynamic organizational environment. Strengths, weaknesses, opportunities, and threats (SWOT) analysis is a typical approach of looking at the organizations. SVOR is similar with strengths (S),

Table 2.2 The Place of Strategic Management in the Discipline of Management

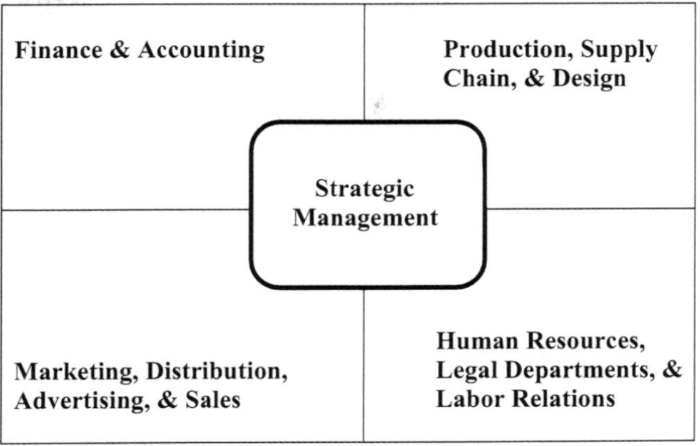

vulnerabilities (V), opportunities (O), and risks (R). Harvard Business school professor Michael Porter developed the five forces as a similar approach: (1) threat of new entrants; (2) existing competition; (3) threat of new goods or services; (4) bargaining power of suppliers; and (5) bargaining power of buyers or consumers (2008, 1980). Strategic management is much larger than lean production in that it may include product strategies in a global market, mergers, or spinoffs that affect the size and opportunities for a company (more agile and focused smaller company versus a larger company with more interconnected resources and skill sets), downsizing and outsourcing, and so forth. Strategic management is often associated with management fads such as core competencies, MBO or OKR; and so forth (Doerr 2018). It is usually associated with CEOs and top level management.

In a sense, lean production should be a clear component of strategic management, but at this moment, one cannot say that this is the case.

Shareholder Value Theory and Black-Boxing Production

In the 1960s, a broad movement developed around worker participation in management and making jobs more interesting to workers. Much of this was connected to socio-technical theory discussed above. This came to be called stakeholder theory in the sense that many people ranging from workers, customers, and the community were stakeholders in the actions that corporations may take (Freeman 1984). As part of a neo-liberal conservative reaction to these ideas, shareholder value strategies emerged out of the crisis of "Financial Managerialism" of the 1960s (Davis and Thompson 1994; Fligstein 2001:84–86, 147–69; Stout 2012). Stakeholder theory became "the enemy" (i.e., a literal "stake in the heart of capitalism"). Intense competition from Japan and low profits in the 1970s led to a distinctive strategy that instructed top managers to pay strong and almost exclusive attention to the people and groups who own their stock (shareholders) and not to multitudinous others (stakeholders). One particularly influential group consisted of institutional shareholders. For instance, CALPERS accumulated such gigantic assets that it could demand the removal of CEO Robert Stempel from office at GM in 1990.

The Development of Shareholder Value and Financialization

Especially coming out of Jack Welch's ideas espoused at GE in the 1980s, shareholder value theory narrowed the constituency of corporations from "stakeholders" to "shareholders". As such, "shareholder value theory" is composed of six parts. First, CEOs and other corporate managers should concentrate on increasing the value of their stock and also in paying dividends to shareholders because shareholders are their most important, and indeed, their "only" legitimate constituency. In other words, owners come first. The other principles relate on how to achieve this goal. Second, it recommends that corporations sell off overvalued and otherwise diversified assets to boost cash reserves and pay dividends. Third, corporations should "assume debt to keep firms disciplined", which keeps the stock's prices and values from being diluted by issuing more shares. While not an original tenet of this model, off-shoring soon became part of removing layers of employees by sending work to low cost production facilities in China and elsewhere. Fourth, corporations should buy up or merge with competitors, which increases market share and creates redundant employees that leads to the next point. Fifth, corporations should "remove layers of management to save money" and layoff or outsource lower level employees to minimize exposure to costly employee benefits and other responsibilities toward workers. And sixth, corporations should offer very high wages and bonuses to CEOs and top executives through stock options in order to align managerial incentives with the value of the company's stock (Fligstein 2001:85; Davis et al. 1994).[15]

Shareholder value approaches mainly involve higher level management strategies and do not directly concern production processes. There is no actual model of quality control, production scheduling, or supplier organization. However, we include this Wall Street-focused theory as being responsible for intensifying "short-term management thinking" focused on stock prices, and thus being antithetical to the "long-term thinking" that characterizes Toyotism. Thus, shareholder value theory's major points do little to specify a production process. Within its framework, production can be severely strained by: (a) the infighting of disruptive mergers, (b) being divested from a parent company, (c) being milked by managers not approving needed maintenance on plant and equipment, and (d) generally ignoring long-term investment in plant, equipment, and employee loyalty (Dobbin and Jong 2010). In some ways, shareholder value is an intense backlash to "stakeholder theories" that tried to show how a much wider group of people including workers and the community could have a positive influence on the corporation. In some ways, it is more of a "division of profits" rather than a "division of labor".

General Electric as a Shareholder Value Corporation

GE has been the poster-corporation for shareholder value. Under former CEO Jack Welch, GE downsized, outsourced and off-shored as much of its production as possible. GE as a conglomerate that produces a wide variety of products from light bulbs, to dishwashers, to air craft engines has a far-flung network of products, suppliers, and Research and Development units. Like its rival, Westinghouse, it faced hard times in the 1970s and 1980s, and CEO Jack Welch brought the firm back into profitability by pursuing a tough strategy of shareholder value principles. As such, Welch cut, merged, and off-shored production until GE was the wunderkind of the 1990s and the new century.

GE began in the late 1890s and often competed with Westinghouse as an appliance and electrical motor manufacturer. It currently manufactures home appliances, consumer electronics, wind turbines, aviation engines, electric motors, lighting, locomotives, software, and certain kinds of weapons. It also produces basic products in terms of water, electricity (nuclear, wind, solar, and gas), wind turbines, gas and oil, and supplies financial, entertainment, and health services. Currently, about half its revenues come from financial services with manufacturing providing the rest. GE's history is complicated because the company has transferred as many operations as it currently has in numerous sales, merger and acquisition activities.

In 1981, Jack Welch became the youngest CEO at GE, and he cleaned house with an emphasis on shareholder value. He was intent on reviving the company by making its many divisions first or second in the industry or by dropping them. His nickname of "neutron Jack" came from cutting more than 100,000 employees or about 25% of the company by 1985 (some by selling businesses and others by straight cuts). Welch took three steps to make GE more profitable: 1) Welch changed the focus of overall operations to financial services, which he viewed as being more profitable, and in the process he cut unprofitable divisions and companies; 2) Welch adopted Six Sigma quality programs in manufacturing in 1995; and 3) in the 1990s, he off-shored and outsourced manufacturing whenever he could.[16] In 1980, before he was CEO, GE had revenues of $27 billion and a market value of $14 billion, and the year before he left GE, those revenues had increased by a factor of five to 130 billion dollars and GE's market value skyrocketed by nearly 30 times to $410 billion. *Fortune* magazine named him the "Manager of the Century" (1999). At the end of his tenure at GE, Welch was widely regarded as a management genius (Slater 1998; Lane 2008; O'Boyle 1998).

Two processes hurt manufacturing employment in the United States but boosted profits for GE. First, the financialization of GE drew resources away from manufacturing, especially as Welch cut basic research within the firm. Second, Jack Welch's "70–70–70" rule was to outsource 70% of production to other firms, off-shore 70% of this outsourcing to low wage countries, and send 70% of this to India (Carmel and Tija 2005:110). His internal comment that GE should "put factories on barges" so that they can be moved around the world in search of the lowest wages available has also become legendary or infamous, depending on your values. Some of GE's manufacturing would be difficult to off-shore, such as military equipment, and especially jet engines (Stevenson 1992). However, a large amount of electronic manufacturing was off-shored to Asia and GE generated record profits on these items—particularly, large appliances (refrigerators, washers, electric ovens, etc.). In 2011, 154,000 of GE's total workforce of 287,000 workers were in China and Asia (Uchitelle 2011a, b).

However, shareholder value theory has little to say about how production is organized on the shop floor. Welch has indicated little concern about teamwork, long-term employee loyalty, and investing for the long-term. Six Sigma is an approach to quality, but it does not actually entail teams and can be quite management-centered, with top people getting their black belts in quality control but shop floor workers contributing little. Downsizing and outsourcing fly in the face of developing employees for the long-term. He instituted a year-by-year policy of cutting the bottom 10% of the company as indicated by performance evaluations. While some view Toyota's practice of getting unproductive workers out of the way—giving them desks by the windows to look at the scenery—as extreme loyalty, cutting the bottom tenth every year is also draconic and detrimental—what Besser (1996) refers to as a "community of fate" that contributes to teamwork. Like shareholder value theory in general, it is hard to glean information about teams or the organization of production based on Welch's shareholder value approach. Welch most likely has little interest in the internal operations of his off-shore subcontractors. While semi-autonomous teams could occasionally be used internally for R&D like Silicon Valley high tech companies, Welch got rid of most of GE's basic research, deeming it "unprofitable". By 2011, GE had more plants off-shore than in the United States (230 to 219) as 53% of its revenues came from overseas manufacturing (Uchitelle 2011a, b). While more evidence would be helpful about teamwork and employee loyalty at GE, one can fairly conclude that GE and shareholder value theories have little concern for these issues. As indicated above, Welch was named the "manager of the century" by *Fortune* magazine, and still has a strong influence on GE and most of corporate America to this day. And shareholder value theory is a mainstay of Wall Street, many institutional investors, and some corporate boards. But it clearly violates many tenets of lean production, especially taking the long view.

The Contradictions Between Management and Industrial Engineering Theories

The iconoclastic economist/sociologist Thorstein Veblen, in *The Theory of the Business Enterprise* (1958/1904), discussed the major conflict in society as that of the owners of businesses and the engineers who work there. Owners have one concern—profit—and this is often bolstered by the stock prices of their firms. Engineers, on the other hand, are more interested in the efficiency of the production process and the quality of the product. In a nutshell, owners like high profits and stock prices, and engineers like good machines and production processes. Often, these two goals operate together. But sometimes they do not. We have just discussed lean production as a process that most managers and engineers embrace, but at the same time, we have many managers and perhaps more stockholders and investors who prefer the shareholder value theory of management, which black-boxes the production process. Cutting-edge lean production requires commitment throughout the organization and legitimacy from the top with long-term results takes considerable attention. A CEO who pursues shareholder value theory is less likely to be committed to a particular type of production process. While we do not want to make lean production versus shareholder value to be a new basis for conflict within industry, we do want to point out that the two differing approaches—one preferred by some managers and the other preferred by engineers—may have entirely different approaches to organizational processes.

The two parts of this chapter mention this same kind of conflict. Industrial and other engineers are often very interested in the most efficient way of producing a generally high quality product. They are invested in their theories and tools. Management leaders may often share these same values, but there is a strong trend among managers who are influenced by rather spare economic theories and a pure interested in profits. They tend to black-box the production process and focus on the short-term generation of profits by whatever means. Perhaps the biggest conflict came in the 1980s when CEOs "milked" various plants, squeezing out the most profits they could generate by not repairing plant equipment. In general, engineers were horrified by this complete neglect of their tools and procedures. Current theory is not as conflict-oriented as this example from the 1980s. However, shareholder value theory, with its often financial manipulations, black-boxes the production process and in the process cares little about the processes that are the engineers' life blood. While shareholder value theory is not against lean production per se; they are focused on the outcomes in terms of profits and share prices, whatever the path might be to them. If lean production works, then fine. If it does not or it largely produces the same result, then they often put pressure on the production process to deliver no matter what the means, and often the short-term solutions are privileged. The longer term process of generating lean processes in industries and services tends to go by the wayside.

Conclusion

Management and industrial relations approaches to lean production are similar in their appreciation and advocacy for lean production. However, it is very clear that lean production has major standing within the field of industrial engineering, not in management. Lean production is a relatively minor area of management, mainly because it is subsumed under strategic management, which is a much larger area, and to some degree, under organizational change. This is unfortunate concerning management. From our perspective, lean production is not a narrow approach to organizational success, but it is a larger view than simply production. As practiced in Toyotism, lean production is a comprehensive and long-term approach to production, innovation, and overall corporate strategy. Again, the term "lean" may be a limiting "narrative" label, but the 14 tenets of lean production are a comprehensive approach to strategic management. Unfortunately, they are not recognized as such in the management literature. When accompanied by the clearly strategic use of "contradictions" used by Toyota (Osono et al. 2008), this yields a complex format for strategic management. In comparison, stockholder value theory is a poor substitute for this more comprehensive strategy that we discuss in this book.

Notes

1. The Japanese usually speak of this model in specific terms. Toyota uses the term TPS. Honda has its own name: Honda Best Practice (HBP) (Nelson et al. 1998/2007). Honda sees Teruyuki Mauro as the father of this process, while Toyota usually points to their founders or Taiichi Ohno. Darius Mehri (2005) notes a Japanese reluctance to theorize concerning their preferences for company specific models.
2. These points contain Liker's 14 points as follows: (1) points 1, 9 and 14; (2) 4, 6, and 8; (3) 11; (4) 11; (5) 5, 10, and 13; (6) 9; and (7) 2, 3, 7 and 12 (2004).
3. How should one interpret the book by Tim Turner and colleagues (2012) who are Toyota workers or managers? This book that praises workers' experiences at Toyota plants around the United States is easy to dismiss propaganda, but on the other hand, it is hard to imagine a book like this about workers' "stories" being published by other auto companies. Of course, Darius Mehri's participant observation in Japan reveals some troubling aspects of Toyotism (2005). Nonetheless, in our interviews, it was relatively hard to elicit criticisms from Toyota workers in the United States who have not been injured or fired.
4. Hans Pruijt and also Steven Vallas refer to lean production as neo-Taylorism. Rather than a new system, they see lean production as highly Tayloristic with two core principles (the one best way and the problem of systematic soldiering) (Pruijt 2003). From our point of view, this fundamentally misses the fact that lean production wants workers to think and Taylorism did not. Further, the job rotation system, flexibility, JIT inventory, long-term job security, and long-term perspective of the

corporation in general are quite different. Consequently, we do not see lean production as neo-Taylorism.

5. Alexey Stakhanov reportedly mined 227 tons of coal in one shift. He became a hero of socialist labor in the USSR and even made the cover of Time Magazine in 1935. The Stakhanovite movement was named after him.
6. The PDCA cycle is not far removed from the concept of time-and-motion studies, which in itself is similar to a rational process of planning and carrying out a plan. One could also say that it fits Neil Smelser's structural differential approach to change.
7. Using the first letter of the first word in each of these categories, Six Sigma methods call the first approach DMAIC and the second one on new products DMADV.
8. Some organizations use yellow belts for employees that have basic training in Six Sigma tools and generally participate in projects, white belts for those locally trained to a small degree, and orange belts for special cases (Six Sigma 2017).
9. Two other engineers are strongly associated with Six Sigma. Genichi Taguchi (1924–2012) has written many books on lean processes, but more from a Six Sigma approach (1992; Taguchi et al. 1999, 2000, 2002, 2004, 2005). He is an industrial engineer but more involved with statistics. He has won many awards, including the Deming Prize for individuals. Subir Chowdhury (1967–present) worked with Genichi Taguchi, and was trained as an engineer in quality control and systems. Chowdhury is especially oriented toward Six Sigma and has written many books on it (2017, 2011, 2002a, 2002b, 2001; Chowdhury and Zimmer 1996; Chowdhury and Taguchi 2016). He has won the Phillip Crosby Medal given by ASQ and the Henry Ford II Distinguished Award for Excellence given by the Society of Automotive Engineers, among many others awards.
10. This does not mean that relations did not previously exist between suppliers and OEM firms in the United States. Salespeople would try to develop strong relations with suppliers (purchasing and sales) through informal means of lunches, dinners with spouses, involving family and friends in weddings, and so forth. This would generate sales through close personal bonds. However, the Japanese methods included formal relations between firms, sharing information and long-term rather than market-based relationships.
11. QR codes (Quick Response codes) are the trademark for a square matrix or two-dimensional barcode. They were designed in 1994 for Denso Wave the auto industry in Japan, and are a machine-readable optical code with more information than a bar code since it is two dimensional rather than a linear strip of lines.
12. As a statistical term, the power law indicates that two variables, no matter what their initial quantities might be, have a correlation with proportional relative change with each other. This is close to saying that although two variables may have very different absolute values, they are still correlated with each other.
13. For recent reviews of the management literature on change see Stouten et al. (2018), and Samuel et al. (2015).
14. There are many approaches to organizational change. For a general overview, see Marshall Poole and Andrew van de Ven's *Handbook of Organizational Change and Innovation* (2004). Institutional theories of organizational change also come out of sociology (Padgett and Powell 2012; Adler 1995:61–83; Cole and Whittaker 2007), and employment and labor relations (Kochan et al. 1997b:325–35; MacDuffie and Pil 1996; Cutcher-Gerschenfeld et al. 2015).
15. This aligns with "principal-agent theory" that is strongly concerned with how an owner (principal) controls their managers (agents). The mechanism by which this is done is by paying CEOs with shares of stock, which increases the CEOs interest in increasing the value of the stock (Jensen and Meckling 1976). However, many CEO got bonuses even though their performance was subpar.

16. Six Sigma is a quality control process that was developed by Motorola in 1986. It seeks to improve the quality of products by removing the causes of defects and minimizing variability in the manufacturing processes. Each Six Sigma project is carried out with a defined sequence of steps with financial targets, and it certifies employees with a green to black belt system. Six Sigma itself refers to six standard deviations from the mean. For further details, see Chowdhury 2001, 2002, 2011, 2012, 2017; Chowdhury and Taguchi 2016; and Murphy and Yates 2009.

3 Social Science, Critical, and Socio-Technical Approaches to Lean Production

The social sciences perspective on the impact of lean production on the workforce covers some positive aspects of lean, such as teamwork, job rotation, job enrichment, and job satisfaction—and many of the negative aspects of lean, such as work intensification, excessive overtime, and the extensive use of temporary workers. The social science literature is not as unitary as industrial engineering or even management, and the diversity of streams of thought is quite evident in viewpoints being positive and negative toward lean production. This chapter will focus on four approaches in the social sciences:

(1) Sociological studies of lean production in automobile plants by Robert Cole, Jeffrey Liker, and our own work
(2) Critical social science, flexible accumulation, and labor process theory of Steve Vallas, Matt Vidal, and the labor process theory approach
(3) Sociotechnical theory as a social psychological approach to teamwork, technology, and lean production of Fred Emery and Eric Trist
(4) Diffusion of theories of lean production throughout the world reviewing David Strang and others

Compared to management, industrial engineering, and employment relations approaches, the social sciences have a more mixed-viewpoint toward the benefits and detriments of lean production. They also have less unity and each of these four approaches can be quite different Schwemmer and Wieczorek 2019).

The Sociological Studies of Lean Production

The first social science studies that observed what was going on within Japanese plants were done by sociologists Robert Cole with *Japanese Blue Collar* (1971) and Ronald Dore (1973) with *British Factory/Japanese Factory*. They opened the eyes of Westerners as to what the organization of work in Japanese plants and the emerging concept of lean production or Japanese manufacturing methods.[1] Dore did a survey research project of two English electric company factories in the UK (Marconi in Bradford and Babcock & Wilcox in Liverpool) matched with two Hitachi factories in Japan (Furusato and Taga) in Japan.[2] He also compared two steel firms—Appley in Frodingham and Yawata in Japan. Dore's work was a quantitative survey research study of about 150 workers in these Japanese and British plants and 300 in the two steel companies concerning recruitment, wages, training, unions, and other aspects of the work environment. (1973:421). His results showed in detail how different the Japanese system was from the British work and employment system. Dore highlights the flexibility of the Japanese factories, but he does not detail a system anything like the TPS. However, he does systematically show the structural and cultural differences between the two countries. He briefly talks of convergence, but largely rejects this assertion.

Dore followed up on this work by going more deeply into culture with his *Taking Japan Seriously: A Confucian Perspective on Leaning Economic Issues* in 1987, and collected these thoughts in *Market Capitalism, Welfare Capitalism: Japan and Germany versus the Anglo-Saxons* (2000). These works emphasized the more institutional and cultural approaches of Japan and Germany, which is a theme we will return to in Chapter 5.

The most influential sociologist who developed the working methods of lean production with a focus on Toyota was clearly Robert E. Cole, who is professor of sociology and business at the University of California, Berkeley. Cole did this in a series of three important books addressing different aspects of the Japanese model. First, Cole started a four-year participant observer and interview study in 1967 of work in two Toyota-linked Japanese factories—Takei Diecast Company as a 2nd tier supplier, and Gujo Auto Parts plant as a 1st tier supplier—and published *Japanese Blue Collar: The Changing Tradition* in 1971. He spent three months working on the shop floor of the diecast company and a month and a half at the auto parts company, along with interviewing 15 workers at each of the two plants. The question that Cole is pursuing is not the methods of lean production for the term did not exist at the time, but rather, "why do Japanese employees work so hard?" But in the process of answering this questions, which is partly cultural, he opened up the processes by which Japanese production is organized and maintained.

With a thorough understanding of how Japanese factory workers think and behave, Cole examined quality control processes in detail in *Work, Mobility and Participation* (1979). Cole defines the QCC movement as being composed of seven parts (1991:135–43,160–73):

(1) First-line leaders being trained in participative techniques and statistical methods
(2) First-line supervisors instructing small groups of workers in their place of work
(3) The spontaneous participation of workers in self-improvement and in solving specific work-related problems

(4) The QCC is an autonomous study group that operates on a continuous basis
(5) The QCC is a group effort with all members participating
(6) The first-line supervisors, engineers, and high-level management provide assistance to the QC circles as needed
(7) The QCC members demonstrate their progress and accomplishments to the public through company awards and national and regional QC Circle conventions (Cole 1989:141)

He presents a detailed process account of how QCCs solved a tire viscosity problem at Bridgestone Tire and a bumper assembly process problem at Toyota Motors Corporation. This included Pareto figures, fishbone (Ishikawa) diagrams, tree analysis, and statistical process control (see Figure 1.1 for these diagrams). Using comparative data from Yokohama, Japan, and Detroit in the United States, he showed how the Japanese methods were clearly superior. Most of this analysis relies on the comparison of rational processes of production, which cite Deming and Juran in their work in Japan since the late 1940s. Much of Cole's work focuses on job mobility, but the major impact of this thoroughly grounded and multi-method study convinced Americans of the superiority of the Japanese methods of production more than the uniqueness of Japanese culture. With the oil crisis of 1979–1980, the efficiency and quality of Japanese automobiles became ever the more evident.[3]

Throughout Robert Cole's work, QCCs and job mobility take the most prominent place. In these two books, it is clear that Cole is aware of JIT training inventory or *Kanban* as it is known in Japanese; however, these two terms do not play a prominent role and neither of them makes the index. Further, Taiichi Ohno is not in the index while Kaoru Ishikawa, the originator of QCCs, is. And while the management and industrial relations approach emphasized JIT over QCCs, exactly the opposite applies in this chapter on social science approaches.

Box 3.1 The Revealers of Lean: Robert Cole and Jeffrey Liker

Robert E. Cole (1940-present) is emeritus from the Haas School of Business at the University of California, Berkeley. He became fluent in Japanese and in 1967, he began his research in two automotive supplier firms in Japan. His 1971 book *Japanese Blue Collar* opened up the whole area of Japanese production methods to the Western world. *Blue Collar* was followed up in 1979 with *Work, Mobility and Participation: A Comparative Study of American and Japanese Industry*, which went into much more depth concerning the management, QCCs, and the nature of the workforce in Japan. His next book, *Strategies for Learning* (1990), examined how Japanese production methods had spread throughout Japan, the United States, and Sweden. It is important to note that his 1971 and 1979 books about the auto industry were one to two decades before the publication of *The Machine that Changed the World* (1990). In 2012, he was awarded the *Eugene Grant Medal from the American Society of Quality*: "For a lifetime commitment to rigorous research into the sociological phenomena that influence quality practices within the work environment and stimulate innovative action through the participation of workers" (University of California n.d.).

Jeffrey K. Liker (1950-present) is professor of industrial engineering at the University of Michigan after getting his PhD in sociology from the University of Massachusetts. After working on life course studies, he switched his focus to lean production methods. He is currently Professor of Industrial and Operations Engineering at the University of Michigan, owner of Liker Lean Advisors, LLC, Partner in The Toyota Way Academy, and Partner in Lean Leadership Institute. Jeffrey Liker has won 11 Shingo Prizes for Research Excellence and *The Toyota Way* also won the Institute of Industrial Engineers Book of the Year Award in 2005 and the Sloan Industry Studies Book of the Year in 2007. In 2012, he was inducted into the Association of Manufacturing Excellence Hall of Fame. Liker is best known for his Toyota Way Series of books. His breakthrough work was *The Toyota Way* (2004) that lays out the underlying philosophy and principles of lean production in the TPS at Toyota's highly efficient and quality-obsessed plants in the United States. The uniqueness of his book is the close association and insider like account of Toyota's methods. Few Americans have gotten such access to the Toyota Corporation. He subsequently followed this up with many books in the *Toyota Way* series on product development, talent, culture, leadership, crisis, and training.

Second, three sociologists examined Japanese production methods further. Terry Besser (1993) from Iowa State University did a detailed observational study of the Toyota Georgetown plant (TMMK) that makes Camrys while she was at the University of Kentucky. While examining many aspects of lean production similar to Robert Cole, she tackled the main impetus or cultural glue that held the plant together. She emphasized three nested communities of fate: the individual in the team, the team within the larger plant grouping, and the Toyota company within the economic system of the United States and world. She emphasized in particular the parallelism of each of the "communities of fate", which are different from typical American plants (Besser 1993, 1996).

James Lincoln and Arne Kalleberg (1985, 1990, 1996; Lincoln 1987; Kalleberg 2011; Kalleberg and Lincoln 1988; Kalleberg and Reynolds 2003) develop some survey research studies of work values and commitment in the United States and Japan. They find mixed evidence for structure and culture in worker commitment. Lincoln and Kalleberg in *Culture, Control, and Commitment* (1990) state that labor commitment and motivation are deeply imprinted in Japan, but Japanese workers are not more committed to their companies or satisfied with their jobs. They conclude that part of any Japanese advantage in labor commitment and motivation comes from the social structure of the Japanese workplace that has tight-knit work groups and strong relations to superiors in the organization. QCCs with a certain amount of participatory decision-making, enterprise unions, and some welfare and health services reduce alienation and increase commitment in Japan. If these practices would be used by American companies, company commitment would also rise. While they do not minimize cultural and historical influences on the Japanese employment system, their findings support a type of "welfare corporatism" as an organizational form that maximizes commitment. As such, they do not conclude that this type of corporatism is exclusively Japanese, which is somewhat similar to Dore's comments about Germany (2000).

Terry Besser (1993) disputes many of their claims saying that much of the community of fate cannot be developed in large scale surveys because much of the culture of work is organizationally specific, especially to Toyota and Honda. Some of what Besser says can also be supported by Mike Rother's book *Toyota Kata* (2010; see also Rother et al. 2003) where he describes the unspoken and underlying "improvement katas" and "coaching or mentoring katas" of the Toyota Corporation.[4] A *kata* consists of a group of underlying cultural norms and values that workers themselves most often cannot express. But they become important in the development of Besser's "communities of fate". As a result, general surveys of labor values and behavior do not capture these *katas* or deep-seated community values. Cole et al. (1993) counter her claims, which is a strong point since they cover both sides of the methodological debate. However, Besser's work does not focus directly on Japan and is limited to one Toyota plant in Kentucky. However, in as much as this is a methodological dispute, it exceeds the scope of the present book.

Beyond the culture and commitment approach, Lincoln and colleagues have also studied the nature of the *kieretsu* in Japanese automobile and electronics industries using network analysis (Lincoln et al. 2017, Lincoln and Gerlach 2004, Lincoln et al. 1996, Lincoln and Shimotani 2010). They find that the impact of networks of cooperative and non-competing firms were very helpful to the growth of major OEM industries from 1985 to 1991 (and undoubtedly much earlier); however, in the post-bubble economy of Japan from 1992 to 1998, the *marugakae* or "all-encompassing embrace" of the *Kieretsu* lost its advantages in research and development. Further, the *keiretsu* alliances had a negative impact on forming new and creative network structures (Lincoln et al. 2017). So, the once *Kieretsu* moved into the new computer age, it became a disadvantage rather than a dominant *zaibatsu* that Chalmers Johnson seemed to describe (Johnson 1982).

A third sociological viewpoint comes from Jeffrey Liker, who gained open access to the Toyota Motor Company in the United States, and especially the Toyota Motors Manufacturing Kentucky (TMMK) plant. His international best-seller—*The Toyota Way* (2004)—develops much more of the theory and principles that drive Toyota toward quality and efficiency through 14 principles. Liker's approach to the philosophy and principles that drive Toyota's quality and efficient culture are much more comprehensively covered with his 14 points (though updated a bit different from Deming's 14 points). This was followed up with *The Toyota Way Field book* (Liker and Meier 2005) that focuses more in implementation, *The Toyota Product Development System* (Morgan and Liker 2006) on design and innovation; *Toyota Talent: Developing Your People* (Liker and Meier 2007) on HR (HR) and recruiting exceptional employees; *The Toyota Culture: The Heart and Soul of the Toyota Way* (Liker and Hoseus 2008); *The Toyota Way to Lean Leadership* (Liker and Convis 2011); *The Toyota Way to Continuous Improvement* (Liker and Franz 2012); and *Toyota Under Fire: Lessons for Turning Crisis into Opportunity* (Liker and Ogden 2011); and *The Toyota Way to Service Excellence* (Liker and Ross 2016). While it may seem that like other sociologists, Liker has not written on JIT, this is not the case as he has nine articles on the topic, and *The Toyota Way* certainly includes JIT issues, but JIT is not the main focus of his work.

Liker has won many awards in the industrial area.[5] With this impressive array of work on lean production, Liker is clearly at the forefront of research in lean production. However, these books with McGraw-Hill have a major strength and a major weakness. The strength is the insider information that comes from his access especially co-authoring with Toyota employees in writing the books. The weakness is that he is so closely associated with Toyota that it is often difficult for him to express critical views toward their operations. Nonetheless, these works constitute a major delineation of the TPS in more detail than most other sources.

Fourth, sociology's contribution to educating students in lean production has been limited at the undergraduate level. Sociology of work textbooks either do not mention lean production or they have a brief mention of post-Fordism with a vague discussion of Japanese methods. One largely looks in vain for a meaningful discussion of lean production in American textbooks on work (Hodson and Sullivan 2012; Vallas et al. 2009; Volti 2008; Grint 2005; Grint and Nixon 2015). When we visited the late Randy Hodson at Ohio State University, who has written extensively on work, he told us that neither a qualitative nor a quantitative dissertation or other study had ever been done in sociology at the Marysville Honda plant, which was the first Japanese transplant in the United States.[6] Organizational theory textbooks, which tend to be more sociological organizational behavior texts, have a similar lacunae. The only sociological textbook on work that has a sophisticated discussion of lean production is Stephen Edgell's *The Sociology of Work* that does an excellent job in discussing lean production in the context of other models including the German and US models (2011:87–99).

Finally, sociology is often focused on class, race, and gender in organizational studies. Early studies of Japanese production processes by Robert Cole has shown significant discrimination towards women in the workplace with the expectation that they will leave to start a family by the time they are 30. Darius Mehri's (2005) more recent work on Toyota shows that women are scarce in the production process, but that Africans, Italians, and Asians are actually employed at Toyota. Since the 1970s, a few women have been hired, but they are still an extreme minority. Darina Lepadatu and Thomas Janoski (2011) focused directly on diversity in lean production at *Kaizen* Motors (a *Fortune*-500 Japanese auto manufacturer in the United States). Their qualitative study of 120 workers and use of the company survey of more than 3,000 employees, found that women have made significant strides into the workforce and that they play an important role in terms of quality assurance and safety. Clearly, Japanese transplants in the United States can employ large numbers of women and avoid most (not all) problems concerning discrimination in the workforce. African-Americans have also been hired in numbers equal to their percentage in the area, but older workers find it quite difficult to keep up with the physical pace of the plant. We can see some of this diversity spreading to Toyota at TMMK where Will James, an African-American became president of TMMK in 2010, and Susan Elkington succeeded him when he retired in 2018.

Critical Social Science: Post-Fordism, Flexible Accumulation, and Labor Process Theory

In a second group of social science scholarship, there are three schools that are quite critical of lean production: post-Fordism, flexible accumulation, and labor process theory.

Post-Fordism. The simplest model is to put "post" in front of the old model, so post-Fordism was the first innovative conception of the division of labor to emerge. Coming out of post-modern theory with Fredric Jameson's *Post-modernism or the Cultural Logic of Late Capitalism* (1991), it was developed more specifically as a production process by David Harvey in *The Condition of Post-Modernity* (1989:141–72). Harvey was delineating the impact of the division of labor on cultural systems. Although he avoids using the specific term "post-Fordism", his use of post-modernism is quite similar to post-Fordism despite preferring the term of flexible accumulation.[7] Post-Fordism was the term initially implied in the early development of this theory. Daniel Bell's work on the post-industrial society with the strong movement from manufacturing production to service provisions (1973) was among the first discussions of this shift.

Harvey's work laid post-Fordism largely at the feet of the accumulation crisis of modern capitalism made worse by the oil crises of 1974 and

1980. This energy crisis led to a decline of capital accumulation in the West (though it rose in the Middle East), a rise in inflation and unemployment, and a major reaction by Western capitalists to regain their edge. Therefore, new labor models have emerged to address the past flaws of both Taylorism and Fordism.

The technical aspects of work in Post-Fordism are somewhat general. First, there is a massive shift from manufacturing to service work (i.e., post-industrialism). Second, in the area of manufacturing, mass production is replaced by more specialized production processes, which provide more unique designs and higher quality. Third, specialization requires shorter production runs and even sometimes batch processing, but in either case they are produced by flexible labor arrangements. Fourth, new technologies involving computerization and robotics also make flexible production possible and profitable. Fifth, production is controlled through flexible technical and human systems using computer technology and instantaneous communications. And sixth, large organizations with bureaucratic characteristics are changed into more responsive, flexible, and usually flat organizations with minimal hierarchy. Also associated with post-Fordism is a shift in consumerism such that people now prefer more interesting and unique items. Emblematic of this was the shift from modernist architecture with its straight and shiny lines indicating total functionality (and somewhat contrary to the theory—complete flexibility within the building) and post-modern architecture with buildings with more interesting nooks, crannies, and sometimes surprising juxtapositions of different styles. Post-Fordism went along with the view that all of society was changing in the direction of post-modernity, distancing itself from Fordism with its boredom with standardization in production, imitative products, and cookie-cutter urban landscapes.

Flexible Accumulation. Although somewhat present in post-Fordism, flexible accumulation emerged as a model much more focused on the production process. It has some of the benefits of coming later with more focus on the current globalization of production and outsourcing. Steven Vallas, in a number of journal articles, has elaborated theories of flexible accumulation and applied them in a more detailed way to the production process itself. Flexible accumulation as a system of capitalism has five features. First, the structures of vertical hierarchy or authority are externalized outside firm boundaries and to some degree flattened or delayered into more flexible ways of accumulating capital. Second, the delayering is accomplished by out-sourcing (and by implication, off-shoring) of work through subcontracted relationships with other firms. Third, the workings of individual firms themselves are segmented into professional or specialized employees who are privileged and work in teams. Fourth, at the same time, blue collar permanent workers are still subject to work under standardized procedures in more traditional roles, and temporary workers are used as buffers to the regular workforce and provide a disciplining factor by showing that permanent workers could easily be replaced. This conditional and bifurcated treatment of blue collar workers along with privileged attention to professional and technical workers creates a complex arrangement of privilege for some and subjection for others. This is an especially important instance of flexibility. And fifth, flexible accumulation has innovations such as supply-chain management (JIT), computer-assisted design (CAD/CAM), and other more flexible processes (Vallas 1999, 2001, 2006a, b).

From the technical aspects of post-Fordism and flexible accumulation, we go to the social and political aspects of these models. First, trade unions that are characterized by bureaucratic rules and rigidities of Fordism are no longer used to represent the interests of the post-Fordist work force that is highly differentiated into types of trained employees and professionals with various levels of employment security. A core labor force is then buffered by a temporary work force that is not represented by any group. Second, decentralized individualistic or collective bargaining replaces more centralized bargaining procedures that in the flexible accumulation view only produces conflict and resentment. Third, these workers, employees, and professions develop different identities and as a result demand more differentiated arrays of cultural enhancements, material products, and diverse lifestyles (e.g., professionals driving a Volvo and blue collar workers driving a four-wheel drive pick-up truck). Fourth, the bureaucratic welfare state can no longer meet the needs of this newly diverse (de-massified) population, and "more flexible institutions are required" (Clarke 1990:73–74; Gartman 1998; Lipietz 1997). And finally, these needs will be fulfilled by more flexible organizations and politics to supply these goods and services in the way that they are needed (Vallas 1999; Vidal 2011; Clarke 1990). Most often, these services are divided into high- and low-paying segments. In large part, post-Fordism fits well into the neo-liberal political view of cutting the welfare state and its associated union-busting institutions.

The combination of new production processes and the new lifestyles of post-Fordism, with multiple and complicated identities, presents a highly complex and comprehensive approach to the division of labor. Nonetheless, post-Fordism has a number of weaknesses. Production is said to be flexible in a number of different types of organizations, but not much else is said about the labor process other than production runs are shorter and may use new technologies (Vallas 1999:74–75, 95). The points about unions no longer being needed and decentralized collective bargaining hardly describe the techniques of union-busting or the hiring of temporary workers, much less off-shoring to China and beyond. The welfare state no longer meets workers' needs, yet it is largely still there along with 401k and 403b plans that have nearly replaced defined benefit plans. All in all, post-Fordism and flexible accumulation leaves one feeling a bit empty. There is nothing approaching the Occam's razor of scientific management.[8]

The addition of "accumulation" to "flexible" indicates that this is a capitalist model that extracts surplus value, especially from the semi-permanent and temporary blue collar workers. This goes back to Claus Offe (1986) and

James O'Connor's distinction (1973) of a legitimation and accumulation crisis in capitalism (see also Berg and Janoski 2005:62, 76–78). The accumulation crisis occurred with the oil shocks and the rise of inflation, but flexible accumulation largely fixes this crisis for capital. Vallas does not seem to follow this theory; nonetheless, he indicates that the beneficial aspects of the system for workers are largely an illusion. The main approach is to accumulate greater and greater amounts of capital for profits, further investment, and distribution through dividends or increasing stock value.

The flexible accumulation approach outlined by Vallas (1999) certainly captures the "flexibility of production", but it does not capture the partially positive aspects of blue collar workers involved with teamwork. This means that flexible accumulation would not apply very well to the TPS. In fact, Vallas tends to avoid discussions of Japanese production and Toyotism, and tends to develop more of a theory that emerges from the natural development of Western capitalist firms.[9] So while flexible accumulation is a useful model that is researched in considerable theoretical and empirical depth, it does not examine the good points of TPS or more international issues of off-shoring. As a result, the critical models of the East in terms of Japan and China are underplayed. All in all, flexible accumulation theory tries to ground its theory in the somewhat natural development of Western capitalist processes without specifying exogenous influences.

Labor Process Theory

Derived from Marx and revived by Harry Braverman in 1974, labor process theory has been around the longest of these three critical veins of social science. It has its own international conferences, but is most evident in the United Kingdom. According to Marx, the labor process is where labor is materialized from use values into exchange values. Labor is an interaction between the working person and the natural world that consciously makes an objective product. Based on this, there are three elements of the labor process: (1) the human activity at work is a useful and productive activity; (2) there are clear objects that are worked upon; and (3) there are tools that are used in the process of work (Marx 1976:284). According to Marx, work is a process at its most elemental or shop-floor level in which man and nature works together. Humans begin and control their reactions between themselves, other humans, and natural objects. While interacting with tools on raw materials and other people in the changing world, humans simultaneously change their own nature. The labor process is purposeful activity that produces use values but at the same time under the auspices of management or owners, the worker produces exchange values that can be traded often in a market. In the process of exchange, surplus value is created through the labor process. The labor process exists in all societies, and the organization and control of a labor process varies by the mode of production and the type of society (Knights and Willmott 1990).

Labor process theory critiques rational approaches to production that are controlled by owners and managers, especially scientific management or Taylorism. Since the labor process is embedded in Marxian theories of surplus value, worker resistance and bargaining with management through the labor process produces a repeated patterns of behavior. Some labor process theories try to explain the bargaining power of workers in society. Labor process theory was developed further into a set of interventions that critique lean production due to its exploitation of workers' energy and time in the labor process. In *Labor and Monopoly Capital: Degradation of Work in the Twentieth Century* (1974), Harry Braverman seeks to update Marx's critiques of the capitalist labor process through an attack on the deskilling that has spread throughout industrial society, although Braverman's primary focus is the degradation of work in the 20th century that comes with increasing management control. This degradation and deskilling then reshape occupational and class structures toward lower pay and benefits. Thus, labor process theory looks at how people work, how managers try to control their work, what skills people gain or lose, what resistance workers exhibit, and what the outcome result for workers in terms of their quality of life.

Box 3.2 The Labor Process Theory and Harry Braverman

Labor process theory is a micro-interactionist theory of the shop-floor organization of work under capitalism. According to Karl Marx, the labor process refers to how labor is an interaction between the worker and his/her work environment in a process of making useful products. The labor process consists of: (1) the work itself on the shop-floor, (2) the materials of work such as raw materials, and (3) the technologies that facilitate the process of work. Workers start, regulate, control and end the labor process between themselves and their materials. Thus, workers act on materials often with machines, and deal with or resist management, and in the process change themselves. The labor process is rather micro-focused, and financing and marketing of products is less a concern (Burawoy 1979). **Harry Braverman** (1920–1976) popularized the concept of labor process theory, especially with his path-breaking book *Labor and Monopoly Capital* (1974). In his view, labor process theory looks at the skills that people use at work, in what way they are paid at work, and who controls their work. Braverman presents a broad thesis that under capitalism, management deskills or "steals workers' skills", reduces the pleasurable work and makes the job boring, and reduces worker power by controlling skill, cutting their wages by reducing their wages to those of unskilled workers, and increasing the effort that workers must expend. Braverman examines to the class-in-itself or the working class as the subject of management and capitalist brutality rather than looking at social movements that lead to emancipation. Some scholars following in Braverman's footsteps have criticized his "deskilling" thesis as not being universal; and others, have attended to working class resistance to Fordism. The key element of labor process theory is the micro-analyses of management control and worker resistance while simultaneously producing a product. Much of management control is achieved by machine power to control workers lives, as on an assembly line. The International Labor Process Conference has existed since 1983 and its international conference have become quite popular among academic social scientists in the areas of work and organization.

When focused on lean production, labor process theory makes a very close examination of the activities and contexts of work, often referred to as "the shop floor" (Stewart et al. 2016; Smith 2015; Smith and Thompson 1999). Clear and focused attention goes toward the speed of work, the skills that are developed or lost, the hours and speed of the work, the amount of participation in the work process, the level of autonomy on the job, and the pay and benefits that workers may receive. In general, labor process theory is critical of speed and hours of work, which is often referred to as "lean and mean production" and an overall labor speed-up. Following in Braverman's deskilling footsteps, others have criticized his deskilling thesis as not universal, and have looked at worker resistance to lean production. While lean production with functioning teamwork may have greater skills than Fordism, the work is still repetitious and often on an assembly line. Antilla et al. (2018) point out that the Scandinavian auto firms took a clear step back from their socio-technical roots when they adopted lean production. Worker autonomy clearly decreased with lean production's use of standardization. Thus, a key element of the labor process theory is an analysis of the local systems of production involving workers and management in the control or resistance of work.

A critical aspect of the theory is how much power both labor and management may hold (Vidal 2020; Smith 2015; Stewart et al. 2016; Wright 2011; Mehri 2005; Zeitlin 1998; Zeitlin and Tolliday 1992; Rinehard et al. 1997; Kamata 1982). However, in some instances, resistance may reinforce management control as in attempts to "get over on the system" actually lead to workers buying into the very system they oppose (Burawoy 1979). As might be expected, the labor process approach is very critical of lean production, and often sees Toyotism as a failed model (Pardi 2005, 2007; Vidal and Tigges 2007; but see Smith and Vidal forthcoming).

While not using labor process theory, Lepadatu and Janoski indicate that "team intensification" causes workers to put work in a lean production environment at a higher level in their ranking of identities. Using symbolic interactionist theory, they showed how work in a lean environment takes so much time and effort that family matters were often placed in second place (2011:6–9). This largely fits with the "management by stress" critique.[10] Further, this process is even stronger for temporary workers who are paid less and have job insecurity (Lepadatu and Janoski 2018).

Socio-Technical and Social Psychological Approaches to Lean and Teamwork

Socio-technical theory coined by Eric Trist is a social psychological theory that has existed since the Kurt Lewin came to the United States in 1933. Many of this theory's findings have presaged the more recent development of teamwork, especially in Silicon Valley. Generally, this approach is not critical, but it is currently focused on improving organizational performance through teamwork and other techniques. After Lewin, Fred Emery, Eric Trist, and Einar Thorsrud are important figures in this theory as is the Tavistock Institute in London (Mohr and Amelsvoort 2016; Weisbord 2012). Socio-technical theory recognizes the interaction between people and technology in the workplace. The term also refers to the interaction between society's complex organizational structures and human behavior. In this sense, society itself, and most of its substructures, are complex socio-technical systems.

Sociotechnical systems regard the social aspects of workers in society and technical aspects of organizational structure and processes. The technical includes both physical technology and organizational procedures and related knowledge. Sociotechnical refers to the inherent connections of *social* and *technical* aspects of an organization. "Technical" refers to both structures and techniques and the "social" concerns how people are organized together. Sociotechnical theory therefore is about *joint optimization* of both technical performance and the quality of employee's experiences. Sociotechnical theory proposes a number of different ways of achieving joint optimization. They are usually based on designing different kinds of organization that leads to productivity and wellbeing. As such, sociotechnical theory is founded on two main principles (Sussman 1976; Greene 1973; Taylor and Felten 1992; Weisbord 2012; Pasmore 1994):

(1) The social interaction of social and technical factors creates the conditions for successful organizational performance. But this interaction has both direct effects and also indirect results that are often unpredictable

due to complexities and unknowns. Both types of interaction occur when socio and technical elements are combined.

(2) The optimization of either the social or the technical alone (i.e., neglecting one or the other) leads to unpredictable outcomes that are problems for organizational performance.

Therefore, sociotechnical theory is about the *joint optimization* of the social system and the technical system so that they work together. Sociotechnical theory proposes a number of different ways of achieving this joint optimization of social and technical teams and communities.

Eric Trist and Ken Bamforth developed the principles of socio-technical theory in an article in *Human Relations* (1951). They based their case study on the paradoxical observation that despite improved technology (read Tayloristic methods) of the long-wall method of mining coal, productivity was falling, and absenteeism was increasing. The shift in technology to the short-wall method with organically functioning teams (read "semi-autonomous teams") increased productivity, decreased absenteeism, and produced highly functional teams. This example provided the results that put socio-technical theory on the theoretical map. Their subsequent analysis introduced the terms "socio" and "technical" and developed the following three principles (Trist 1981, 1997; Trist et al. 2013; Trist and Murray 1993, 1997; Pasmore et al. 2019; Weisbord 2012; Walker 2015; Walker et al. 2010; Pasmore 1995; Sussman 1976). This involves three principles.

First, concerning autonomy at work, sociotechnical theory shifted the focus of the study of work from the individual as the unit of analysis to both teams and the technological context of work. In effect, this was a shift from psychology to sociology. Management has a less controlling role and gives teams a form of "responsible autonomy". Workers in semi-autonomous teams participate in decision-making with greater involvement, motivation and creativity. Participatory decision-making creates a sense of team cohesion (the Zeigarnik effect). As a result, semi-autonomous teams interact with each other to create awareness and understanding of each worker's role. This is combined with a context of technology that also helps to shape how team members interact with each other. Knowledge-sharing between team members throughout the organization prevents bureaucratic delays and rigidity and depressed motivation. This sharing of knowledge between different team members, groups and divisions in the organization prevents "silos" that limit cooperation.

The second point concerns adaptability and flexibility through Participation. Most organizations have both intense complexity inside their boundaries and global uncertainty outside their borders. While organizations try to reduce both types of uncertainty, they and their employees must adapt to change. This can be done in two ways. First, the organization can redesign the organizations structure and functions to deal with the complexity and change. Second, leaders can relinquish their internal control to the workers closest to the changes. This creates knowledge through employee discussion and increases their motivation and acceptance of flexibility (Lawler 2006). Consequently, these complexities come under the vision and skills of the workers closest to the changes and challenges. For example, the semi-autonomous teams deal the coal seams that vary each day and even hour in Trist and Bamford's (1951) the short-wall method.

And third, teamwork and holistic work are important. Rather than divided work as in Taylorism, the holistic task becomes the playing field of the team. As a result, larger tasks become more meaningful. The team approaches these tasks with flexibility, adaptability and responsible autonomy. With this independency, each team member recognizes the significance of other team members' many different skills. This cannot take place unless upper-level management and industrial engineers give up their control of workers tasks. In sum, the team must decide on their solutions in consultation with each other and other teams in the organization (Cherns 1976:786).

The group-based form of organization design proposed by sociotechnical theory combined with new technological possibilities (such as the Internet) provide a response to this often forgotten issue, one that contributes significantly to joint optimization.

Sociotechnical Systems. Sociotechnical systems theory is a mixture of joint optimization and general systems theory. It recognizes team and organizational boundaries and interactions with a system (and its sub-systems), but also relations between the organization and the environment. Sociotechnical systems describe an interlinked mix of people, technology, and their environment, and encounter it with specific processes and structures. There are two major organizational components that can be involved.

First, semi-autonomous or self-managed teams are an alternative to rigid role descriptions and traditional assembly line methods. Employees are organized into small teams with tasks such as assembling products, designing new processes, or coming up with policies or procedures. These teams are self managed and are independent of one another. They are based on two creative processes (Weisbord 2012:296–310; Lawler 2006; Huber and Brown 1991). First, there is job enrichment as a complexity process of giving the employee a wider and higher level scope of responsibility with increased decision-making authority. This is related to job enlargement that means increasing the scope by extending the range of its job duties and moral responsibility. This avoids over-specialization, but one needs to be careful about overwork. A third point is job rotation, where an individual is moved through a number of different jobs over the day or weeks. This provides exposure to different skills, pressures, and a view of the entire operation. Job rotation requires a certain amount of training for new jobs, and this plays a prominent role in lean production.

These self-managed teams create a desire and willingness to do something well. A motivated person can be reaching for both short-term and long-term goals. Motivation is different from emotions and personality, and it needs to be increased by both intrinsic (e.g., on-the job recognition) and extrinsic (e.g., pay and benefit) factors.

Second, process improvement in organizational development is a series of actions taken to identify and improve existing processes within an organization to meet new goals and objectives. These actions often follow a specific activity to create successful results. There are environmental scans and two types of committees. The environmental scan draws people from all levels of the organization (i.e., a vertical slice) in order to get the widest viewpoint on issues. Sometimes this is called a "future search". It involves things going on within the organization and processes developing in the market, and wider changes in society and the economy (Weisbord 2012:258, 281–95). The steering committee takes information from the environmental scan to develop new programs or change old ones to create change in the organization or suggest new products for the market. The steering committee explores an area of concern and then develops a plan to implement the changes. It then develops other committees that will implement these plans (Weisbord 2012:312–15). Working closer to the actual work, the work design committee selects their members from all levels in the organization in order to analyze how a new or reformulated task is to be accomplished. This includes manual and mental activities, task durations and *takt* times, task allocation, task complexity, environmental conditions, clothing and equipment, and how this task or job fits into other tasks and jobs. This information can then be used for teamwork processes, selecting personnel and training, tools or equipment including automation, and even compensation and promotion. One main design feature is how the task fits into the overall process flow of work in a particular union and connections to other units, and this includes job enrichment, enlargement, and rotations (Weisbord 2012:311–36).

Deliberations can take place in other types of committees, but these are three committees useful for developing new products or improving the processes of existing processes. The most important aspect of these committees is that they include a vertical slice of all levels of the organization as in cross-functional committees (Pava 1983, 1986; Pasmore 1988, 2001, 2015; Pasmore and Minot 1994; Austrom and Ordowich 2018; Geels 2004).

These two socio-technical processes are not very far from lean production. First, lean systems use teamwork. However, QCCs and other types of teams are more controlled than semi-autonomous teams in that they mainly focus on production related issues. So, semi-autonomous teams have more freedom of maneuver. In many ways, these types of socio-technical teams are at work at Apple and Google, where wide open creativity is more at issue. Second, the various types of committees can be found in lean production firms. For instance, the term "vertical slice" comes from Laurie Graham's study of Subaru-Isuzu (1995). And Toyota innovation and development teams follow the environmental scan and steering committee models. Job redesign may be more up to engineers in lean production, but it is not unheard of these engineers consulting the workers who will be affected.

Box 3.3 Socio-Technical Theory

Socio-technical school developed from the more gestalt or field theory that **Kurt Lewin** (1890-1947) brought to the United States in 1933, fleeing Nazi persecution in Germany. Lewin was a social psychologist who endorsed the American symbolic interactionist school of social constructionism, but was more widely oriented toward force-field analysis and "action research", which looked at how employees and soldiers behaved in an interaction of social factors and technical constraints. An early and seminal influence on STS is the **Zeigarnik effect,** which states that people are more highly motivated and perceptive when they participate in the design and development of their own tasks (Weisbord 2012). Lewin's most important equation states that behavior is a function of the person and the environment, which served as a basis of the social and the technical being combined together. He met **Eric Trist** (1909-1993) at the **Tavistock Institute** in London in 1933, and they became lifelong friends, though Lewin died in 1947. Trist had helped form the Tavistock Institute with Rockefeller grants, which became connected to the British National Health Service. His study with Ken Bamford of the shift from long-wall (Tayloristic) coal mining with short-wall (team-based) coal mining led to the organizational research that socio-technical theory is known for redesigning work systems and establishing semi-autonomous teams. In 1947, just before Lewin's death, they established the journal *Human Relations,* which publishes a large amount of socio-technical theory and practical studies. **Fred Emery** (1925-1997) was born in a small town in Western Australia and became a psychologist. He joined the staff of the Tavistock Institute in 1957 and worked with Eric Trist as a UNESCO Research Fellow. He published *The Characteristics of Sociotechnical Systems* in 1959 and later worked with **Einar Thorsrud** from Norway. Emery and Trist then published many articles and books together promoting the socio-technical theory of jointly optimizing the social and technical, rather than letting technical characteristics of new production methods simply ruling the design of work.

The socio-technical approach with its semi-autonomous work teams and cross-functional and hierarchical committees became popular with organizational design consultants like Marvin Weisbord (2012) and theorists like Gerald Susman (1976). They strongly influenced the HR professions and the **National Training Institute** claims to be the world leader in organizational learning and development with many programs and services. Many of its ideas have been absorbed in industrial psychology, institutional sociology, and management theory so that it is not as popular as it had been earlier. It presaged many of the ideas in QCCs and small creative teams, but it may have given employees more freedom than most managers wanted to give (Whitworth and de Moor 2009).

The Diffusion and Differentiation of Lean Production

Lean production methods have both diffused internally within firms and externally to many different industries and countries, and lean production has also differentiated itself due to cultural differences and variations in industrial forces like unions and structures involving the state and dominant industries from manufacturing to services to rural processing and farming. For instance, the relatively few countries with strong auto industries may take advantage of lean methods faster and more extensively, while countries with more rural and service-oriented economies may be slower. Shah and Ward (2007) indicate that lean production is now the most prevalent model in manufacturing, and Krafcik (1988) indicates that lean production has "triumphed". This section will deal with diffusion first and then look at differentiation.

In sociology, Robert Cole's third major work—*Strategies for Learning: Small-Group Activities in American, Japanese, and Swedish industry* (1989)—focuses on the diffusion of quality circles and self-managed teams in the United States. He demonstrates how Japanese production methods were disseminated in the United States through two different means: one firm-based and the other professionally based. The first was the ASQ, which was strongly connected to industrial engineers and the legacy of Deming and Juran as statisticians. The second was the International Association for Quality Circles (IAQC) that was much more firm and policy entrepreneurially oriented. This largely parallels the differences between industrial engineers and firm managers but in their professional associations. The ASQ had the liabilities of age and a tendency toward Six Sigma, and the IAQC had liabilities of newness and a somewhat faddish notion. Nonetheless, the idea of QCCs was spread throughout the United States (Cole 1989). Eventually, the IAQC became the AQP, and then it died out as QCCs were superseded by more general approaches to teamwork (1989:205–50).

Also, in sociology, David Strang and co-authors deal extensively in the diffusion of lean production. Strang and Macy (2001) show that reaching excellence through efficiency can be spread through adaptive emulation sometimes called "benchmarking". One example they use is the QCC that took off in the early 1980s. Freund and Epstein (1982) found that 14% of small firms with more than 100 employees and 43% of firms with more than 10,000 employees had used QCCs (Strang and Macy 2001:150). They then found later that many QCCs had disbanded and only about 16% of firms had them. Nonetheless, they are promoted by a large number of consultants as part of lean production, who themselves had dropped from 469 in 1983 to 60 in 1994 (151). However, the opposite can be found with JIT inventory, which has spread to most businesses with more and more efficient supply-chain management systems especially as computers have replaced the "*kanban*" card as the means of implementing JIT systems.

Anttila, Oinas and Mustosmäki (2018; forthcoming) show in a survey of 35 European countries (The European Working Conditions Survey) that lean production is used in about 43% of workplaces, and the "learning approach" (i.e., socio-technical theory) is used in about 29%. In the US (American Survey of Working Conditions) lean is present in about 46% of workplaces and the learning approaches in about 21%. The least lean countries are in Southern and Eastern Europe (forthcoming: figure 2). This is the first comprehensive survey data showing the strength of lean production as the new dominant division of labor. In a survey of 500 factories across 30 countries for the Japanese Multinational Enterprises Study Group (JMNESG), Tetsuo Abo (2015) shows that most plants in the non-Japanese world are about 60% towards the Japanese model in Japan, but that Korea and Taiwan, and North America (in the 1989 rather than 2001) surveys were the closest.

Industries. Lean production has spread to many different industries. It started in the auto industry and manufacturing. Since then, it has spread into the service sector (Bamber et al. 2014), healthcare (Stanton et al. 2014), call centers (Doellgast 2010); higher education (Sunder and Antony 2018) legal, electronics manufacturing, software development, etc.

These managerial styles can, of course, change over time also on the basis of product market conditions. Particularly, when product market competition is high and demand is uncertain, companies tend to experience limited profit margins. One reaction to this situation is to lower labor costs, which entails the adoption of poor organizational practices and unitarist perspectives (Edwards et al. 2006, 2009), paving the way for intensity-based lean systems. Profit pressures can equally affect management styles and approaches. When these pressures become high, managers and firms tend to privilege the coercive and speed up aspects of lean production limiting or neglecting employee participation (Adler 2012).

Other scholars have emphasized the role institutions play to support or constrain different union, workers, and managerial strategies over lean production adoption. Social actors and institutions are seen in constant interaction among them (Doellgast 2010). At the same time, institutional factors give social actors leeway to strategically decide whether and how they should be activated (Hauptmeier 2012).

The many studies done on the diffusion cannot be cited here, but significant diffusion has been made into the medical services and hospital industries (Ng et al. 2015; Morris and Wilkinson 1995). The reader may be referred to a number of handbooks that have chapters on individual industries and less in the oil and gas production industries (Netland and Powell 2017; Defeo and Juran 2016; Juran and Gryna 2011; and Pyzdek and Keller 2012, 2014).

Countries. Occupational structures, union and employee associations, educational institutions, and many other factors can explain how lean production spreads in different countries. Research based on case studies has demonstrates that worker participation can be limited (Durand et al. 1999; Blanpain 2008). Quantitative research has confirmed that the application of the enabling aspects of lean production highlighted by case studies can be limited (Amosse and Coutrout 2011; Gallie et al. 2012). More recently, Anttila et al. (2018) have found out that lean production has increasingly diffused in Nordic countries with more work standardization and quality control. Their study is significant since it uses a large organizational data set for the European Union. So in effect, in looking at Scandinavia, they also show extensive diffusion of lean principles throughout Europe. However, the previous Scandinavian approach using a socio-technical approach of richer job content and semi-autonomous teams has faded as a result of the introduction of lean. The Volvo Kalmar plant had even abandoned the assembly line using *takt* times of many hours. However, this experiment, as promising as it seemed to be, had to be abandoned. The use of lean with more work standardization and less employee autonomy was likewise noticed in the research project carried out by 21 scholars that was based on case studies in the auto industry involving seven different national socio-economic systems such as the United States, Australia, Germany, Sweden, Japan, South Korea, and China (Blanpain 2008). However, one must not forget that lean teamwork and participatory principles, while less than socio-technical theory, are still a major improvement in autonomy and participation over a Fordist system.

Nonetheless, the diffusion of lean principles does not triumph everywhere. It clearly is somewhat hampered in more authoritarian countries. Jürgens and Krzywdzinski (2016; Schulzenko 2017) show that lean is severely underdeveloped in Russia, and considerable adjustments were made in Brazil and India. And as we previously discussed, lean teams and human resource policies in China are largely hampered by Communist Party direction, especially their habit of throwing more workers at problems. As with industries above, the many studies done on countries cannot be cited here, but significant diffusion has been made into Europe and Asia. We again refer to handbooks on lean (Defeo and Juran 2016; Juran and Gryna 2011; Juran and de Feo 2010; Pyzdek and Keller 2012, 2014; Netland and Powell (2017), and Janoski and Lepadatu (forthcoming).

Conclusion

The four elements of the social science approaches have strengths of looking at teams, inequalities, and problems in the lean production approach. However, they generally have little to say about JIT inventory, new product development, or the higher quality achieved by lean production plans and companies. While the social science approach seems to be lacking in the aforementioned areas, the management and industrial engineering disciplines have large lacunae in their approach exactly where the social science approach excels. Consequently, the disciplines discussed in both chapters are certainly compatible in their niche filling roles, and although these two approaches—management and industrial engineering and the social sciences and critical approaches—tend to ignore each other's work, this is unfortunate and avoidable. However, each discipline tends to have its own blinders; management and industrial engineering approach to lean production sees the efficiency and production aspects, and the social science approach focuses more on the workers and their protection.

Notes

1. Three anthropological works were also important. James Abegglen wrote *The Japanese Factory* in 1958, and *Management and the Worker* in 1973. However, his work has gone in the management direction, especially after he has worked as the Tokyo representative of the Boston Consulting group. Thomas Rohlen wrote about a Japanese bank in 1974 and this was a well-done study. However, the anthropological approach focuses on the uniqueness of culture, which is certainly true for much of Japan. But the approach of this book is to focus on lean production, which was started in Japan, and how it has spread in the world. Finally, Ezra Vogel, a Parsonian who graduated from the Harvard Program in Social Relations, was nominally a sociologist, but his approach is really much more anthropological in its approach. He wrote *Japan's New Middle Class* in 1963 and *Japan as Number One* in 1980 followed by *Is Japan Still Number One?* (2000). These works were not directly concerned with lean production.
2. At the time, Ronald Dore (1925 to 2018) was a professor of sociology at the University of Sussex having been previously a professor at the LSE. He learned Japanese during World War II and initially wrote about *Life in a Tokyo Ward* (1958) and later *Shinohata, A Portrait of a Japanese Village* (1978). Since he did his graduate work from 1947 to 1950 at the University of London, he was impressed by the reigning theorist of the time—Talcott Parsons and functionalist theory. However, he was attuned to the influences of unionism in the UK and the "second union" breakup of the 1950s union offensive in Japan (1974:322–35).
3. Robert Cole's book *Strategies for Success* will be discussed in the section on diffusion that follows.
4. For those familiar with the martial arts, the term "*kata*" is Japanese for the English term "forms", which are detailed choreographies of punches, blocks, and kicks that are memorized and performed at each testing level from white to black belt. They create an underlying sense of sequences that become a basis for free sparing so that they do not have to be consciously thought about. They are an underlying pattern that often emerges without thought (i.e., tacit knowledge).
5. Jeffrey Liker has won 11 "Shingo Prizes for Research Excellence", and *The Toyota Way* won the 2005 "Institute of Industrial Engineers Book of the Year Award", and the 2007 "Sloan Industry Studies Book of the Year". Liker was inducted into the "Association of Manufacturing Excellence Hall of Fame" in 2012.
6. We did find a PhD in HR dissertation done in the business department at Ohio State University that dealt with specific hiring issues at the Marysville Plant. However, it did not deal much with lean production.

7. Although Harvey does not use the term "post-Fordism", we still put him in this category for two reasons. First, his major statement on the topic comes from developments in the mid-1980s which focused on the early development of the concept. And second, his work is strongly associated with post-modernity, even though he is continuously seeking rapprochement between post-modernity and Marxism. As we will see with our explanation of the related concept of flexible accumulation, Steven Vallas has less concern with post-modern concepts or reconstructing Marxism. Instead, he is squarely focused on the development of the production process and the division of labor.
8. Vallas goes into much more detail in his critique of post-Fordism. He specifically criticizes this model for not recognizing a tension between power and efficiency, its multiplicity of market adaptations, and its failure to recognize the dualism inherent in the workforce (1999:74–76; 2003).
9. For instance, "the two distinct strains of flexibility theory" that Vallas identifies are "post-hierarchical models of work" which go back to the McGregor and Hertzberg studies of the 1950s and 1960s, and "flexible specialization of Piore and Sabel that we will discuss under our last model" (1986:70–73; Sabel and Zeitlin 2002). It is interesting that Vallas avoids citing Robert Cole, Jeffrey Liker, or Terry Besser though he does mention Paul Adler on NUMMI and Laurie Graham on Subaru-Isuzu. However, this avoidance of Toyota and discussion of Japan in general, helps to distinguish this approach from the lean production view.
10. *Karoshi* is the Japanese for "death due to overwork", which has been legally recognized in Japan and Toyota has lost cases involving it. *Karojisatsu* is suicide because of overwork. There have been over 1,000 compensated cases in Japan for *karoshi* from 2002 to 2011 (ILO 2013).

4 The Labor and Employee Relations and Human Resources Perspectives

Balancing Employer and Worker Interests

The industrial relations and human resources (HR) perspectives on lean production try to balance the management and the worker approaches to lean production. Both often simultaneously focus on efficiency and worker protection (pay and benefits, safety, job security, retirement, etc.). In the first area, labor unions often play an important role in the industrial relations approach to lean systems, but as unionization in the United States has declined precipitously with Southern industrialization, industrial relations have morphed into considering non-union shops and HR approaches. Even its main association—the Industrial and Labor Relations Association (ILRA)—has been renamed the LERA. In the second area, the HR professions, who are housed in HR departments and its academic discipline—the Society for HR Management (SHRM)—have grown in their scope but from a management perspective rather than labor. But HR constituencies—workers, women, minorities including African-Americans and Hispanics—push them in many ways toward the employee motivation and protection.

Over time, these two groups are starting to overlap in a number of different ways, especially as unionization continues its decline. Closely related to the HR approach to employee relations is the social psychology discipline housed in most psychology departments. This approach has focused on teamwork from a largely psychological approach to group processes, unlike socio-technical theory, which examines the intrinsic groupness of teamwork. Most importantly, these approaches emphasize aspects of what the previous two chapters lack—mechanisms for protecting and motivating workers and employees generally, and more specifically doing these functions in a lean production environment. So, this chapter first considers the labor and employment relations discipline, and then goes to the HR approach.

Labor and Employment Relations Approach

Industrial Relations and Labor Relations departments in corporations developed out of the strikes of the industrial revolution being avoided by national legislation to legalize trade unions. In order to deal with collective bargaining, corporations developed industrial relations departments with negotiators, economists, and other experts to deal with unions. Unions developed many of these same personnel with negotiators and economists. Departments of Industrial relations developed with some of the most prominent being the New York School for Industrial and Labor Relations (NYSILR) at Cornell University, the Sloan School at MIT, and similar departments throughout the Midwest. The southern United States was noted as a very difficult place to organized unions, and the institutionalization of industrial relations was much smaller so a few departments exist there. With the decline of unions in the United States and other countries, industrial relations started to shift toward human resource departments; however, distinct schools and corporate departments still exist in states that were heavily unionized.

Lean production came relatively late to labor and employment relations. Their attention is often focused on labor unions, collective bargaining, grievance procedures, arbitration, and other processes including strikes. It was not until the mid-1990s that labor and employment relations scholars started looking at lean production. Two approaches can be distinguished. Using the main association's initials, the first one is LERA-1, which is the labor approach to lean production as it started to be introduced into unionized shops especially in the automobile industry. The second approach, which we call LERA-2, is more focused on high performance firms without unions, and how this approach can provide for the well-being of employees.

LERA—1. The first labor and employment relations (LERA) approach was led by Thomas Kochan, John Paul MacDuffie, Russell Lansbury, and Fritz Pil. The LERA approach centers on employment practices established jointly by management, labor and the government (Kochan et al. 1997a) to produce institutional systems of organization that enact lean processes or not. The key is the staffing, training and paying of employees and managers to enact specific forms of work organization to cooperate in making economic and social success (see figure 1.1 in Locke et al. 1995). LERA scholars refer to this as "high involvement work practices" and HR policies (Kochan et al. 1997a:12). MacDuffie and Pil (1995) define lean production with a high involvement work practices index (composed of teams, involvement in groups, job rotation, useful suggestion systems, and quality control) and a complementary HR practices index (composed of hiring criteria, new training, advanced training, performance pay at the group or team level, and smaller status differentials). Both the LERA and *The Machine that Changed the World* perspectives have come out of the international automobile studies at MIT, yet the LERA approach is much more cautious than Womack, Jones and Roos" exuberance. Their empirical findings taken around 1992–1993 cite the index of high involvement work practices as being 32.9 in the US, 78.4 in Japan and 56.7 in Europe. And while the US percentage of teams is only 23, Europe is actually higher than Japan at 80 to 70 (MacDuffie and Pil 1995:17). In comparison, the HR policies and practices index is more mixed at 28.7 for the US and Canada, and 59.3 and 42.5 for both Japan and Europe (1997a:21). These data all come from their International

Assembly Plant Study (MacDuffie 1991, 1994, 1996; MacDuffie et al. 1996; MacDuffie and Pil 1997; and Pil and MacDuffie 1996).

However, with both the labor and employment relations approach and *The Machine that Changed the World* (MCW) scholars coming out of MIT, one would think that they have a lot in common, and to some extent, they do. Nonetheless, the labor and employment relations approach is critical of the "Machine" or MCW group. Kochan et al. say that the MCW approach has "elevated [the issue] to the level of folklore" (1997a:3–4), and that Womack et al. (1990) say that it is the one best way to organize production. The LERA view does not agree with this and they question the weak focus that Womack et al. have on "employment practices and direct attention to the critics who cite the possibilities of a work speedup and excessive overtime" (1997a:4). Kochan et al. (1997a) question the near determinism of Womack, Jones and Roos approach, and see a variety of responses form from "flexible production", which is not only flexible in the workplace, but also flexible from plant to plant, company to company, and country to country. In other words, the LERA-1 approach sees more variation both in the application of flexible methods, and in their focus on employment practices, which are bargained and not dictated by management. Their approach is much more thorough in its international and comparative focus that looks beyond Japan and the US to Germany, Canada, the United Kingdom, continental Europe, South America, Korea, and South Africa.

In a sense, Kochan et al. (1997b, c) is a counter-point that questions the "universally applicable system" of *The Machine that Changed the World* (1990:301), and a clear protest against the seeming management and industrial engineering domination of the whole process. The result of lean production is not a "duplication" but rather a complex variation of different strategies that mix a variety of employment practices that result in different effects in different contexts (1997b:303). In addition, there is a "larger range of strategic characteristics to firms, unions, and governments" (Kochan 1997c:303; Signorelli 2019) that produce these results. Throughout their approach, unions play a stronger role than in *The Machine that Changed the World.* In their book indexes, there are only three mentions of unions in Womack et al. 1990 versus 22 mentions in Kochan et al. 1997a. But unions, which *The Machine that Changed the World* gives short shrift, are often a stronger presence in Europe and a moderate presence in the US. Unions even exist in Japan, but are a different form as company unions with national federations in Japan. They are weaker on grievances and working conditions, but still have an impact on wages since they represent lower level supervisors and white collar workers (Chalmers 1989:173–96; Cole 1971).

Throughout the LERA-1 approach, unions play a strong role, but unions are declining, which is a fact that LERA-1 struggles with. They are investing much more in the analysis of HR departments. Again, in the book indexes, Kochan et al. (1997a) mention human resource departments 23 times, and Womack et al. (1990) do not mention them at all. HR departments and professional associations are playing a role in the "employment practices" environment, which to LERA-1 is the most important aspect of lean production. While LERA-1 clearly point to theoretical and practical weaknesses of the Womack et al. (1990) approach, the declining influence of unions and the lack of response by government in the United States is a clear problem that they need to clear up. The following section on HR management will address some of these issues but in a different way.

Box 4.1 The Two Approaches to Lean of the Sloan School of Management at MIT

Ironically located in the Sloan School of Management, which is named after the former CEO of GM, a major research focus from two different directions had a strong influence on lean production in the United States and the non-Japanese world. First, the Sloan School of management received a multi-million dollar grant to study lean production. The blockbuster book by **James Womack**, **Daniel Jones,** and **Daniel Roos'** *The Machine that Changed the World*. Their backgrounds are somewhat puzzling with Womack being a PhD in political science, Jones holding a Bachelor of Arts degree in economics, and Roos being a PhD in civil engineering. Nonetheless, they began a major exposition of lean production with *The Machine that Changed the World*. This was followed by many other books on lean production by Womack and Jones, and a variety of consulting firms and consultations with major corporations. Their influence on spreading lean production throughout the world cannot be overestimated.

Second, the Sloan School of Management also had a major influence in labor and management relations. **Thomas Kochan** and coauthors published *After Lean Production: Evolving Employment Practices in the World Auto Industry* (1997a). Three other professors who graduated from MIT demonstrate a labor focused approach toward lean production from the Womack, Jones, and Roos approach. **John Paul MacDuffie** at the Wharton School at the University of Pennsylvania and **Fritz Pil** at the University of Pittsburgh have followed up on these studies but with an industrial relations and labor and management relations perspective. Their research examines how labor relations and employee viewpoints interact with the diffusion of lean or flexible production as an alternative to mass production. Especially important to this is how HR systems and work organization impact on economic performance. Their emphasis is on collaborative problem-solving inside firms and throughout industry in managing human and social capital. Critically important to this is the interaction of strategic choices and industry structure as strategies and technologies change. **Joel Cutcher-Gershenfeld** from Brandeis University, who comes from a family of labor arbitrators with a PhD from MIT, has a similar background as MacDuffie and Pil. However, he is more focused on high performance work systems and especially how one builds consensus between labor and management within a lean production environment with extensive job flexibility. His focus is on large-scale systems change, high performance work systems, negotiation and dispute resolution, cyberinfrastructure, and especially new technologies that require higher skills (2015, 1998 and Gerschenfeld et al. 2017).

LERA—2. The second LERA approach is more focused on industries other than automobiles and especially high technology firms. Joel Cutcher-Gerschenfeld and Aileen Appelbaum are prominent in this approach. In *Knowledge-Driven Work*, Joel Cutcher-Gerschenfeld and thirteen co-authors (1998) study the cross-cultural diffusion of ideas about the organization of work between the US and Japan. These diffused ideas, linked to lean production and the knowledge of the workforce, have become a major source of competitive advantage in the global economic system. Looking at eight Japanese-affiliated manufacturing facilities, they demonstrate the complex processes through which the employment relation interacts with innovative ideas from both countries. They trace the flow of ideas from Japan to the US, and the "reverse diffusion" of innovations returning to Japan.

They look at six factors of work: (1) the cross-national diffusion of employment practices; (2) teamwork within production systems; (3) continuous quality control or *kaizen* and how employees are involved; (4) the development of employment security; (5) how human resource management systems work; and (6) the overall functioning of labor-management relations. First, there examination of teamwork shows that teamwork processes are incredibly important, but they are fragile. Second, their typology of teams shows the conflict between self-management and the necessities for interacting with other teams and units in the organization. Third, their examination of *kaizen* shows that employee-driven problem solving counteracts and balances the top-down ideas of management emphases on "re-engineering" and "benchmarking". Fourth, concerning employment security, most US managers believe that job security unnecessarily restrains managerial flexibility, but the Japanese managers at believe that it is essential to the flexibility itself that is part and parcel of effective teamwork and *kaizen*. Fifth, in their view of HR management, management illustrates its connection to competitive advantage even in diverse, older, unionized, and urban work forces. Related to HR is the innovativeness of the lean production system that comes from increased basic and advanced training programs.

Job rotation provides some of this, but effective training is crucial. Sixth, at a higher level, labor-management relations are important to the trust required for effective knowledge-based production. In many areas, especially in the United States, existing labor-management structures remained unchanged producing negative results. This is an area of increased concern that results in role ambiguity and mistrust.

The major theme of their work is "virtual knowledge", which involves the catalytic effect of many new ideas building into a workable program that promotes product development and new work processes that are more effective. They see this combination of explicit and tacit knowledge percolating in teamwork as an essential part of knowledge production in a firm setting. Again, the crucial element is the trust that is necessary to make these semi-autonomous teams work without fear (e.g., loss of job security or job safety). These procedures for collective problem-solving are based on stable teams in larger groups. The authors claim that some corporations who view employment security as a barrier to management flexibility end up destroying these opportunities to create new knowledge in competitive workplaces. However, for the new-age corporations that can create and use digital information in flexible organizational contexts, the rewards of a knowledge-driven workplace using continuous improvements create a competitive advantage for them through both innovation and quality.

Eileen Applebaum and Rosemary Batt (1994) are economists closely associated with LERA. They find two models important after the "breakdown of mass production" (1994:14) The first is a team-based model founded on socio-technical theory and quality engineering with decentralized management and decision-making. The second is an Americanized version of lean production that relies more on managerial and technical experience and hierarchical control. They argue that high-performance teams are more prominent in the high-tech section where wages and benefits are much higher (we discuss much of this model in Chapter 3 on socio-technical theory and in Chapter 8 on Nike, Apple, and Google).

More recently, the three Gerschenfeld brothers—one a LERA scholar, another a physicist, and the third a computer specialist—in *The Third Digital Revolution*, take flexibility to another level with many new technologies. Gerschenfeld et al. (2017) see three revolutions in technology over the last half century: the computing the communications and the fabrication revolutions. The first is computer-aided design and digital computation via computers to produce existing products from automobiles to the writing of news articles, fiction, and academic work in articles and books. The second digital revolution is in communications via smartphones linking people to each other and the Internet of things via computer chips in various devices, talking to each other from cars, to refrigerators interacting with external data services like the cloud. The third digital revolution is fabrication in the form of 3D printing and additive or accretion production of physical products, as opposed to the subtractive or cutting technologies, that create new products from the base up with little or no waste as compared to cutting technologies that produce waste even as efficiently applied as possible. For instance, a 3D printer only uses the resin it needs to build-up a product while a laser cutter completes a pattern and leaves material behind that is cut off or drilling that produces the waste of various types of filings (or sawdust in construction).

There are many very specific consequences of the third digital revolution since 3D printers or additive fabrication appears to be very inexpensive and relatively easy to learn. Cutcher-Gerschenfeld et al. (2018) sees the creation of digital fabrication labs (Fab-Labs) that are accessible to the general public as aided by the internet of knowledge with blue prints and extensive instruction (one negative application of this concerns the fabrication of guns). These technologies can be combined with laser cutters to produce sophisticated and decentralized production methods. For instance, one can easily create a car part, a dental crown, a computer item, or a cupboard hinge in a few minutes using a 3D printer, rather than going through a corporation that makes it and a distribution channel that sells it.[1] The jury is still out on whether this community-oriented production process, with its extreme decentralization, will work. For instance, the second industrial revolution was portrayed as starting in garages (Apple and Microsoft), but those humble beginnings were soon replaced by massive corporate behemoths that were created by acquisitions, highly complex programming, and dominant design and marketing schemes that are near monopolies. A garage or otherwise decentralized Fab-Lab are not going to seriously compete with the Googles and the Apples, and even if they do succeed with a major innovation, they will be bought out by the Internet or Foxconn behemoth. Still, much is possible in terms of a new niche in the division of labor.

Perhaps more likely than either scenario is that corporations will find an immediate application for additive technologies to supercharge their JIT training inventory practices. This will have an immediate impact on suppliers who previously designed and produced these component parts—an integral part of lean production. Much of their activities will be subsumed by the original-equipment assemblers. Suppliers will be forced to come up with new and probably highly design oriented products that are not so easily reproduced by additive production, which of course, they will probably use themselves.

The Human Resources Approach

HR departments select evaluate, supervise, train, and discipline employees in an organization. It evolved from the more basic personnel department to a vision of what HR departments as a much broader organization with many more duties. Specific areas that have increased the scope of HR are labor law, employment standards, employee benefits, employee grievances, background checks, drug testing, some aspects of payroll, and maintaining data bases on employees. They share hiring with the managers of the area and sometimes the teams in which prospective employees are intended to enter. One of the biggest areas of expansion has been discrimination due to gender, race, religion, and national origin.

The first personnel management department started at National Cash Register in 1900 in response to union unrest, and FMC followed in 1915 due to labor turnover. In the 1920s, personnel administration focused mostly on the aspects of hiring, training, and paying employees. However, they did not focus on any employment conflicts, employee performance, and discrimination. In terms of seminal research, HR departments were boosted by the Hawthorne Studies at the Western Electric company in the 1920s. This created the "human relations school" of organizational studies.[2] Considered the father of human relations school, Elton Mayo from Harvard and other found that employees work harder when their supervisors pay attention to them (the Hawthorne effect). HR departments started to take over policies that paid attention to communicating and cooperating with workers. This evolved into paying more attention to how prospective employees respond to changes in the labor market. Douglas McGregor as the X and Y theorist formulated a more person-oriented view of dealing with employees. Other well-known HR people are Richard A. Swanson (1942-present), Kurt Lewin (1890–1947), Lillian Gilbreth (1878–1972), and Channing R. Dooley (1878–1956). HR did not have a strong identity or theory, and many considered the departments from the 1920s to the 1970s as being relatively powerless and focused on record keeping and some benefits. When the NLRA passed and the NLRB sanctioned unions in the 1930s, Industrial Relations Departments and unions were the most important aspect of employee relations within the firm.

The civil rights movement and the decline of unions changed this. By 1964 and 1965, the civil rights movement led to workplace legislation about discrimination, and many lawsuits were filed against corporations. The Economic Opportunity Act of 1964 required that companies pay more attention to discrimination in the firm in hiring, promotions, layoffs, and firings. As firms became less and less unionized, these kinds of issues assumed more and more importance as personnel morphed into HR Departments (Dobbin 2009). Personnel Departments morphed into HR Departments that began their rise especially connected with training for employees and advice for managers about how to handle discrimination (Dobbin 2009). Now HR management is incorporated into most business schools and separate schools for labor and management relations that were connected to unions began to decline or reorient themselves to HR (note the change in name of the IRRA to the LERA, which is more HR friendly). The Society for HR Management (SHRM) then began to rise and become large and powerful. With the rise of lean production in a non-union environment, HR Departments began to take over grievance and conflict management functions but not wage bargaining.

By the late 1980s, HR departments emerged as more powerful since they controlled major areas of uncertainty, and they even made claims for breaking into the more dominant area of strategic management (Kaufman 2004). In fact, the overall value-added component to strategic management of performance is a critical claim of the new approach to HR management (Baron and Kreps 1999; Ulrich and Brockbank 2005; Leatherbarrow and Rees 2017). In the United States, government regulation developed beyond those regulations concerned with unionization. As a result, instead of simply hiring, firing, paying, and keeping records, the HR Departments started to deal with the employment relationship as a whole (Calveley et al. 2017), flexible working arrangement including transfers, training, job rotation and temporary employment (Hutchinson 2017), justice in the workplace including gender, race, ethnicity, and foreign recruitment (Neugebauer 2017), and the general demands of a globalized workplace (Rees and Smith 2017). As unionization continued to decline, HR Departments tended to take on union-like functions, including not just firing but the adjudication of the fairness of layoffs, retirements, and firings. For instance, Toyota developed a three-step process led by the HR Department concerning firings (Besser 1996). Much of these justice issues were propelled by the impact of law suits, and HRM interacted with the legal departments of their firms. These lawsuits could be very costly in terms of money and corporate reputation in the media.

In terms of organizational change, HR professionals most often rely on John Kotter's eight-step process from *Leading Change* (1996):

(1) Establish a sense of urgency
(2) Create coalition to guide and defend it
(3) Develop a vision with an associated strategy
(4) Communicate this vision of change to the organization
(5) Empower your employees for action
(6) Generating short-term goals that can be achieved
(7) Consolidating achievements and producing further changes
(8) Anchor your new behaviors and vision the organizational culture

Kotter's theory is popular in HR, and all three managers we interviewed at FMC mentioned it.

Currently, HR directors sit on company executive teams because of the firm's interests in planning and developing their employee base. The factors affecting HR planning are organizational structure, plans for growth, demographic changes, sensitivity courses for race and gender, environmental concerns, and overall talent management. In many lean production organizations, teams play an increasing role in HR management. The SHRM and HR professionals in general are pushing for HR to be a major element of strategic management. HR Departments also began to get involved in training for teamwork, training for new jobs and promotions, and overall compensation and promotion policies. Often involved with promotions, they handled transfers to new locations, housing, and finding the appropriate fit in a community—especially for foreign employees coming to the home country or home employees going to another country. This involved not only location but also immigration policies, visas, and solving cultural conflicts and disputes abroad. The management of increasing diversity became solidly in HR domain (Noe et al. 2017).

As HR expanded into the strategic realm, how did this connect to lean production? In a non-union environment, HR departments became the arbiter of the employment relationship, an area previously claimed by Industrial Relations Departments. In academia, the Industrial and Labor Relations Association (ILRA) changed its name in the 1990s to the LERA dropping "industrial" as the vestige of the previous era when IR was dominant with collective bargaining and grievance procedures. This was to purposely include HR functions.

In the lean production environment without unions, HR often become more of a protective or justice oriented department most connected to the handling of disputes, which was formerly held by union grievance procedures. Managers of production would often say that HR departments are "coddling" employees or "taking their side". While HR professionals are not union-like advocates, they have to balance production or other managers' desires to fire or lay off employees with their new-found functions of (a) developing a high performance work system with employees who are highly skilled and treated well, and (b) firing or laying off employees due to management concerns in the production process. This also involves counseling programs for alcohol dependence, drug abuse, and other often emotional issues that intrude on the workplace. In the field, the SHRM professional association offers many programs for dealing with these problems, but one has to differentiate the prescriptive claims from the evidence-based findings (Miller 2018; Lepadatu and Janoski 2011).

Lean production revolves more around high-performance workers with flexibility and often higher levels of training than mass production or bureaucratic environments. Workers are expected to participate in teams, produce high quality goods in process improvements (*kaizen*), and develop innovative and new products. HR is charged with finding these employees, promising them a high-quality work environment, and integrating these demands into a system that functions well and delivers on its promises.

Human Resources and Temporary Employment

Based on our analysis, temporary employment is a structural aspect of lean systems that directly involves HR departments. Although it is part of the lean production model, it contradicts other lean principles of a long-term philosophy, equality in teamwork and job rotation, and loyalty toward employees.

The JIT workforce is a human buffer for sick or injured team members, and a buffer against temporary increases in production volume that then disappear in the future leaving the company with too many permanent employee, which is when the temps are let go during times of economic recession. Although it is hard to generalize from our data collected from several major transplants and suppliers in the United States to the entire population of lean organizations, history shows that the Japanese practices often become benchmarks in the auto industry and the organizational models developed in the auto industry ("the industry of industries" in the United States) tend to spread to other industries. This means that temporary work is (or may become) a core structural aspect of lean production. The lean, loyal and long-term principles of mature lean production to protect permanent workers are based on the disloyal and short-term use of temporary workers.

Human resource scholars and sociologists study temporary employment. Sociologists are generally quite critical of it (Lepadatu and Janoski 2018; Kalleberg 2000, 2007, 2018; Rogers 1995, 2000; Smith and Neuwirth 2008). But it is HR professionals who have to deal with the problem in two ways. First, this is particularly important because some firms including Toyota (TMMK) have made temporary work a standard practice in hiring for the corporation. In this case, temps are hired by a subcontracting outside firm and they are then reviewed by supervisors and HR as to whether they would make a good permanent employee. But no matter how good they are, hiring them as permanent employees depends on the firm's needs at the time. If the economy is entering a recession, many good employees will not be hired on as permanent. Second, HR practitioners in the firm have to deal with the problems brought about by temporary workers. As Lepadatu and Janoski (2011, 2018), temps have a negative impact on teamwork because they "feel" that they are temporary and do not want to invest in teamwork or the team members do not want to listen to them. Also, temps may work "too hard" in order to get hired and this causes permanent workers to resent them for promoting a work speed-up. In either case, HR professionals generally "support and justify" the use of temporary workers, and temps are less likely to go to HR to solve their problems (Nollen and Axel 1996; Pearce 1993).

This raises the question as to pushing the envelope to make everyone a temporary worker (Hyman 2018; Levenson 2000). Certainly, this would increase the flexibility of firms in their use of the labor force. This model was somewhat presaged by the labor market models of upstream markets with firms increasing training and moving to higher value added products, downstream markets with firms avoiding training and moving to easily made products, and white water markets where employment where employees moved from firm to firm as their skills and emerging problems demanded (Hall 1996; Janoski, Luke, and Oliver 2014). The white water model was especially appropriate for Silicon Valley, where firms sought highly skilled programmers for special projects, and then those employees moved to the next firm for new projects—a consultant type of model where the employee was a self-employed entrepreneur. Free-agency in professional sports is similar with Lebron James or Kevin Durant moving from team to team. But the white water model doesn't work very well for most employees who are not labor market "stars" as in Silicon Valley or the NBA.

However, this is where the use of the word "lean" might be inaccurate. The Toyotism model does not recommend the unlimited or maximizing use of "lean" in cutting the labor force. It firmly embraces a permanent and loyal labor force for the long-term future of the firm. Again, lean production uses permanent employees who are then buffered by temporary employees. For the most part, HR as a discipline embraces this model, yet with its own internal contradictions.

The Interests of Human Resources Departments in the United States and the United Kingdom

There are some considerable differences in HR research, HR departments, and their predecessors in industrial relations departments. Although Industrial Relations departments were most often hard bargainers with unions, they did understand unions (i.e., know thy enemy) and as a result in comparison to other corporate departments, industrial relations would sometimes be looked at by other managers as being too friendly with the enemy. Since the 1960s and 1970s, industrial relations departments have faded with the decline of unions.

However, HR departments and the SHRM in the United States do have relatively highly paid management jobs for college graduates who are more interested in the people aspect of corporate work. In an environment where anti-discrimination laws have become tighter and more complicated, HR has control of an area where uncertainty abounds and multimillion dollar lawsuits can seriously impact the corporation. And their pay can be commensurate. So, while American human resource professionals may be seen as too liberal or careful, they do not want to kill the goose that laid the golden egg in terms of pay and status. US research in HR at universities reflects this.

On the other hand, HR departments at universities in the United Kingdom are more radical. Business schools in universities in general in the United Kingdom are more radical than the United States with many critical social scientists being in management departments. In fact, many labor process theorists are actually in business departments which is definitely not the case in the United States. In the United States, HR scholars and practitioners would like to see how HR could be brought together with strategic management and gain greater power within the corporation (Kornelakis and Veliziotis 2019; Delery and Doty 1996; Huselid 1995). In the United Kingdom, HR researchers look more at the "shop floor workers" in response to quality control and management (Glover 2000, 2010; Glover and Noon 2005; Glover et al. 2014; Butler et al. 2013; Butler et al. 2011). There is an overlap between the two countries on many issues like turnover, productivity, and diversity.

Thus, Bridgit Miller asks "is there a conflict of interest in HR Roles" (2018). Much of the differences between HR in the two countries is due to a bit more focus on gender and race in the United States, and a long-standing interest in class in the United Kingdom. In terms of lean production, the US version will err in the direction of productivity and personnel issues (individual complaints and grievances), while the British approach will look more from the shop-floor up concerning how lean production is perceived in a more group orientation.

Conclusion

These two approaches—labor and management relations and HR—show the way for a more diverse and employment centered approach to lean production. Their strength is their focus on employment relations and how workers' interests are represented while management objectives are also recognized with equal importance. This puts much of their analysis at the organizational and institutional level. Management and industrial engineering theories do not have this type of focus. Sociology and social science approaches also focus on workers and employment relations, but often they are less concerned with management goals of profit and firm strategy. Each discipline has different roots but they converge on the employment relations process. LERA comes out of the industrial relations tradition of collective bargaining and union formation, but as unions have declined, they have embraced more employment relations, which is largely in HR area. HR comes out of management theory, but since it is so focused on hiring and firing employees, it has become more active because of two factors. First, employment discrimination due to race and gender has become part of its bailiwick, and considering equity within the corporation has become important in avoiding lawsuits. Second, it plays a role in solving a number of teamwork and job rotation issues within the lean production framework. In fact, many team members rotate into HR itself. Extensive training/retraining and diversity programs then become part of HR. Yet at the same time, HR is a creature of management.

However, this focus on employment relations is also among their weakness. Both labor and employment relations and HR are not very focused on the actual production process. Sociology and labor process theories focus on this aspect of lean production in a much more micro-interactionist way. LERA is much more institutionally focused as is much institutional theory in the social sciences, but this "shop-floor" focus is not one of their strengths. Similarly, HR focuses on the employment relation, but they have to work within a management framework to whom they are ultimately bound. But their association (SHRM) is strongly engaged in a wide array of training, teamwork, and other programs that provide a menu of possible programs. But again, each discipline has its strengths and blind-spots, and LERA and HR are clearly strong on the employment relation.

Bruce Kaufman asks in two chapters of *Theoretical Perspectives on Work and the Employment Relationship* (2004) whether a more integrative theory of HR management and an extension of employment relations (i.e., a new industrial relations) might be possible (2004). As an example, one can look to the Ford-UAW cooperation on flexibility that Cutcher-Gerschenfeld et al. (2015) describe. Gregor Gall and Jane Holgate talk about "rethinking industrial relations" (2018), and Miguel Lucio and Robert MacKenzie (2017) look to the state for additional regulation. These two approaches indicate a contradiction in these two fields. Both see the protection of workers to be part of their mission, but one leans towards management and the other toward workers. Fusing the two is a tall order, so one should keep in mind the subtle differences between these two fields and how that might affect the results that their research puts forward. These two fields are small. Industrial relations has departments in about 30% of schools in the United States, and has fewer and fewer departments in corporations. HR are a subfield of management in academia. Nonetheless, they make valuable contributions to the study of associates and employs in the study of lean production.

However, what is really needed from these two disciplines is a new approach to the protection and advancement of labor other than past dependence on a declining labor movement carrying the albatross of the NLRB around its neck, and a move away from the timidity in the face of management from HR researchers and professionals. Working together, they could forge a new vision of labor's role in the economy. Such an approach might be what Thomas Kochan and Lee Dyers call for as a "new social contract" (2017).

Notes

1. Cutcher-Gerschenfeld et al. (2018) refer to people or communities owning "the means of production". While their Fab-Lab survey of 179 leaders was impressive, one wonders if the barrier of "bringing these products to scale" will make them profitable vis-à-vis the corporations who will be intent on bringing them to mass markets themselves.
2. The human relations school has been criticized for not being able to connect employee job satisfaction to job performance. Hence, it started to be called "cow sociology" after the Borden Company's commercial that claimed that contented cows produced better milk. None the less, HR continues in the tradition of work satisfaction and trust.

5 The Wider Sweep of Global Lean Production

Diversified Quality Production, Models of Production, and State-Led Capitalism

Germany and France have two specific approaches to lean production and industrial relations that are different from the disciplinary approaches listed earlier. The *Modell Deutschland* approach describing the German industrial relations system has morphed into the more specific "Diversified Quality Production model" that incorporates aspects of lean production and other approaches (Sorge and Streeck 2018, 1988; Streeck 1991; Herrigel 2015; Streeck 2014; Herrigel 2015). The French regulation school of institutional economics and sociology has developed more specific view of productive models that places lean production in a larger context of institutional economics (Boyer and Freyssenet 2002, 2016). This French approach includes additional aspects of marketing, capital accumulation, and government relations. The two models differ in that DQP is specifically a model of Germany, albeit in including global German manufacturing and services, whereas the French Productive Models approach is about many different national models. Nonetheless, these two approaches have distinctive angles on lean production in its world expansion. A third, totally different approach is the state-led capitalism that is evident in China and, to some extent, Singapore. This approach is critical of Western capitalism and extols the virtues of Asian economic growth that has a much stronger state element in its expansion.

Diversified Quality Production in Germany

German social scientists have a long tradition of institutional analysis going from Max Weber to economics with *Volkswirschaft* or institutional economics. Arndt Sorge and Wolfgang Streeck first put forth their model of German production in 1988, but it had long been recognized since the 1950s Economic Miracle that institutions and quality production were different in Germany (Sorge and Streeck 1988, 2018; Köhler 2019). This chapter is about viewpoints related to lean production that are quite diverse, but from a common source in Europe rather than the United States or Asia. The first is DQP, which comes from Germany but is also a description of the German model of production. The second is the Productive Models approach, which comes from France, but it is not necessarily a description of French production methods. Instead, it comes from the regulation school of production but is expanded to include the Japanese methods of production in specific models. Each approach gives a relatively unique perspective on lean production.

Components of DQP. The Components of the diversified quality production system in Germany consists of four main parts. First, the German system is an advanced training system for the whole workforce, not just college educated graduates. As a result, in addition to an excellent system of higher education, a vocational education and training system that is based on early training in specialized fields within specific firms, and examinations assuring competence when students graduate at the age of 17 or 18. This is opposed to other systems that provide a more general form of education somewhat modeled on a watered down college education system. The training system, sometimes called the dual system, provides apprenticeship training and specific schooling that produced more highly skilled workers than found in other countries (Thelen 2004; Janoski 1990; Sorge and Streeck 2018:590). This training is more extensive than that found in lean production systems.

The highly trained workers provide the basis for a much more flexible and talent rich organizational system. It allows organizational structures that provide overlapping and enriched work roles in departments, occupations or jobs. The high skill levels developed in one above provide the basis for this design principle of interpenetration and overlapping expertise rather than the segmentation of skill levels within boundaries. One can refer to this as a process of sharing different types of information coming from diverse sources of expertise.

Second, the German organizational system is more decentralized than most with a high degree of worker participation at different levels. These institutional structures are legitimized by state legislation. This institutional system allows for decentralized authority in three ways. First, this system requires fewer supervisors since workers can generally work their way through problems on their own. Second, the organization is organized in such a way that information flows vertically and horizontally in an unobstructed way. This requires a great deal of decentralization that at the same time can be controlled and channeled by the organization. Third, the characteristics of these multidirectional flows are the works councils at the plant level and codetermination in corporate boards. This amount of decentralization requires a large amount of trust between labor and management at all levels. The wage determination system is not done within the firm in most instances, but is rather conducted in state-wide collective bargaining. There is a lasting social peace between management and workers that allows enough trust to develop new production processes and new productions. This does not mean that there are strikes or various demands made by the unions but that the resolution of these conflicts is expected to be done on a relatively routine basis. It is noteworthy that workers have a bit less ability to stay out on strike for long periods of time because a whole industry (not just a company) will go on strike for a whole region of the country. It is very expensive for the union to absorb these strike costs. Also, it is expensive for the employer to afford the shutdown of a whole industry. As a result of these high stakes, solutions to problems are generally achieved in a relatively short amount of time. This is rather different from American strikes of one company at a time.

Third, a strong form of active labor market policies, which may fluctuate over decades, provides support for workers and firms to engage in this highly institutionalized system. The state supports the system in two ways. First, laws require that when laying off workers, firms must provide three months' notice during which the *Bundesantalt für Arbeit* or federal employment service works with the firm and works councils to create a social plan that will reintegrate workers into the labor force (Janoski 1990). This may be by finding them jobs in other industries, or it may be in terms of providing for retraining for laid off workers and the otherwise unemployed (Janoski 1990). Second, the federal employment service also provides advanced training for those not threatened by unemployment but who need or the company needs new and more advanced skills. Thus, the employment service has a reputation not just for the unemployed but also for social mobility and advancement in terms of promotions and pay increases.

And fourth, Germany has a system of co-determination that is applied in two ways. The first way is with legislated works councils, which began as early as the 1920s but in their present form come from national legislation in the 1950s. German firms above the size of 15 may have a works council with worker representatives elected and the council meeting with management to co-determine certain HR and technology policies (i.e., the works council can veto management's proposals). On other issues, they have advisory rights to be informed and to register their opinions (but not stop the policy). The second way is with codetermination on corporate boards, where workers and employees can elect representatives to be members of the board of directors (called a supervisory board). There are three laws on these boards, but most of them give workers and employees slightly less than half the representatives. These two policies form a pattern of participation and economic democracy that make Germany relatively unique. But the EU has also enacted "European Works Councils; that have also spread to other countries in the EU (Krzywdzinski 2011).

Box 5.1 The German Model and Diversified Quality Production

The German model had a difficult time with Fordism but found a modicum of success with it. Similarly, it encounters lean production with a full embrace of some concepts and a rather tenuous distance on others. The German system consists of many parts, but the most obvious are codetermination and works councils. The three major **codetermination** laws were passed by the German government with the first one on coal and steel firms in the 1950s, and other firms in the 1970. Each law is somewhat different, but they mandate that employee and worker representatives must be members of the German board of directors of the firm. These representatives are from 40% to 50% of these boards, and would be an anathema to US corporate leaders. But they work well in general and there are strict rules about board members of any sort revealing company secrets, which if they do, they can be prosecuted and jailed. **Works councils** operate on a different level on the shop floor. Each firm above a certain, rather small size allows its employees to elect a works council representative with larger firms having many representatives. These works councils have powers to reject certain decisions related to staffing, discipline, new technology, and other processes. They have rights to be informed on many other issues. Germany does have trade unions who negotiate wages for industries on a regional level, but works councils have nothing to do with these negotiations as a clear line is drawn between employment rights on the shop floor (works councils) and wage levels (trade unions. However, union representatives often gain positions on works councils. The director of HR at a major VW plant and member of the VW global supervisory board is a worker who made his way up from the shop floor. When asked what it was like being on the corporate board, he said somewhat tongue in cheek: "It was like going over to the dark side". But clearly, he could manage both roles, and there are also considerable legal penalties for workers revealing any confidential information about company strategy.

The German model is wider than codetermination and work councils and involves the structure of the very large apprenticeship system and levels of cooperation between trade union federations, employer federations, and the state. Intellectually, it comes out of the German *Volkswirtschaft* tradition of institutional analysis, which has no equivalent in the Anglo-Saxon world. Arndt Sorge and Wolfgang Streeck have been at the forefront of discussing how the German model works and to what extent lean production principles have been employed in what they call "diversified lean production". The two most prominent advocates of the DQP model are **Arndt Sorge** (1945-present) who is the director of the Research Unit in Internationalization and Organization at the Wissenschaftszentrum in Berlin (WZB); and **Wolfgang Streeck** (1946-present) who is the Director of the Max Planck Institute for the Study of Societies in Cologne.

Changes in DQP. What we have just presented is the original model that relied on Arndt Sorge and Wolfgang Streeck work in the 1970s and 1980s. However, with the fall of communism, the unification of Germany, and the rise of neo-liberalism there have been major changes in the model. The former East German wages and regulations were equalized with West Germany, and the result was that investment avoided East Germany and went to Poland and Czechoslovakia where wages and regulations were much lower. In response to lower growth and budget deficits, the Hartz reforms in the welfare state and labor market took effect in 2002 and 2003. Peter Hartz, a HR director at VW advised Social Democratic Chancellor Gerhardt Schröder on reforming the welfare state. His recommendations reigned in German unemployment compensation through the reforms called Hartz I through Hartz IV. The committee devised thirteen "innovation modules", which recommended changes to the German labor market system. These were then gradually put into practice: The measures of Hartz I–III were undertaken between January 1, 2003, and 2004, while Hartz IV was implemented on January 1, 2005. The "Hartz Committee" was founded on February 22, 2002. Hartz I and II provided support for vocational training, mini-jobs, entrepreneurial grants, and more job centers. Hartz III restructured the job centers and changed the BAA into the *Bundesagentur für Arbeit* (Federal Labor agency). Hartz IV consolidated different types of unemployment benefits and set them at a lower level of *Sozialhilfe*, which is more of a welfare rather than earned benefit, and requirements for receiving the benefits were stiffened. This created a rather strong incentive to get back to work after a layoff. Some claim that this caused the resulting boom.

Thus, they make three points. First, they show why and in which ways DQP was more heterogeneous than we had originally understood. Second, based on political and economic changes in Germany, they show that the DQP Mark I regime with 1980s characteristics turned in to DQP Mark II processes under neoliberalism. As a result, major "complementarities" disappeared between the two models especially concerning the connections between the production modes, industrial relations and economic regulation. Regulation shows the largest change, but business strategies and production organization show more continuity. This is why Germany maintained stronger economic performance after the mid-2000s. Their strongest theoretical point is that the complementarities emphasized in political economy are historically relative and limited, so that they are not fixed configurations.

The Nordic Model of DQP. The German model has many affinities to the Scandinavian system that is similar in many ways and different in a few instances. In Sweden, there is a similar path of DQP to Germany. In the Sweden I model, corporatist bargaining between strong trade union federations, employer federations, and the state produced a low unemployment and high welfare state approach to social policy, and an emphasis on keeping wages down for corporations to sell high technology products in a global market place (Berggren 1992, 1993; Pil and Fujimoto 2007). However, in the Sweden II model, wages and unemployment were allowed to rise with some restrictions on the welfare state. However, Sweden is still recognized as having the strongest welfare state in the world, except perhaps by is cousin Norway which also has extensive oil revenues. The education system is not so much based on the apprenticeship system as in Germany, but education is excellent but more targeted for national needs that the US would ever be. Finland, a former colony of Sweden, is also renowned for the excellence of its educational system (*The Economist* 2013).

Evaluating DQP. The diversified quality production system has its promoters and a few detractors. First, Wolfgang Streeck is one of its foremost promoters, and he states that although DQP has been changed by lean production and also neoliberalism that it remains a unique model that has some influence beyond Germany with powerful German multinational automobiles, chemical, tool and die, and pharmaceutical firms. Sorge and Streeck in 2018 reassert their DQP model by bringing it up-to-date on many of the changes over the 30-year history of their concept. They discuss the unification of Germany with a rise in unemployment and a decline in unionization. Although unionization has decline, it is still high in the automotive sector, and the unemployment rate since 2011 has been lower than the US by almost 3.0% in 2011 but has narrowed to 0.7% difference from the US. This is remarkable given that massive tax cuts for the US correspond with much higher taxes in Germany. Sorge and Streeck indicate that as DQP has evolved with changes in lean production and neoliberalism; however, most of its main features remain intact though modified (2018:610–11). However, the model has change. There is higher quality and more competitiveness, while the model is more unstable and may change further (606–607). However, despite the instability, it remains unique with works councils and codetermination intact with government legislation. Lucio Baccaro (2018) says that DQP has changed a lot, but that the theory has moved from a firm strategy to a larger political economy strategy (see also Baccaro and Howell 2017). Virginia Doellgast largely concurs (2018). Fossati (2018) adds that political regimes with particular institutions can shape firm's choices in terms of teams and cooperation. This makes DQP take a concerted theoretical shift to a larger theory. In large part, these are highly friendly critiques (Jürgens 2004).

Martin Krzywdzinski (forthcoming) and others says that lean production and neoliberalism has largely overwhelmed the traditional institutions of Germany. He discusses the development of the "humanization of work" in the 1970s, and then "group work" from the IG Metall union that was a response to lean production. A number of group work projects were carried out in the 1980s and early 1990s. But he calls these efforts as insufficient. One of their objectives was to extend *takt* times and Mercedes-Benz in the 1990s still tried to do this. But this paled in comparison to the onslaught of lean production and the successful Christian Democratic Union (CDU) governments. Krzywdzinki argues that "Lean production wins out" (7) and states that it became dominant. Many German firms used "holistic production systems" instead of lean production, but the conception was the same. As a result, four things were solidified:

(1) Teamwork became important, but German approaches supported the election of team leaders and quality control and maintenance to be integrated into the teams
(2) Team participation in *kaizen* activities were included in plat collective bargaining agreements
(3) The supervisor or *meister* was redefined into more goal setting, participation, and coaching
(4) New forms monitoring performance replaced piecework regulation with agreements

The Auto5000 project at Volkswagen in 1999 involved the company and the works council to find an alternative to lean production. Under this plan, 3,500 unemployed workers were to produce a new Touran SUV in a new factory in Wolfsburg. The plan was embedded in the IG Metall collective agreement with VW. Pay was 20% lower but the works council had stronger rights to control the project. Team sizes ranged from seven to 17 employees with job rotation, multi skilling, elected team leaders, and problem solving. Added to this was a development and training consultant (Krzywdzinski forthcoming; Eurofound 2001). Krzywdzinki does not see this as an alternative to lean but rather an actual implementation of lean influenced by German institutions. The company, trade unions, work councils, and academics see it as a success (Schumann et al. 2006), but it failed because of the 20% wage cut. By 2009, Auto5000 was integrated into the VW collective agreement and team-based improvement became quite limited. But this is not a surprise. After 20 years (1999–2009), why would the workers in Auto5000 accept wages that were 20% lower and even then when they were inside the VW complex surrounded by higher paid workers? The company answer might be—well you were unemployed and we gave you a job. But that wears thin after 20 years. Contrary to the plan, how well would Toyota have done in Georgetown if they offered wages 20% lower than American auto companies? We would think—not well.

Since 2010, German auto companies have been busy with what they call *Industrie 4.0,* which integrates the internet of things into automobiles. This makes automobiles much more complex and involving extensive computerization. The traditional "machine whisperer" (Kryzwdinski forthcoming) has become a data processor and programmer. This change is still in progress and its impact on teamwork is uncertain.

From German lean issues, we move to German plants in Mexico, the US and the BRICs. In general, DQP in German transplants in the rest of the world is still partial in many ways because their institutions and laws are quite different from Germany. US labor law and the NLRB have cause problems in implementing works councils at the Chattanooga, VW plant.

The attempted works councils were declared illegal unless union participation was present. The non-union union opposed this. While German transplants often assure that sufficient training facilities are present, this still does not simulate the German experience where training is advanced and available to everyone (Janoski 1990). In India trade unions are quite strong, but in China, they are dominated by the state and rarely protect workers. Nonetheless, VW has implemented many of the elements of the German system in Brazil, Russia, India and China (the BRICs). The least success has been in Russia, but the other three countries have adopted most of the system (Jürgen's and Krzywdzinski 2016). However, going back to the Auto5000 case, if lowering wages and increasing flexibility and overtime is the criteria for lean production, then the BRICs certainly demonstrate these criteria in spades compared to high wage and highly regulated Germany.

However, looking at auto companies hides many of the strengths of the DQP system in Germany. Much of the German system is engaged in a form of batch production with few assembly line technologies attached to it. These are major energy and tool and die projects that are done by Siemens AG, Trumpf and Schuler, and many other firms. For instance, Siemens has been in existence since 1847 and it produces industry, energy, healthcare and infrastructure products. It is the largest industrial manufacturing company in Europe. It is now based in Munich and has over 350,000 employees. As an example of batch production, Siemens and partners won the *Power Magazine* "Plant of the Year Award" for the Egypt Megapower Generation Project that involved twelve 445-ton gas turbines for three power generating plants that created 4.8 Gigawatts of electrical energy. The turbines had to be transported by 192-wheel vehicles in Germany, and after being transported by sea, a 40-axle trailer in Egypt. The project was completed in two years and 3 and a half months. It has taken Egypt from 25% blackouts to electrical energy exporting (Larson, Patel and Proctor 2019). These turbines are not created on assembly lines with complete standardization but instead require highly skilled workers operating according to precise plans. Trumpf in Ditzingen, Germany is in a different but similar batch-oriented industry with 13,420 employees in 2018 and they produce laser cutters and machines that make machines in the tool and die industry (Trumpf 2019; Dougherty 2009; Neary 1992). And Schuler Presses (since 1839 but now part of the Andritz AG) makes stamping and cutting technology. The new Audi factory in Puebla, Mexico had a number of the gigantic presses, but what was notable was how they worked. At TMMK, the stamping presses shook the concrete floor and our bodies like an earthquake, while at the new Audi plant, the Schuler presses were cushioned and at first we thought they weren't really working. While there are companies who do similar businesses around the world, Germany is unique in having a larger number of them because of the strong vocational education and engineering system we mentioned earlier. It makes Germany predisposed to this highly skilled and somewhat unique approach to production. This doesn't mean that it does not do lean production, but it has to deal with quality in an entirely different way. Like building a bridge, you need to be right the first time. *Takt* times are more like three to six months rather than 70 to 120 seconds. Taking into account this more batch processing firms gives a more rounded out picture of diversified lean production in Germany.

Productive Models Approach coming out of the French Regulation School

The Parisian or French "regulation approach" was a theory that developed in the 1980s and 1990s in an institutional and Marxian vein to explain the system of capitalism at that time. It came out of a critique of Althusserian Marxism, which was more heavily deterministic and often relied on "the last instance" of Marxian results when faced with momentary changes. Alain Lipietz (1987) said "we are the rebel sons of Althussier" (1993). More early works of Michel Aglietta (1979), Robert Boyer (1990), and Lipietz (1987, 1997) and others in the French regulation school.[1]

The general argument somewhat reversed the Marxian sub- and superstructure to show that instead of one capitalism in constant reiteration of itself as technology and society change, there were multiple forms of capitalism that were quite different and that this was especially so because of the different institutions in a plethora of nation-states that both affect the development of capitalism. These different strains of capitalism then became known as the "productive Models" approach of Boyer and Freyssenet (2002). But we first develop the initial regulation school because one cannot just leap to the specific models of Boyer and Freyssenet.

The regulation school originated in French economics (Jessop 2001a:ix), but it is not a description of one country, but rather a theory that classifies types of capitalism in a number of countries. It combines institutional economics, evolutionary economics, and a new form of political economy that is a midpoint between neo-classical and Marxist economics of social-political economy (perhaps more Weberian but without the driving rationality) (Albritton 1995:431). The changes in the new political economy are based on the social action of class and other struggles (Jessop 2001a:ix), and is hence not deterministic or based on "one model of capitalism".

The regulation school approach is based on three components (Jessop 2001b):

(1) **The Industrial Paradigm** is a shop floor and meso-management system of production composed of technical and social components. Mass production or Fordism is an application in the firm (rather than the whole economy) is a frequent example.
(2) **The Accumulation Regime** is a long-term pattern of production and consumption in which money circulates through various parts of

the economy. The regime of accumulation goes beyond Marxism to include the thought of John Maynard Keynes, Michał Kalecki, Joseph Schumpeter, and especially Michael Polanyi.

(3) **The Mode of Regulation** departs from Marxian theory to take into account the institutions, organizations, and supply chains of various industries (Jessop 2001b). Here, the regulation school explicitly takes into account the state as it affects economic institutions, corporations and community networks in how they modify and constrain the previous two elements of the industrial paradigm and accumulation. This has a strong connection to the "varieties of capitalism school" but it sees more variety and puts class and other struggles at the forefront of change (Hall and Soskice 2001; Marsden 1999).[2]

These broad categories are then developed in four ways. First, the industrial paradigm has two components. Wages are set by labor markets in hiring and controlling workers, which has a lot to do with the lifestyles and expectations of workers. Second, the capitalist enterprise itself exists with particular forms of finance capital in markets with extensive competition with other firms in a negative network of finding the right price and quality niche. Firms must generate profits from their positive network of suppliers, and if a sub-contractor, a network of industrial (usually OEM) buyers. Third, moving from firm finance to the more general accumulation regime, the circulation of money is necessary for bank loans and stock markets that provide a source of funds for corporate investment, but also provide a consumer and employee loans. While credit mechanisms and stock markets provide for financial investments for corporations, they also provide a means of grading firms on their future prospects through credit ratings and stock prices, which further guide general investment in the market. This also affects the future of employees in firms as their firm's evaluations go up, down, or stay the same.

Fourth, the state and institutions provide guidance through opportunities to develop or constraints to tighten for firms. These institutional effects and regulation can offer tax write-offs for homes for employees, and encourage businesses to pursue certain investment, or avoid them (e.g., speculation, fraud, or exploitation). It is important to note that's non-state institutions—neo-corporatist bargaining, voluntary association assistance, collective bargaining by unions—have an important role. In the US, examples are university and school accreditation associations, professional practice associations (e.g., AMA, ANA for Medical licensing). They also include private governments (e.g., the NFL, NBA, MLB, NCAA, the Olympic Committee, and FIFA in sports) that govern themselves with episodic interventions by lawsuits. It is the unique emphasis of the regulation school that the different combinations of these state and non-state organizations create forces that mold capitalism.

Fifth, the global fields of nation-states and capitalism produce an array of global institutions to regulate society and the economy. To start the process, they are international forms of trade-based or norms of protectionism and tariffs and liberal (increasingly neo-liberal) forms of exchange. A vast array of global and regional organizations has been developed to arrange on territorial regions and world sites (e.g., the WTO, World Bank, GATT agreements, NAFTA/USMC trade agreements, the Eurozone and EU, the UN and bi-lateral agreements between nation-states). Meanwhile the most global institution of all is financial. Firms, banks, and stock markets who measure success or failure in the world economy.

The regulation school examines two areas pertinent to our discussion: the decline of Fordism, and the rise of new models. First, the decline of Fordism was explained by a distinctly worker-led approach. Fordism declined because it alienated workers with the standardization of mass production, it promotes deskilling of workers which is dispiriting, and it created worker dissatisfaction contributing to conflict at work that in turn produced declining productivity, lower quality, higher costs, and declining profits (Edgell 2011; Aglietta 1978, 1979). This in turn produced declining incomes that resulted in lower tax revenues, which then promotes cuts in the welfare state as a threat to the Fordist compromise between labor, employers and the state. In other words, the rise of the neo-liberal political paradigm. However, the theory tends to discount the more consumerist explanation that says the massified standardization of Fordism in products and life-styles produced a demand for greater variety of products in stores and dealerships. Along with the introduction of computer technology, led to the ability to produce more diverse products from cars to food (Edgell 2011:100–1). For instance, the mass produced Budweiser beer in the United States eventually gave way to micro-breweries making unique but more expensive beer then led Annheiser-Busch to become more flexible and produce a wide variety of beers from Michelob Ultra for the health conscious to Amber beers that are heavier with more taste.[3] They also purchased or created ten different craft breweries. While the Regulationists might concede that consumer tastes changed, they find the profitability crisis coming from worker dissatisfaction to be the dominant force.

Second, the regulation school looks at the rise of post-Fordist production methods with Boyer and Freyssenet's "productive models" theory they developed starting around 1996. This involves a more specific theory of production processes that involves a wider range of materials initially focused on a firm's profit strategy comprised of a compromise between product policy, productive organization and employment relations. This is the then embedded in product and labor markets and then connected to a national income distribution and growth system at the nation-state and international regime level (Boyer 1990, 2004; Boyer and Freyssenet 2000, 2002, 2016; Freyssenet 2003, 2009; Boyer et al. 1998). It emphasizes the development of profits through different strategies of the automobile industry and develops the following models:

(1) The Taylorian model based on time and motion studies to find the one best way
(2) The Woolardian model that involves mass production but also an early form of JIT inventory
(3) The Fordian model based standardized production and on volume achieved through mass production
(4) The Sloanian model based on volume as the other approaches but emphasizing product diversity (instead of any Model T that you want in any color as long as it is black versus the diversity of multi-colored Chevrolets, Buicks, Oldsmobiles, and Cadillacs)[4]

    But more relevant for lean production, they introduce two productive models from Japan.

(5) The Toyotan model based on permanent cost reductions at a constant volume, which is similar to Toyotism but with a somewhat different profitability focus
(6) The Hondian model based on product or technical innovations and flexibility

Their productive models approach finds more variation in the typically labeled Fordist era, and then finds variation in the present stage of lean production (Boyer 1990; Boyer and Freyssenet 2002, 2016; Boyer and Drache 1996; Freyssenet 2003; McMillan 1996). Both of the two new models are based on flexibility in the production process, but their marketing strategies differ.

The productive models approach is much broader that typical lean production models. It has a generous portion of political economy with institutions and questions of finance and accumulation of capital. Brenner and Glick (2001) and Albritton (1995) and probably world systems theory criticize the productive models approach because it does not have an overall concept of capitalism. However, we would say that this is exactly their point. It is not useful to go after nomothetic conceptions of capitalism that fit poorly in many countries. Looking at its lean production aspects, one can find much to disagree with especially concerning "contradictions".

However, the productive models approach has a facet quite different from other analyses of lean production. It looks at two factors: (1) the pertinence of the profit strategy in relationship to the nation-states national income and distribution and economic growth patterns; and (2) the nature of the state and enterprise governance compromise that allows the firm to be successful. This involves aspects beyond production. First, the corporations embedding in a certain political and economic system. And second, the design, innovation, and marketing side of the productive process, especially at how profits are produced. Thus, the profit strategy is paralleled by the product market and labor market, embedded at the national level in the enterprise governance compromise with a certain product, and more largely dealing with national income and the international regime. What is the distinctive characteristic of each company that drives success beyond production? According to Boyer, Toyota is based on "the permanent reductions in costs" strategy based on impeccable quality at relatively low costs produced at a constant volume (Boyer and Freyssenet 2002:77–88, 2016). While Corollas and Camry can be criticized on the basis of conservative style, the Prius and hybrid cars relying partly on batteries is a major innovation that Honda also developed but did not effectively implement like Toyota did. Also, Toyota's strong emphasis on teamwork and *kaizen* seems to be unmatched by Honda. But Boyer and Freyssenet emphasize a somewhat mundane approach based on costs and quality.

Boyer and Freyssenet see Honda pursuing an "innovation and flexibility strategy" that are based on quick reactions and flexibility to changing conditions to produce new innovations (2000:89–100). While Toyota is firmly focused on automobiles, Honda sees itself differently. It is a power company that currently focuses on engines in a variety of forms—automobile, motorcycle, lawnmower, outboard motor, generator, and jet aircraft engines. Boyer sees Honda as a distinct innovator taking more risks than Toyota. Certainly, Honda has participated in Formula 1 racing for a longer period of time with three manufacturer championships (1962 compared to Toyota starting 40 years later in 2002) but Toyota has a stronger presence in NASCAR, which Honda avoids. In some ways, the Honda strategy is somewhat of a "David and Goliath" approach. Toyota has been the big "generalist" corporation, while Honda is the smaller or "niche" firm that has to take a new innovation or design and be first into the market with "conceptually innovative models" (2000:402–4).

Box 5.2 The Regulation School: Robert Boyer and Michel Freyssenet

The Productive Models school comes out of **Regulation Theory**, which is a neo- or institutional-Marxist theory not about simple government regulation but rather about the rules of the capitalist system that shape production and business in general. **Robert Boyer** (1943-present) is the most prominent writer in this tradition. He is a French economist with a combination of a Marxist, institutionalist, and rather independent orientation toward the world of work and how it is organized. He was born in 1943 during World War II in Nice and now does his research in Paris as a professor and member of a number of high powered research groups. Boyer examines the regularities and changes of economies over the long-term. And he is especially interested in how macro-economic and institutional developments vary across countries at the same point in time, especially in comparing France and the United States. The first iteration of his analysis of institutional change is *Régulation Theory*, which targets the permanent transformations of capitalism and its crises (it is not related the English-speaking field of government regulation). Instead it focuses on important controls and changes in institutional and historical macro-economics, especially wage and worker interaction, innovation systems, monetary and financial systems, and the integration of national economies in the global economy. The second iteration is the **productive models theory**, which presents a variety of different approaches to efficiency, quality and success.

Michel Freyssenet (1941-present) received his PhD in Sociology. Initially, he worked on formulating plans for developing regional development plan for the Ministry of Works Public and Habitat in Algiers for the program of sociological studies in planning the development of the main cities in Algeria. He has worked for CRNS and in 1982 co-founded and co-coordinated with Patrick Fridenson the **GERPISA** (*Groupe d'études et de Recherche Permanent sur L'industrie et les Salariés de l'Automobile*) network that has done yearly conferences and major research on the automotive sector throughout the world. He was the co-director from 1992 to 2006. GERPISA maintains an international network of researchers in the automotive industry, which is closely related to lean production. Its conference this year is "Paradigm Shift? The Automotive Industry in Transition—the 26th International Colloquium of GERPISA".

While the productive models approach has a point in looking at a larger range of strategies and contexts, which aligns it more with the strategic management approach, we see the two companies as more similar than different (as discussed in chapter 6).

State-led Capitalism in China and Singapore

Lu Zhang (2015a, b) reports that lean production developed in the Chinese auto sector with x, y and z. However, Chinese firms tend to rely on a greats deal of excess labor since it has long been quite cheap. The main use of lean production is in JIT, but much labor has made it work.

But aside from lean production, China represents a non-liberal model, which is called state-led capitalism. Ian Bremmer in *The End of the Free Market: Who wins the War between States and Corporations* (2010, but also see 2009 and 2018; Huang 2017; Woolridge 2012) describes China as the primary driver for the rise of state-led capitalism as a challenge to the free market economies of the developed world, particularly in the aftermath of the financial crisis of 2008. Bremmer draws a broad definition of state capitalism. In the state and actually party-led system, governments use various kinds of state-owned companies to manage the exploitation of resources that they consider the state's crown jewels and to create and maintain large numbers of jobs. They use select privately owned companies to dominate certain economic sectors (Baumol 2007). The state uses sovereign wealth funds to invest their extra cash in ways that maximize the state's profits. In all three cases, the state is using markets to create wealth that can be directed as political officials see fit. And in all three cases, the ultimate motive is not economic (maximizing growth) but political (maximizing the state's power and the leadership's chances of survival). This is a form of capitalism but one in which the state acts as the dominant economic player and uses markets primarily for political gain. Following on Bremmer, Paul Aligica and Vlad Tarko (2013) further develop the theory that state capitalism in countries like modern day China and Russia is an example of a "rent seeking" society. They argue that following the realization that the centrally planned socialist systems could not effectively compete with capitalist economies, formerly Communist Party political elites are trying to engineer a limited form of economic liberalization that increases efficiency while still allowing them to maintain political control and power.

Nial Ferguson (2012) warns against "an unhelpful oversimplification to divide the world into "market capitalist" and "state capitalist" camps. The reality is that most countries are arranged along a spectrum where both the intent and the extent of state intervention in the economy vary" (get quote source). Ferguson then states that the real conflict is not between a state-capitalist China and a market-capitalist America and Western Europe. Instead, it is with all three regions in attempting to find the right balance between the private economy and the state in trying to regulate and share the wealth of capitalism. In other words, the real struggle is redistribution.

Julan Du and Chenggang Xu (2005) find the socialist market economy of state capitalist system directly opposed to a market socialist system because financial markets in China are different from socialist markets. The state profits are retained by enterprises rather than being distributed among the citizenry in terms of social dividends. They find state capitalism to be neither market socialism nor a long-term form of capitalism. Xi Jinping has rolled back increasing freedoms of the previous premiers/presidents (Economy 2018; Mann 2007), and has affected the development of lean production and semi-autonomous teams in China (Epstein 2010). Ian Bremmer does not believe that China will ever develop a free enterprise system (2009, 2010, 2018). However, as Chinese wages rise and they emphasize internal innovation, and as technology transfer from the West declines, China will lose competitive advantage of low price labor and gain only partially form innovative teams and research. Free speech is generally a requirement for this. The result will be a need for industrial espionage in the face of slower growth (Musacchio and Lazzarini 2018; Musacchio 2015; Scissors 2009; OECD 2009). On the other hand, Florian Butollo (2015) finds that state-led or private firms may profit from state guidance, but average workers (e.g., in his study garment and LED lighting in the Pearl River Delta) will not benefit much from their profits as wage increases will rise only a little over the cost of living. Consequently, in either the higher wage or stagnant wage model, Chinese growth will be increasingly limited.

Singapore has developed a growth model uniquely situated for their city-state. It combines urban-planning with a form of state capitalism. Bremmer (2010) calls it "a system in which the state functions as the leading economic actor and uses markets primarily for political gain" (get page number). The approach, long led by leader Lee Yen Ku, uses academic studies, government reports, and corporate information to direct the economy towards the use of real estate but also the educational thrust toward new and high value-added sectors of the economy. The biotechnology and pharmaceutical industries have been the main thrust of this policy that is often supported by educational programs and government assistance to the commercial ventures that are approved. It consists of both state-owned and private enterprises. It has been very successful economically but less so in terms of political democracy (Shatkin 2011). Unlike Hong Kong, it has not engendered much political opposition.

Conclusion

This chapter has covered lean production in a diverse array of methods in different countries outside the Anglo-Saxon orbit. Clearly lean production has been a Japanese and English-speaking country phenomenon. However, it has spread to many other countries as well (Jürgen and Krzywdzinski 2016). This chapter sheds light on three diverse views of similar types of models, both in the plants and in theory.

First, diversified quality production model in Germany and in German corporations throughout the world use JIT and a few other aspects of lean production (Jürgens 1997; Jürgen and Krzywdzinski 2016). However, the DQP model differs significantly from lean production in that it involves much more participation and representation of labor through works councils and co-determination (Sorge and Streeck 2018). With the decline of unionism with lean production, this is an important drawback of lean production that DLP seems to solve.

Second, the regulation approach is not specific model of any one country, but more of a theory that describes lean and other forms of production. It does make a strong statement that lean production is not a single force or "one best system (Boyer and Freyssenet 2000; 2002). Instead interaction with institutions and the state make a big difference in the implementation of lean production.

Third, the discussion of the Chinese system and lean production shows that lean production and diversified quality production both have limitations in a non-democratic country with authoritarian control of the economy. We say this despite the claims that the Confucian ethnic is especially useful in team work and employment systems (Bellah 1957; Vogel 1971, 2001). This shows that lean production and DQP have some basis in democratic systems since they involve economic and production rights in their approach to work. Further, Daron Acemogulu and Richard Robinson in *Why Nations Fail* (2012) argue that authoritarian systems will ultimately fail in the area of prosperity and poverty because the institutions that they create do not sustain the freedom need to innovate and prosper. This is why the core economic powers in the world system tend to allow free speech, something that China is wont to do.

These three models are not generally included in discussions of lean production but they will connect back to the larger discussion of innovation, social control, and institutions that we synthesize about political economy and lean production in chapter 10.

Notes

1. Further works on the French regulation school and post-Fordism include Aglietta (1979), Boyer (1993), Boyer and Saillard (2001), Jessop (1990), Jessop and Sum (2006) and Peck (1996).
2. There may be come confusion with the term "regulation." It is used in a much broader sense than simply the state or private government regulating the economy. It more accurately refers to "social structures" of institutions with the corresponding "agency" involved with social struggles for change or new attempt by capitalists to increase profits. In some ways it is more of an institutional bargaining theory that attempts to resolve the break between structure and agency (Jessop 2001a).
3. This might be seen as evidence for a resurgence of Italian flexible specialization model (Sabel 1984; Sabel and Zeitlin 2002); however, since large corporations like Anheiser-Busch can easily adapt to this flexibility and specialization with lean production methods, we do not find this almost batch production method to be a major model. Neither does Edgell (2011:202)
4. Frank G. Woollard (1883–1957), was a British engineer who worked for Morris Motors in the UK for nearly three decades in design, production, and management. He was a pioneer in what is today called JIT inventory using it in the 1920s well before the Japanese and again in the 1940s and 1950s. He helped Morris get 34% of the British automobile market. Although Woollard (2009) wrote many papers on flow production, he remains a largely forgotten developer of lean management. In passing, Henry Ford also conceived of JIT production, but was unable to implement it (Levinson 2002). However, his approach was to vertically integrate the firm and provide all of his own parts (e.g., he had his own steel mill at the Ford Rouge plant).

Part II

Models of Lean Production

6 The Leaders in the Field of Lean Production

Toyota and Honda from Japan to the World

The major characteristics of the Toyota System of Production, the closest model to the ideal type of lean production, are most prominently displayed in the Toyota and Honda Corporations. Toyota has the greatest emphasis on teams, while Honda has somewhat less, but both of these corporations implement lean production much more than any other firms. These two automotive companies have been highly successful and constitute a full implementation of lean production (see Table 6.1 for revenues and employees).

Toyota System of Production

The automaker that held the number one spot in global vehicle sales for four years in a row (until 2016) can trace its origins back to the textile industry. Born at the end of nineteenth century and named one of the most important inventors in Japanese history, Saitchi Toyoda had humble beginnings on a farm and later on, founded a spinning and weaving company (Chambers 2008). After many years of trial and error on the assembly line in the textile industry, Kiichiro Toyoda and later on, Eiji Toyoda transferred the flow production method to the automobile factory. Historical accounts describe that JIT techniques were practiced at Toyota since 1935. In the end, Taiichi Ohno helped establish the Toyota System of Production (TPS), built the foundation of the Toyota Way and the basic tenets of the Just-in-Time Production (Ohno 1988a; Fujimoto 1999). Throughout Toyota's history, the elimination of waste, generated by excessive inventory, unnecessary steps in the production flow, or defective products in some cases, was a central philosophical principle that the TPS was built on. From the human side, another Toyota's guiding principle was to avoid layoffs as much as possible (Toyota Global 2018; Grønning 1997).

Although the term "lean production" is not largely used by the Japanese themselves, lean production with its symbolic Toyota System of Production is a system of production centered on minimization of waste, continuous improvement (*kaizen*), and obsession with quality. The complexity of the TPS is best represented in the image that we call the Toyota House (University of Kentucky Lean Systems Program 2018). TPS is grounded on certain

Table 6.1 Revenue (in $US) and Employees at Toyota, Honda and VW

Organization Outcomes	*Toyota*	*Honda*	*VW*
Revenues in 2018	$ 254.6 billion	$ 129.2 billion	$ 259.9 billion
Net income	$ 22.46 billion	$ 9.53 billion	$ 11.827 billion
All Employees in 2018 (% Temps)	369,124 (20%)[2]	215,638(20%)[2]	656,000 (unknown %)
2017 (% Temps)	365,445 (20%)[2]	199,638–2014 (20%)[2]	634,000 (unknown %)
Revenue per employee (ee)[1]	$ 689,586/ee	$ 598,980/ee	$ 396,301/ee
Entry level pay in 2018	$ 15–17/hour	$ 16.66/hour	$ 16.00/hour
CEO pay in 2018	$ 2.84 million/year	$ 0.7 million/year	$ 19.7 million/year[3]

Notes

1 Revenue per employee is simply revenues divided by the number of employees.

2 The number of temporary employees at Toyota and Honda varies from none in times of recession to about 20% in more improved economic times.

3 Chief Competitive Officer Didier Leroy, who is French, earned $9.4 million at Toyota, which was the highest wage ever recorded in that company.

philosophies or taken-for-granted assumptions on human nature: customers go first, respect for humanity, and elimination of waste. For example, in Toyota's statement of their philosophy, they say: "each person fulfilling his or her duties to the utmost can generate great power when gathered together", and "if each person makes the most sincere effort in his assigned position, the entire company can achieve great things" (Toyota Global 2018). The human strategies again are the foundation which complemented with technical strategies guarantee the success of the TPS. The model puts people first with giving particular attention to customers (delivering the highest quality at the lowest cost and in the shortest time), employees (investing in work satisfaction, consistent income, and job security) and the company (focused on market flexibility and generating profit from cost reduction). Again, flexible and motivated employees are at the center of TPS's powerhouse, but the complementarity between the social and technical aspects are crucial for Toyota's success. The strategic technical pillars of the TPS model are *kaizen* (continuous improvement), *jidoka* (automation with a human touch or the ability to stop the line to prevent the passing of defects), and JIT (JIT production: making only what is needed, when is needed, and in the amount needed).

New industrial and organizational terminology emerged as being quintessential elements of the TPS. The "Five Ss" refer to the legendary obsession with cleanliness and orderly organization in the Toyota factories: *seiri*—sifting, *seiton*—sorting, *seis*—sweeping, *seiketsu*—spick and span, and *shitsuke*—worksite discipline. There are three different types of waste- also called the "Three M's" that are strictly monitored in lean organizations: *muda*—elements of production that add no value to the product and only raise costs, such as overproduction, waiting or correcting defective products, *mura*—irregularities or unevenness that comes when production volumes move temporarily up or down, or when workers that workloads that move up and down, and *muri*—when machines or workers are overburden by the tasks at hand. *Takt* time, which is a German word, can refer to the time it takes to produce one vehicle, but most often indicates the number of seconds it takes for a worker to do one portion of a repetitive job (e.g., to install one piston), which also dictates the number of minutes it takes for the assembly line to move from one station to another. When workers identify a problem down the line, they may push a button or pull the *andon* cord, which is a device of visual alerts indicating when a workstation has a problem. Another term that is used only at Toyota, and not by other automakers, is *kanban*, an electronic communication billboard that displays information on lead-time and cycle time, supporting essentially efficiency and the JIT production. Last, but not the least, one of the terms involving the human aspects of lean is called *nemawashi*, a process that involves collecting information and seeking input from workers impacted by a proposal or a change of policy, in other words, consensus decision-making (TMMK 2007; Hino 2006).

In their *Extreme Toyota* study of 220 workers from 11 countries, Emi Osono et al. (2008) concluded that the TPS alone cannot explain the constant Toyota success over the years. In addition to this hard innovation, Toyota is built on a soft innovation that comes from its corporate culture where workers are challenged to engage in problem-solving and suggestion systems on a daily basis. While the TPS and the organizational culture balance and reinforce each other, they are grappled by constant contradictions and paradoxes (Takeuchi et al. 2008).

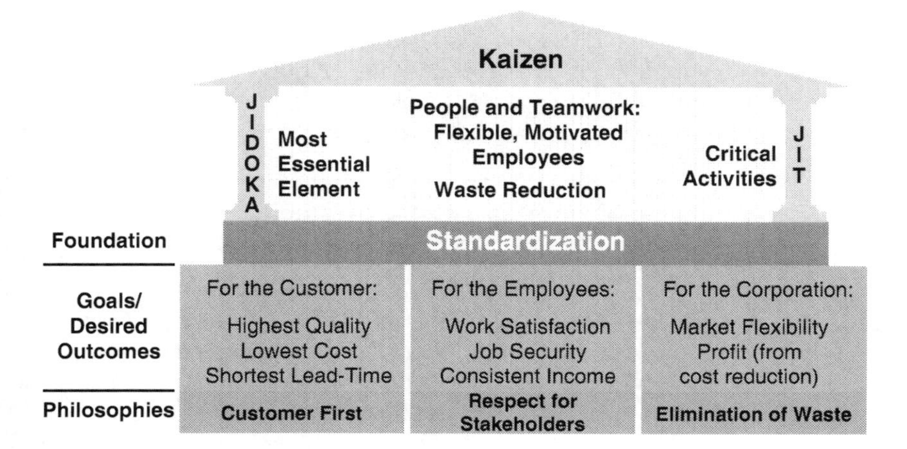

Figure 6.1 The Strategic Technical Pillars of the TPS Model: Jikoka (intelligent automation) and Just-in-Time-Inventory.

Source: This is a small modification of the University of Kentucky figure (University of Kentucky 2018, see also Liker 2004:33.

In contrast with the traditional Fordist practices, Toyota treats its workforce as knowledge workers and invests a great deal in increasing the capabilities of its HR. Osono et al. (2008) and Takeuchi et al. (2008) describe a culture of contradictions at Toyota:

(1) Toyota fosters a culture of tensions and contradictory viewpoints in order to come up with effective solutions and generate innovative ideas to get ahead of its competitors
(2) It moves slowly and steadily, yet it takes occasionally big leaps, such as launching the hybrid Prius, promoted earlier than other rivals
(3) Although Toyota experiences record sales and growth, its mantra is continuous change and improvement: "never be satisfied", "there has to be a better way", etc.
(4) While being paranoid about the elimination of waste, Toyota employs a higher number of office employees (not assembly line workers) that spend a longer amount of time in meetings than its competitors
(5) Toyota is very cost conscious, almost stingy in some areas (e.g., saving electricity during lunch time), but very extravagant in key areas (e.g., investment in manufacturing facilities, HR, or 170 million dollars a year for the Formula One races)
(6) Internal communication is simple, yet very complicated (presentations should be simple and logical, yet the company promotes myriad of horizontal and vertical networks to keep the employees informed and connected to the Toyota community)
(7) Whereas Toyota is described as being hierarchical and centralized, the employees are encouraged to provide constructive criticism and challenge the hierarchy and leadership

These principles are well described in Terry Besser's (1996) study of the Toyota Georgetown Plant in Kentucky (see Figure 6.1 for a symbolic description of the strategic and the technical pillars of lean production, and to Box 6.1 for a short descrption of how it works).

Box 6.1 Team Toyota: What is it like to Work at the Largest Toyota Factory?

Through participant observation and interviews, sociologist Terry Besser (1996) described how the Toyota Way was implemented in the largest Toyota factory in the world in Georgetown, KY, which opened production in 1988. Attracted with generous tax subsidies and cheap land, Toyota settled in Kentucky to avoid workers with union sympathies and prior auto experience. In an area with high unemployment and few job opportunities, Toyota was quickly seen as a God-send for their community in the outskirts of Lexington, KY. Workers coming to work on the first day of the plant were shocked to discover that the factory was designed with no restrooms for women on the shop floor based on the assumption that women will be office ladies, not assembly line workers. It turned out that about 25% of Toyota's assembly line workers were women. Japanese culture and civilization courses were meant to break the ice and shorten the cultural distance between American workers and the Japanese managers, which were similarly shocked to have women enrolled in samurai sword fighting courses and men in the Japanese tea ritual classes. The Japanese managers were surprised to see the female workers fighting back in an assertive way when they were not treated like their male co-workers. On the assembly line, workers are empowered to take action and stop the assembly line with the *andon* cord if they see a problem. With a goal of zero defects in mind, workers are aware that it is too difficult to fix defective parts at the end of assembly. However, any idle time when the assembly line is not moving is counted by electronic monitors and turns into overtime to help workers meet the production goals. A typical day at work involves one to two hours of overtime. Overall, the associates seem to be enthusiastic to belong to Team Toyota which extends from their small group of five to eight people, to the larger work group of 20, the Production Division, and then the entire corporate team.

The metaphor that best describes the Toyota Corporation is the green tomato—a company that sees itself as young, fresh, imperfect, and still growing. By contrast, the red tomato has accomplished its mission and is ready to be consumed at the end of its life cycle (Takeuchi 2013). Three forces of integration counterbalance three forces of expansion in order to preserve the values and identity of the company against rapid and sometimes risky ventures. According to Takeuchi et al. (2008), the three forces of expansion include: setting impossible goals (e.g., zero defects or making each and every customer happy) that push the employees to innovate and constantly improve; local customization (Toyota offers more customized models for each market than any of its competitors; the Made in Japan label was replaced by Made by Toyota since new models are launched on almost every continent); experimentation (Toyota encourages its workers to constantly experiment and "had built a culture remarkably tolerant of failure").

The forces of expansion are complemented by three forces of integration that stabilize the company's change and growth. First, there are values from the founders expressed in stories and lessons of wisdom from the Toyota founders to contemporary executives. These are shared and revolve around four simple beliefs of optimism, teamwork, experience, and loyalty. Typical phrases are "tomorrow will be better than today", "everybody should win", "see for yourself" (*genchi genbutsu*), and other phrases about long-term employment that include investment in on-the-job-training, problem-solving, job rotation, mentorship, and so forth. Even during the most extreme economic crises, Toyota's mantra seems to be "cut all costs, but do not cut people" since people are seen as reservoirs of knowledge necessary for innovation.

Is Toyota Losing its Way?

An August 2009 YouTube recording of a California family who lost their lives when their Lexus crashed due to unintended acceleration had received more than 250,000 views and generated a public hysteria among the Toyota owners. What was proved to be an incorrect installation of the mats at the dealership pushed Toyota into the most challenging crisis in its history (Liker 2011). Toyota's recall crisis (2009–2011), when more than 20 million vehicles were recalled worldwide, seemed to shake the taken for granted assumption that quality is embedded in Toyota's DNA. The recalls involved unintended acceleration, pedal entrapment, and braking problems caused by software errors. President Akio Toyoda sees the year 2003 as a turning point where the rapid growth in sales might have affected quality as Toyota's number one priority. However, the studies commissioned by the National Highway Traffic Safety Administration revealed that the quality problems at Toyota amount to 1/ 100,000 vehicles a year, which prompted Toyota specialist Jeffrey Liker to attribute the recall scandal not to quality issues, but to ineffective corporate communication, sensational media, and government regulators reacting to political pressure to support the US auto makers (Cole 2011a, b). Powerful executives pushing strongly to acquire 15% of the world market have neglected early warnings on quality problems coming from the lower ranks, while the main competitors came from behind with improved quality measures. Nevertheless, faced with enormous financial and reputations costs, Toyota is currently implementing serious countermeasures, such as slowing down the development process of new models and expanding rapid quality response teams around the globe.

The recall crisis also happened during the economic recession of the last decade. The JIT works perfectly under normal conditions, but natural disasters and economic crises have put it to the test (Nelson et al. 1998/2007). Toyota Europe lost one billion euros in 2008 alone and 8,500 jobs were lost between 2008 and 2010 (half administrative jobs and half production jobs). Toyota coped in many ways with the recession. The assembly plant in Burnaston in the United Kingdom had closed production for four weeks and then entirely eliminated an entire shift (Hertwig 2015). Operation budgets were cut with 10% and the company had withdrawn its participation in the Formula 1 racing activities (Nelson et al. 1998/2007).

But it did not take long for Toyota to bounce back. In 2017, Toyota inaugurated its new North American headquarters in Plano, TX just outside of Dallas. This large business campus cost one billion dollars and employed 4,000 employees working in open spaces with removable walls. The new headquarters consolidate staff from other operations in California, Kentucky and New York and provide 12 eateries, a fitness center, pharmacy, and even a rock climbing wall. Although the company has no plans to build more car factories, it had pledged to a 10 billion dollar investment over the next five years in North America, where it currently produces more than two million vehicles per year (Carlisle 2017). According to the Toyota North America CEO Jim Lentz, this strategic move will help the Toyota associates "collaborate better, innovate faster, respond quickly to changes in the market, and most important, make more timely decisions in response to our customers" (Toyota Corporate News 2017).

The Honda Approach to Production

When Soichiro Honda clipped on an engine to his bike, he just wanted to take care of his mobility needs and test his inventive abilities. Stubborn and independent, Honda was a poor student at school and learned to value hands-on innovation, not the traditional schooling (Falloon 2005). What followed was the pursuit of a dream and a moment that "redefined" the US motorcycle industry (Pascale 1996). Founded by Soichiro Honda in 1948, Honda Motor Manufacturing is one of the youngest and most diversified "mobility companies" of the world. Under their logo—"Honda: The Power of Dreams"—the company is selling almost four times more motorcycles than cars, but also a wide range of other products, such as scooters, outboard motors, jet engines and aircraft, as well as lawn equipment, home generators, and even personal robots (however, manufacturing robots in their plants are purchased from suppliers). In fact, Honda defines itself primarily as a producer of engines, which can be applied in many different industries—a bit different from Toyota, which centers itself on automobiles. Toyota has not manufactured motorcycles, and neither have VW, GM, Ford, Chrysler, and Daimler-Benz. Honda's strategies are already working towards a clearly set corporate vision for the year 2030, which is to "lead the advancement of mobility and enable people everywhere in the world to improve their daily lives" (Global Honda 2018).

In their concept of continuous improvement and quality, Honda refers to *Monozukuri*—the art of making things, but this acquires a greater mean with other terms. *Kotosukurim* means making new experiences based on brand storytelling about the art of making things and this pushes *Monozukuri* into the area of continuous improvements of productions process and new productions. *Hitosukuri* refers to making people or workers who are clearly flexible and able to create new production processes and suggest new productions. These three terms are central to communicating this vision (Global Honda 2018; Ballé et al. 2019).[1] While building on existing strengths, Honda will focus on the following initiatives in the future: (1) Mobility (robotics, artificial intelligence, energy solutions etc.); (2) Accommodate the characteristics of diverse societies by putting "people at the core"; (3) Environment and Safety, striving to become number one in the efforts to create a carbon-free and collision-free society (Global Honda 2018; Sato 2006). With the most frequently used words of "dreams" and "joy of mobility", Honda's vision for the future also embeds "storytelling" and "art" more so than engineering and manufacturing. It is interesting to note that at his retirement in 1973, Honda attributed his company's success to "dreams" and "youthfulness", and prevented his sons from joining the company (Falloon 2005).

The youngest and most versatile automotive manufacturer and the largest engine producer in the world, Honda has never posted a loss in its history. On the contrary, a dollar invested in Honda in 1987 has a return of about $800 in 2014 (Rothfender 2014). With more than 215,000 employees worldwide, Honda sells only 12.5% of its products in Japan, 4.5% in Europe, 24.5% in other Asian countries while its largest market is North America (52.5%). In fact, Honda takes great pride from being the first Japanese auto manufacturer to start production in the United States, first with motorcycles in 1979 and then with cars in 1982. When Honda first set foot in the United States, motorcycling had a bad image being associated with the rowdy, black-leathered teenage troublemakers. "You meet the nicest people on a Honda" campaign was hailed as one of the most successful advertising companies in history and managed to shift the cultural trend against motorcycling. A newcomer on the American market, Honda made their lightweight bikes accessible to everyone and beat the rival Harley Davidson against all odds (Pascale 1996).

Grounded on Respect for Individual as a basic fundamental belief, Honda President Takahiro Hachigo describes Honda's organizational culture as including three key points of initiative, equality and trust; and three joys of buying, selling, and creating (Global Honda 2018). With strong aspects of Asian culture, Honda philosophy is thriving for harmony in the workplace and a harmonious flow of work. But the Honda way also strongly encourages intellectual curiosity, the pursuit of research, and personal initiative in solving problems that may occur on the production floor. The white overalls, worn by all the employees regardless of their status, rank, or seniority from the recent hires to the president, are a powerful symbol of equality at Honda (see Figure 6.2 below). The white color stands for "cleanliness" as a reminder that "good products come from clean workplaces" (Global Honda 2018). The white overalls are leveling the playing field when team members get together to solve problems (Cable 2012).

Figure 6.2 The Honda Celebration on Producing 25 Million Automobiles

Quality and safety are one of the company's most strategic priorities. Regarding the relations between employees, open communication and trust are the cornerstone to creating harmonious workplaces. Honda had expressed its commitment to respecting human rights, combating discrimination and enforcing ethical behavior and integrity in accordance with "social norms and common sense" (Honda Code of Conduct 2018).

Honda prides itself with being the first Japanese automaker that opened factories in the United States (1982). For a company with a small family culture, opening factories in a new uncharted territory was a risky move, but expanding their North American market turned to be their most ingenious business strategy yet (Rothfender 2014). The factory in Marysville, just 42 miles from Columbus, with currently more than 4,200 employees, was the first Japanese transplant factor to be opened in the United States in 1982 (motorcycle production had started in Ohio in 1979). Honda North America later expanded to include 12 manufacturing plants, most of them in Ohio (East Liberty Assembly Plant in 1989, Ana Engine Plant in 1985, and the Alabama plant in 2001); and Honda of Indiana in 2008. Honda also has plants in Japan, Belgium, Brazil, Indonesia, India, China, Thailand, Argentina, Taiwan, and the Philippines.

Honda loves the anonymity: its factories are usually away from the spotlight, with no visible signs or major advertisement on the highways. (See Box 6.2 for the COO of Honda, North America's experiences in working for Honda.)

Box 6.2 Interview with Tom Shoupe, Chief Operation Officer, Honda North America and former President of Honda Motor-Alabama

I joined Honda in 1988, so I've been here for 30 years. One thing that was very attractive to me at Honda was the kind of the viewpoint of the company and the magic about how Honda had started, kind of going against all odds to be a company that could make cars. And the fact that Honda had, I thought, a truly global viewpoint at the time in the 1980s, and I was very intrigued by the people I met in the company, the passion level they had for what they were doing. I was also a Honda owner, so I had familiarity with the product, but there was a uniqueness I felt about Honda I didn't see at other companies at the time.

By the time I got at the Marysville plant in Ohio, there was motorcycle and auto production, so it was a real exciting time. It was the center of a lot of social and political interest in the United States at the time too because it was expanding so rapidly, and there was a lot of debate about foreign companies doing that kind of business in the United States. It was a very exciting atmosphere that attracted me to this place! As for the vision for the future, my vision is to create a company that for the long-term is able to keep those same things that attracted me and others, the passion, the excitement about providing value to the customer, which is why we exist, and making sure that that's passed on to future generations is a really key part of how we develop. Some people call that culture. I'm not sure if that's culture or not, but my dream would be to have a next generation that's capable to satisfying the customer at that level, not kind of waiting to react to customers, but being in front of this. Our founders had that kind of spirit: they won't wait to react; they would have to go forward and reach out and do things that were considered impossible. So how to use that heritage and that spirit to build that kind of company, that's what we're working towards, I think . . . every organization or person makes mistakes, but it depends on how you respond to those mistakes. We always feel like we've been at the cutting edge of production, and the challenge is how to stay true to those values with the external change, but also the rapid growth that we've had. I think those kinds of things become more difficult as you get larger as a company. For example:

I worked in Corporate Planning and Production Control, but when I got back from my visit to Japan to learn about new model development, I've been given that opportunity to work in manufacturing, and to me that was a radically different kind of environment.

I mean it's one thing to be working in the business side of the manufacturing company and another thing to be on the floor to be with the people where products are made. So, I though it wasn't all glory and wonderful experiences every day if you don't have some frustration. I don't think you grow either, so, you know, it causes you to really think of what I could have done better in this situation or why did this issue happen. The deeper you think about that the more it causes you to grow, you know?

In the late 1980s, people in the United States would ask me different questions about working for the Japanese. A lot of people assumed that all the Japanese were the same, that Honda was like Toyota and Toyota was like Nissan. My experience quickly taught me that's not true at all. The Honda start was so radically different. We had a very passionate and magnetic type of leader who started this company with his own hands. He was engine guy and we were an engine company. We like to think we still are an engine company whereas when other companies started, their expertise was assembly and that's a fundamental difference if you know the industry. There's a different approach in thinking. Our culture is more of a give and take. Ideas do not come only from the highest-level people; we encourage ideas to come from anywhere. People can make major strategies, you know, and there's just some subtle but fundamental differences in the way our companies (Honda and Toyota) think and approach the work. When I joined the company, I was told: what do you want to do? What's your dream? Why did you come here? I've never been asked that before. We haven't worked in every company in the world, but Honda was to me very unique. Their culture and approach were very natural even though it's a Japanese company. It seemed to have a human, friendly, natural approach.

During the 2010–2012 economic downturn, we've made a lot of changes and adjustments. Probably the biggest thing we learned was about our supply base and our character and constitution if you will. We had to reduce production without impacting our associates. We kept everybody, and actually in Alabama's case sadly right after the tsunami, about a month later we had the terrible tornados that came through here. Our people really wanted to do something to help because the damage was so extensive, and when we had the downtime, one of the things we came up with was we basically paid people to clean up after the tornadoes. Our cash was a little tight and we gave some money to relief efforts, but what helped more than anything was our people. Busloads of Honda associates in white uniforms went in different communities and just helped people. It was a great example of how we take a bad situation and try to look at how can we make the best out of this. That's what we did in our business: we pushed ourselves and tried to figure out how can we recover as quickly as possible and do the right things to support the people. Then, we had the Thailand flooding which was the third major natural disaster to affect us in one year, and we had to back off again and kind of recharge things, and we were able to apply what we learned from our tsunami crisis. We are back in pretty good shape, but we learned a lot of lessons and it gives you an indication of how innovative the supply chains are in the world, and how it's a truly global supply now. How even really small components of a part that you put in a vehicle can really affect your whole business.

At Honda, we have engineers who don't have a degree, but who have experience and are highly skilled. We need a blend or balance of both type of engineers (with degrees or no degrees), especially as start to grow a little older as a company. We need to keep that line alive.

The competition around us has intensified dramatically. One of our biggest challenges for next years to come is to keep our ability to stay true to our reputation, being an innovative company. Like I said, as you grow large sometimes it's harder to keep the focus on that kind of thing. How do you stay ahead and innovate, and give people things that are excited about? How do you anticipate what they want and how they can use it? I think for us as a manufacturing plant, we have to stay focused on high quality. That's probably the biggest challenge that we have! Date of interview: August 25th, 2012.

Unlike the American car manufacturers that are located primarily in urban settings, the Japanese automakers have strategically set their new factories in rural regions. The availability of cheap land, ease of access for suppliers, lack of traffic jams during the JIT delivery of parts, availability of workforce eager to work and lower unionization rates made the rural areas to seem more suited for auto production. However, in their early phase of settlement into the United States, the Japanese car makers have been criticized for a systematic discrimination against minority and women hiring (Holusha 1988). Black workers, mostly living in urban areas, were historically hired in large numbers by American car companies, and the perceived lack of access to jobs in the Japanese factories, which were settled 30–40 miles from the big cities, angered many of them. The Honda factory in Marysville had only 2.8% black workers in an area with 10.5% African Americans. This prompted a federal investigation and Honda had to settle a 6 million dollars racial and sexual discrimination lawsuit (Cole and Deskins 1988; Lepadatu and Janoski 2011).

Since these early scandals, both Toyota and Honda are trying to actively support and recruit employees from the Historical Black Colleges and Universities (HBCUs). Honda is providing an annual scholarship fund for students pursuing training in the field of engineering, supply chain and manufacturing at HBCUs, as well as encouraging HBCUs students to apply for internships and co-op positions to gain more industrial exposure before applying for jobs (Ohio Honda 2018).

One of the features that both Toyota and Honda have in common is a systematic attempt to avoid unionization. In fact, the UAW was not successful at unionizing any of the Japanese plants in the United States. The Japanese car makers appear to have a consistent strategy of moving to Southern states with strong right-to-work legislation and lower rates of union workers. The state of Alabama had offered 158 million dollars in incentives for the opening of the Lincoln plant, but required that the company hires workers from the entire state, whereas in Indiana and Ohio were more restrictive. Honda had imposed residency requirements in their hiring process. The residency requirement to be hired at the Greensburg plant in Indiana was to live within an hour drive to the plant, so that workers can arrive at work on time during bad weather conditions. This included only 20 of the 97 counties of the state, which also meant mostly rural white workers and not workers from Indianapolis badly affected by the American plants' layoffs (Boudette 2007). Honda refused to hire workers who have offered to move into the hiring zone claiming this would slow down their "aggressive launch schedule". This refusal to hire the workers willing to relocate within the proximity of the plant fueled speculations that the hiring zones are an insidious strategy of keeping the union sympathizers away from the plant.

In their corporate communication with the employees, Honda Alabama portrays unions as a threat to their team culture. The company points out the challenges and even failures of the heavily unionized American car manufacturers that had to lay off workers and close plants, and concludes that they do not provide more job security and job growth than Honda. Since the company has not laid off a single worker in more than 30 years, it does not justify, in Honda's view, the need for a union (Kiesiel 2007). Also, the Honda employees being all called "associates" and not antagonistically divided between workers and managers is symbolic for uniting the workforce in a single community or team (Hertwig 2015). As such, better and more transparent communication between workers and managers, and not unions, is seen as the panacea to problems and discontent in the workplace. Honda had adopted a European Work Council in 1995 as process for communication and consultation with employees from its many European factories (Hertwig 2015). Similarly, the Toyota Employee Forum was established in 1996 as a labor-management council.

Honda opened its first business operation in Los Angeles due to its large Japanese community, mild weather suitable to motorcycling, and growing population (Pascale 1996). Its start-up loan was so small that the first three executives shared an apartment with two of them sleeping on the floor to save costs (Pascale 1996). Recently, Honda had decided to move its corporate offices from Los Angeles to Marysville, OH, where their oldest factory in the North America is located. The move is symbolic of Honda philosophy of "seeing with your own eyes" the problems on the floor, but also to speed up decision-making and cross-connect wider operations in automotive design, technology, manufacturing, but also motorcycles, power engines, and other products (Eisenstein 2013).

The Honda Way

Whereas armies of engineers, managers, consultants, and academics teach and preach the value of the Toyota System of Production and lean systems in general, the Honda Way is shrouded in mystery. Officially, the Honda Way is called the Honda Best Practice System, consisting of best position in competition, best productivity in improving process, best product in quality and design, best price in cost and maintenance, and best partners in terms of supplier and dealer relationships Nelson et al. 1998/2007:25). A number of participant observation studies and ethnographies have been conducted at most of the Japanese transplants in the United States, but Honda is not among them. While the Accord had become the 2018 North American car (Global Honda 2018) and Honda cars in general are ubiquitous on the US highways, little is known of the differences between Honda and Toyota.

The Honda Way seems to be grounded on a unique organizational culture and a strong ethic of efficiency and economy (Chapelle 2009; Mintzberg 1996). It is based on "a religion of doing more with less", but also on a rebellious and risk tasking attitude, and a "success against all odds" culture. When the founder Soichiro Honda was not granted permission to expand and diversify his company from motorcycles to cars in Japan in order not to compete with the existing Japanese auto manufacturing giants, he had rebelled against the pre-established order and oriented the company towards car production for the North American market, which is its largest market to the current day. Honda's adventures in North America in the late 1960s, before the contemporary trend of companies to expand global overseas operations, was a risky move for various reasons.

The small Honda cars were not popular among Americans who preferred sedans, trucks and SUVs. However, the oil crisis of the 1970s shifted the American consumer preferences to smaller cars, so it was perfect timing for the Civic to become successful. Japanese companies also tapped in a closed-knit system of suppliers (*keiretsu*), which Honda could not rely on in North America, but it still did not stop it from pursuing its "go it alone" attitude (Chapelle 2009). American auto workers were also not used to job rotation and training for multiple jobs at the same time, and anti-Japanese sentiments were high at a time when Ford, GM, and Chrysler were laying off workers because of the fierce competition with Japanese companies (Chapelle

2009). Honda also went against the grain and resisted the temptation to join mega mergers in the auto industry. When many of the mergers of the time, such as Daimler and Chrysler proved to be problematic, Honda preferred to take the risk of "flying solo" (Chapelle 2009). To ease critics, Honda is creating cars with more than 90% parts produced in the United States and even exports more than 100,000 cars made in the United States to Japan and other markets in a move to smooth US-Japanese relations (Chapelle 2009).

Equality is deeply embedded in Honda's social structure since its inception. The founder Soichiro Honda strongly believed that "an egalitarian culture is a problem-solving culture" (Cable 2012). The associates, not even the executives, do not have individual offices. Single status at Honda refers to equal opportunity: all the employees benefitting from the same pension plan, sick days, vacation plan, canteen, lockers, and uniforms (Mair 1998). Weekly and sometimes ad hoc *waigaya* meetings where team members regardless of rank and position can share their voice, recommendations and concerns. They are encouraged to leave their titles at the door (Cable 2012). Some of these *waigayas* last for years, with associates meeting daily for five minutes until they finish a new innovation project. Although these meetings are time consuming, noisy and sometimes chaotic, they are seen as some of the most basic principles of the Honda Way: the employees' involvement in the consensus democratic decision making. Honda teams are a bit larger than Toyota with about 12 team members on average, while Toyota aims at eight or so (Interviews). Of course, this varies a great deal with some units having small or no teams (a paint shop section with fewer employees on duty at one time).

Honda is a global corporation that is decentralized at the regional level. Described as a global local corporation, Honda is believed to have transferred its production system entirely intact and adapted some of its "softer" aspects related to HR and employee relations to the local environment (Mair 1998). Each of the six world regional headquarters have the autonomy to design and produce their own cars for their own region. Individual plants also have the responsibility to run their own business (Cable 2012). At Honda, the associates are considered the experts of the manufacturing process and the most knowledgeable to plan new product launches. Honda uses a flexible integrated manufacturing process with various model of cars produced on the same assembly line with no stops whatsoever between one model and the next. Its JIT philosophy also required higher reliance on parts produced by local suppliers and not imported from Japan.

Since it opened its first business office in the United States in 1959, Honda prides itself with not laying off a single worker. Even when the economic recession hit in 2008–2009 at the same time with the devastating earthquakes in Japan and Thailand, and production slowed down significantly, the associates poured their energy in training, equipment repair, continuous improvement projects, or community service (Cable 2012). In Europe, 1,300 workers were laid off, mostly through a voluntary release program, in the middle of the economic recession (Hertwig 2015). Previous research, however, shows that the lifetime employment of permanent associates is guaranteed by the hiring of a JIT workforce of temporary workers, up to 20% of the assembly line workers at Honda that may easily be let go in hard economic times (Lepadatu and Janoski 2018).

On the global scale, Honda's business strategy seems to be focusing on high volumes of low cost products, precision engineering especially on engines, redefining the market ("you meet the nicest people on a Honda"), aggressive pricing and advertising (Pascale 1996). The Honda story reveals, in fact, more "miscalculation, serendipity and organizational learning" than the pursuit of a grand strategy (Pascale 1996). Thus, the so-called Honda Effect refers to the process in which an organization uses mistakes, miscalculations, and serendipitous events to reinvent itself (Pascale 1996). Organizational learning from every single person involved in the production, advertising and sale process and a bottom-up approach to leadership with managers and executives in very humble roles of facilitators seems to be the cornerstone of the Honda miracle. As such, Honda success can be defined as the incremental adjustment or adaptation to unpredictable events (Pascale 1996) or "strategy evolving with the organization through good bottom-up communications rather than being planned rigidly in advance" (Mair 1999:30). In other words, a triumph of learning over planning: "Honda managers did almost every conceivable mistake until the market hit them over the head with the right formula" (Mintzberg 1987:70). Chuck Ernest, in charge with overseeing the Honda factory in Lincoln, AL, noted that "No one will blame you for making a mistake if you tried something new; in fact, you may be promoted for that. However, you could fall out of favor if you are afraid to stray from what worked before—no matter how well it worked". (Rothfender 2014:4). The goal is to constantly challenge the status quo. In fact, from early on, Soichiro Honda's philosophy was that success is 99% failure. In conclusion, the very few journalists who toured its plants describe Honda as having an "unorthodox DNA", built on skepticism and curiosity, paradox and contradiction, demonstrating a strategic capability to manage apparently contradicting concepts and practices, and reconciling organizational learning with pure luck (Rothfender 2014; Mair 1999).

In the management literature, Honda is labeled as a learning organization where engineers learn from experience (Mair 1998). For instance, Tom Shoupe, COO at Honda North America, is a Political Science graduate who had specialized in manufacturing and industrial engineering at the workplace and during several long-term visits to Japan. Indeed, Honda started as a high autonomy organization with a flat hierarchy where individualism was encouraged, and bosses got their hands dirty with the workers (Mair 1998). The JIT production requires that only 0.3 day of parts (five hours) are stored in the factory (Mair 1998). Due to its structural flexibility, Honda received the name of "flexifactory" (Mair 1998). This varies over time and plant; for instance, the Marysville plant now has about a day and a half inventory (36 hours). Still, these are short time periods for inventory (Interview, April 4, 2019)

Honda as "The Power of Dreams" company uses a broad concept of teamwork that refers to giving employees the most responsibility possible and a broad emphasis on cooperation loosely defined rather than a specific type of team structure (Mair 1998). The concept of teamwork extends to the entire company- team Honda-, not just the primary group of 5–10 co-workers; for instance, even office workers are trained to fill in the production line during emergencies or illnesses of assembly line workers (Mair 1998).

Honda sees employee involvement as intimately linked to job security. Quality circles were set up seven years after the opening of the Honda plant in the United Kingdom to allow workers to learn the production system. The employees were encouraged to learn statistical analysis (histogram, scatter diagram, charts, etc.) and "got a theme, form a team" became a slogan at the Swindon factory in the United Kingdom. The hiring strategy was to seek "good basic people who can grow with Honda" rather than "stars who can land with their feet running" (Mair 1998).

The Honda UK plant organization reveals that the Japanese associates are more in a helping role and that there is local control in factory. The local managers are in charge with HR, employee relations, as well as management training and development. Employees earned a 5% bonus for perfect attendance at Honda UK and could receive meal or clothes vouchers for special contributions. Honda UK was one of the very few automobile factories without union representation in Europe. Citing the numerous opportunities to voice concerns in associate committees, Honda UK managers would see as their own personal failure if their workers would unionize (Mair 1998). Because Honda did not hire workers with prior auto experience and did not adopt the suggestion system in its first seven years, Mair (1998) concludes that the Honda model of production is closer to Theory X than Y.

Whereas very few Honda associates mention the word "lean" to describe their company, the Toyota insiders begin and end every conversation about their company with the TPS. As a result, Rothfender (2014) concludes that lean principles are technical tools that Honda uses to stay competitive among the other auto giants, but lean is not necessarily the entire Honda identity. Honda had added the element of craftsmanship or craftspersonship to manufacturing by means of mass production (Rothfender 2012).

The word *gemba* defines the continuous improvement philosophy of Soichiro Honda. *Gemba* refers to the actual place, *gembitsu* to the actual part, and *gengitsu* to the actual situation. All the five senses must be involved in assessing the environmental factors that create a production problem: (1) hearing—"are the sound of machines okay?" (2) smell—"no toxic smells allowed", (3) sight—"products move downstream easily", (4) touch—the machines should not be greasy and dirty, (5) taste—"no toxic tastes in the mouth" (Nelson et al. 1998/2007).

The Honda approach can be summarized. According to Rothfender (2014), the principles of the Honda way are as follows:

(1) **Embrace Paradox**. Respect for mistakes is engrained in Honda's DNA, and Honda looks at paradox as an opportunity to continuously

reassess the status quo. Honda admits that innovation is a disorderly process, which may involve conflict and dissent, the very reason why it welcomes paradox. Honda does not thrive to imitate or follow other successful companies, but to set their unique, one-of-a-kind business journey.

(2) **Real Place, Real Part, Real Knowledge**. *Sangen shiugi* refers to checking on three aspects before taking a decision: (a) *Genba* (go to the shop floor and see for yourself); (b) *Genbutsu* (use the acquired knowledge to formulate a solution); (c) *Genjitsu* (make decisions based on reality, meaning on data collected from the spot). Whereas *sangen shiugi* is used in other lean production systems, they are most intimately embedded in the Honda Way.

(3) **Respect Individualism**. Honda has an unorthodox hiring process, looking for employees that do not necessarily love cars or have prior auto experience, but are a little bit odd and unconventional, with a fresh look on the nature of work in a car factory. Founder Soichiro Honda is famous for describing the ideal Honda applicant as "people who had been in trouble" (Rothfender 2014:123). The most relevant criteria for employment at Honda are independence, self-reliance, creativity whereas prior auto experience is not necessary.

Honda's affinity for individualism is again an unorthodox approach for a Japanese company, which are known for their collectivistic cultures. This individualistic orientation generates a completely unique concept of teamwork where the individual is the cornerstone of the group. The team is the proper site for the decentralization of innovation. In Honda CEO Takeo Fukui's words, Honda is a "company where each and every person is the main player" (Rothfender 2014). The team exists more as an organizational structure to allow individual ideas and skills to flourish. The associates have no job descriptions, but made believe that each individual job is "everything" in the larger context of the factory. Indeed, Honda's concept of teamwork may refer to three distinct aspects: open offices, absence of status symbols, and joy of work (Nelson et al. 1998/2007).

Box 6.3 A Comparison of Toyota and Honda from an Engineer's Perspective

A former Toyota plant engineer who now works for Honda in Lincoln, AL describes his experiences:"When I was at Toyota, I was asked to create a new assembly line at an existing factory. It was easy; my supervisors in Japan gave me the blueprint, and said, "Here's how we do it, follow the plan": At Honda a few years later, I had to oversee the setting up of the line at the Lincoln engine plant. The experience was so different. The only instruction I got was, "Go to the Anna engine plant in Ohio, study how they do it, talk to the workers and the managers about what they like and don't like, what they would fix and what they would leave unchanged, and then make a better one in Lincoln'. I took this to mean, understand what they've done, and then advance it"

(Rothfender 2014:21).

Honda acknowledges global warming caused by carbon emissions as a significant problem that auto companies must address. (Rothfender 2014). In 1972, Honda became the first automaker to meet the strict US Clean Air Act emissions standards and also introduced the first gasoline electric hybrid vehicle in the United States in 1999 (Rothfender 2014).

Takata Airbag Recall

From 2001 to 2015, one of the largest airbag suppliers—Takata—was found responsible for devices that could explode causing injuries and killing drivers and passengers. The US National Highway Traffic Safety Administration (NHTSA) created the largest safety recall in US history. The airbag would send metal shards throughout the passenger cabin using a propellant based on ammonium nitrate. With high moisture and temperatures, these accidents could happen. As of 2019, there have been 16 deaths in the United States and at least 24 deaths and 300 injuries throughout the world (*Consumer Reports* 2019). While 19 automotive manufacturers were affected, Honda issued recalls on 12 Acura models (with two "high risk"), 20 Honda models (with five "high risk"). Toyota also had 18 models affected, but this was 12 less than Honda with 32, which was the most affected by the recall. Some luxury brands (BMW, Audi, and Mercedes) were also affected, but they had sold fewer cars than Honda. On March 29, 2019, American Honda and the NHATSA confirmed that a defective Takata airbag caused the crash of a 2002 Honda Civic in June 2018.[2] The recalls began in 2013 and continue to this day since car owners resell their vehicles and it is difficult to contact new owners. This means that 87.5% of the Takata airbag deaths occurred in Honda automobiles (American Honda Motor Company 2019). While Honda did not manufacture the airbags, part of lean production is to have good relations with high-quality suppliers. Since the scandal started in 2013, sales have shown moderate growth from 1,525,312 in 2013 to 1,641,429 in 2017; however, quarterly revenues dropped in 2014 and 2015 only to surge again in later years (Macrotrends 2019). However, the airbag problems, while solved at the factory, are not solved since cars with the defective airbags are still on the road (Tabuchi 2014).

Conclusion: Toyota Way Versus the Honda Way

To sum up the comparison of the Toyota and Honda Corporations, they both are grounded on Asian cultural principles and values (Rothfender 2014:9):

(1) Simplicity over complexity
(2) Minimalism over waste
(3) Individual responsibility over corporate mandates
(4) Flat rather than tall bureaucracy
(5) Perpetual change
(6) Decision making based on observed and verifiable facts, not theories and assumptions
(7) Autonomous and ad hoc teams that are accountable to one another
(8) Skepticism about the existing solutions
(9) Unambiguous goals for employers and suppliers
(10) Borrowing from but breaking with the past: innovative discontinuity

From the inside of each corporation, Toyota and Honda have contrasting cultures, but to the outside, these differences pale in comparison to similarities. Womack has used the metaphor of Roman legions to describe the Toyota employees—following orders, walking the line in discipline, and the image of guerilla fighters for Honda associates—trying to anticipate danger at every corner and competing on a chaotic battlefield (Rothfender 2014). As such, Toyota has the reputation of being more hierarchical and centralized than Honda. However, Toyota certainly gives teams considerable latitude and various groups of workers—women, African-American and others—can voice their concerns to upper management (Lepadatu and Janoski 2011; Besser 1996; see also Box 6.1). In fact, the Toyota plant in Georgetown (TMMK) president from 2010 to 2017 was Wilbert James who is African-American, and the current president from 2017 till now is Susan Elkington, a woman. Further along this approach, Toyota teams are smaller at about 6–8 workers, while Honda has about 18 to 20 workers per team.

Honda defines itself as an engine company and, as a result, its products reflect a wider range than Toyota from outboard motors to jet aircraft engines. Honda is defined more through its risk-tasking, innovation, organizational learning, open communication, and local control than the principles of lean production, which are seen as tools, not its core identity (Rothfender 2014). The Honda Racing Corporation, a division of the larger corporation, is involved in formula one and motocross racing, largely because of Soichiro Honda's fascination with racing. This risk-taking is often reflected in the Honda tendency to delegate more of its decision-making to each of its own facilities around the world, whereas the final word for Toyota's major decisions resides ultimately in Japan, (Rothfender 2014). However, Toyota certainly took big risks on the hybrid car. Robert Boyer and Michel Freyssenet, in their "production models" approach to lean production, sees

Honda as using much more of a "innovation and flexibility" strategy, while Toyota relies on a "permanent reductions in costs" strategy (Boyer and Freyssenet 2002). Some view Honda as more stylistically advanced, which may be a matter of taste for racing. And others view Honda as a more of a balanced global and local company making customized products for each region and connected to a global corporation, but relatively independent of central management (Rothfender 2014). While both Honda and Toyota had Japanese and American executives at the beginning, as time goes on, these dual structures give way to more Indigenous managers.

Robert Boyer, in his "productive models" approach to lean production, makes a strong distinction between the two companies. Unable to rely on an extensive supplier network upon arrival in the US, Honda produces more of its own factory equipment than any other manufacturer in any industry (Rothfender 2014).

Both Toyota and Honda have taken criticism concerning quality concerning safety issues; however, they have largely survived in terms of both sales and revenues. Both Toyota and Honda remain at the top of quality ratings whether it is J.D. Power or Consumer Reports. While American automobile manufacturers have improved their quality, there is the "red queen effect" in that they are not chasing a stationary target. [3]The Toyota and Honda corporations are both constantly improving their own quality so the gap between the manufacturers has not narrowed since the American manufacturers have embraced quality production. Toyota and Honda remain at the pinnacle of high-quality manufacturing and design, with Toyota and Honda both putting hybrid vehicles in the showrooms at an early stage. Although each company uses nominally different terminology—TPS or the Toyota production system, and HBP or the Honda best practice system—both companies are the world's gold standard for lean production.

Notes

1. These terms are not foreign to Toyota; however, Toyota refers to this process in the plant as *Kaizen*, while Honda uses the term *Monozukuri*. Workers at Honda, Marysville see *Monozukuri* as process improvement, but it would seem that the three terms—*Monozukuri, Hitozukuri, and Kotozukuri*—might be necessary to get to the same point.
2. Takata filed for chapter 11 bankruptcy in the United States in 2017. It was subsequently sold to a Chinese-owned competitor that operates in Michigan. It is now renamed Joyson Safety Systems. Since the propellant was the major problem, the use of a different propellant apparently solved the problem. However, finding the cars with the old airbags is the difficulty.
3. The red queen effect comes from Lewis Carroll's *Through the Looking-Glass* when the Red Queen says that it takes all the running that you can do to stay in the same place. Another way to look at it is that when companies benchmark against a more successful company, they imitate existing practices, which may improve their performance, while the more successful company has come up with new practices that vault them over their competitors imitations (Barnett 2008).

7 The Emergence of Semi-Lean Production

Ford Motor Company, the Nissan Corporation, and McDonald's

The implementation of lean production some industries has not been as extensive as one might believe. Many corporations have embraced the rhetoric of lean production, but have not fully implemented the model. Most often the JIT inventory approach is implanted within supply-chain management techniques, but the use of teamwork, QCCs and employee involvement clearly lag. In this chapter we cover two automotive companies—FMC and Nissan Corporation—that implement lean principles but in a partial way but for very different reasons. They are enthusiastic about JIT inventory, but tend to drag their feet on questions of teamwork or simply do not implement it at all. Sometimes teamwork is more prevalent at the management level, and often this involves embracing Six-Sigma methods by managers and little for employees. We then look at a third corporation in the food industry—The McDonald's Corporation—where lean principles have had a difficult road to follow, and even further, George Ritzer has put forth the McDonaldization theory in a series of well-received and popular books (Ritzer 1993, 2019) that argue that McDonald's follows the Weberian bureaucracy and Fordist approach to selling fast food. They may have tried lean principles and teamwork, but it simply does not work well with their model. But this approach also involves aspects of emotion work to the service-oriented fast food industry that is somewhat foreign to lean production. Our argument in this chapter does not mean that these companies and corporations have not changed, but it does indicate the full penetration of lean production principles in each corporation has been partial, especially in the area of teamwork and QCCs. The economic and workforce comparisons of these three companies can be seen in Table 7.1.

Lean Production at Ford Motor Company

Like many one industry dominated cities or towns, Detroit tends to be very hierarchical community based on its dominant industry being a cash cow. Developing a career within the Big Three auto companies was a path to upward mobility. Innovation becomes strangled by the dominant industry as executives plot their course to a safe career ending in financial security

Table 7.1 Revenue (in $US) and Employee Statistics at Ford, Nissan and McDonald's

Organization Outcomes	*Ford*	*Nissan*	*McDonald's*
Revenues in 2018	$ 149.6 billion	$ 108.1 billion	$ 21.1 billion
Net income 2018/2017	$ 3.7/7.7 billion	$ 3.13 billion	$ 204.87 billion
All employees in			
2018 (% Temps)	199,000 (5%)	139,000 (20%)	~210,000 (>50%)
2017 (% Temps)	200,000 (5%)	139,000 (20%)	~200,000 (>50%)
Revenue per employee (ee)[1]	$ 751,548/ee	$ 777,942/ee	$ 100,119/ee
Entry level pay in 2018	$ 13.00/hour assembly[2] line, $ 21.00/hour electrician	$ 15.00–17.00/hour entry technician	$ 7.69–13.93/hour $ 22/hour manager
CEO pay in 2018	$ 16.7 million/year	$ 16.9 million/year[3]	$ 15.4 million/year

Notes

1 Revenue per employee is simply revenues divided by the number of employees. They can be impacted by very low prices and the number of part-time and temporary employees.

2 The UAW has agreed to a dual pay structure whereby new employees get a much lower entry wage than employees grandfathered into the previous pay scale.

3 Nissan CEO pay comes from three companies, but was $6.5 million from Nissan, $8.4 million from Renault, and $2 million from Mitsubishi.

and a healthy dose of wealth. This organizational culture was quite the opposite of lean production. Yet crises would occur and auto companies would adopt a new organizational culture for a short time. But then the bureaucratic mass production culture with internal provinces of authority and expertise would creep back in. As a result, lean production would be adopted for a period of time, but then being lost in the shuffle as other interests intervened. Up until the Japanese imports began hitting the US shores in large numbers, Detroit auto producers had a quantity approach to car production, not a quality approach. This created a "lemon culture" whereby a somewhat small percentage of cars would have major defects (i.e., were "lemons") and that the customer would just have to accept it. Customers developed some techniques to deal with this—"Never buy a car made on a Monday or a Friday" because they had a higher probability of being lemons. Ralph Nader and Clarence Ditlow wrote *Lemon Book: Consumer Rights for Car Owners*, which went through four editions (Nader 1965, 1970; Nader and Ditlow 1990, 2007).

The Taurus Project and Quality, 1980 to 2000

Henry Ford II was the CEO of the company for 34 years from 1945 to 1979, and then had largely run out of gas after the beginning of the second oil crisis. Ford's response to the oil crises had been to cut the weight of cars and put underpowered engines into them so that they achieved the best gas mileage possible.[1] Between 1979 and 1982, Ford had incurred $3 billion in losses. In 1979, Philip Caldwell became CEO—the first non-Ford family member to hold this office—for a little over the next five years. He started a new "team" approach to the design and production of automobiles at Ford that eventually resulted in the creation of the Ford Taurus. This included a major effort to revamp Ford's production system toward lean production and to introduce new and exciting models. Ford hired McKinsey Consultants to give them advice on how to do it. These results were partially helpful and over some time, Ford developed a much more lean production system with improved quality for their vehicles. As chairman of the board and CEO, Caldwell approved and directed this effort with the Ford Taurus and Mercury Sable being a risky design with curved lines in an era of boxy cars. It turned out to be one of the biggest successes in automotive history both in terms of quality and design.

The Taurus project, variously named as Signa and RN5, developed a two-pronged approach. First, it introduced and publicized a new quality culture at Ford. In the spring of 1980, Peterson appointed Larry Moore as Corporate Quality Director in an entirely new position. Moore recruited W. Edwards Deming to help set up a new quality culture at Ford (Walton 1986). Deming told Moore that management was responsible for 85% of the quality problems at Ford. Ford soon developed statistical process control over auto design and production. The Ford Taurus was newly developed based on this new statistical approach to quality.[2] Donald Petersen said that Ford was building a "quality culture" and that many of the changes that they would be making were directly taken from Deming's fourteen principles. This new emphasis on quality in the production of the Ford Taurus was reflected in Ford's advertising. The New York advertising firm Wells, Rich, and Greene created a new slogan—"Quality is Job 1"—especially for the new Taurus.[3]

Second, the new car was produced according to a new approach to teamwork. Peterson made the team concept central to his organizational style, and it was clearly part of Deming's approach to production. Team projects were formed around the new Taurus concept car, and the traditional "throw it over the wall" approach to departmental cooperation was abandoned with team leaders plucking employees with expertise for their projects. This was a major aspect of matrix organization and project management, but the American automotive companies, especially Ford, had long adopted a much more bureaucratic form of organization. This organizational culture was a barrier that needed to be overcome.

Third, the Taurus design was to be radically different and a risk for the company (Taub 1991). The 1980 Thunderbird was the impetus for change. It was voted the worst car of the decade based on quality and design (Taub 1991:9). It looked square and boxy as angular designs were the look of the time, but it was a sporty car that had become a two-door limousine (i.e., a contradiction in terms). Its redesign became curvy in 1983 (something like the Lincoln Mark VIII in 1993). This design developed concurrently with the Taurus, which was based on the Audi 5000 and the Ford Sierra, which were rounder (Taub 1991:112). Ford claimed that its more aerodynamic design gave it more miles per gallon, which was important after the second fuel crisis and in the face of the government imposed Corporate Average Fuel Economy standards (CAFÉ). The Taurus's interior was also innovative by being more user-friendly with touch control the left side of the dash curves slightly around the driver to simulate a cockpit. It also marked Ford's decisive move to front-wheel drive.

The Taurus came out in late 1985 as a 1986 model (new models were traditionally introduced in September or October for the next year). The Taurus was very well received by both the public and the press, and it was named the 1986 *Motor Trend* Car of the Year, and sold over 200,000 cars in 1986 and a million by 1989.[4] The Taurus's ultimate success led other companies trying to copy the design like the Chevy Lumina but they were a "years late" and "an effective design short".

However, these reforms were not particularly imprinted into Ford's organizational culture. Four factors led to a decline in sales. First, quality and safety began to decline for the Taurus and other car models but Ford was able to stem the tide of criticism with hotlines and recalls contrary to their earlier "lemon" culture of more or less buyer beware (Taub 1991:252–53). It eventually rose by 1990. Second, Ford was slow to develop new models of the Taurus and they lagged behind Toyota and Honda in developing new and different models. It seemed that Ford could only concentrate on one new model at a time. And third, subsequent CEOs got side-tracked from quality and innovation with other projects. The biggest side-track one was to buy other car or related companies when that money could be used for quality and innovation (e.g., Mazda 1974–2015; Hertz Rent-a-Car in 1987; Jaguar, 1989–2008; Aston Martin 1989–2007), Volvo 1999–2010; Land Rover 2000–2008; FPV of Australia 2002–2014). Only the purchase of Mazda led to some expertise in the area of lean production. Most of the others were "prestige" purchases.

And fourth and perhaps most important is that teamwork at the line level was never strongly implemented. This is due to management interests that diverged from lean production, and a fundamental contradiction between the structure of unions in the United States and the requirements of teamwork. On the one hand, teamwork in the QCC approach is rather demanding and requires a commitment by the company and the workers. On the other hand, the auto union exists to protect the rights, pay, time, and health of workers. And to do this, unions exist within a democratic structure that requires elections and candidates to present a sort of platform that indicates what the union is doing for their members. Often, union candidates compete on better wages or working conditions. As a result, cooperation with the company can be a hard but not impossible sell. For example, at the Mazda plant in Flat Rock, Michigan, the UAW current representatives faced a New Directions slate of candidates who criticized the incumbent union representatives for being too cozy with management. There are largely more pressures on union candidates to oppose management than to cooperate with them. One could say, "If you are cooperating with management, why do we need you?" This view was strengthened by our conversations with a Ford North America union representative, who in retrospect said that he regretted opposing teamwork programs. However, in retrospect, one does not have the pressures to get elected. This pressure is not universal, since Japanese and German unions, for different reasons, did not have the impulse to oppose management efforts to promote teamwork. It is the unique feature of "job control unionism" that tends to be in conflict with flexibility and teamwork on the shop floor (see Box 7.1 for a description of Ford since 1980).

Box 7.1 Ford Motor Company after 1980

FMC was founded by **Henry Ford** (1863–1947) in 1903 in Detroit and then moved to Dearborn, Michigan. The company has been the second largest American automobile manufacturer. However, it has been family owned since its inception even through it has gone public. The family controls the preferred shares that give them effective control. This has been a strength at times, but has led to weaknesses often linked with succession during crises. **Henry Ford II** (1917–1987) came home from the war to take over from his ailing grandfather in 1946, and he revived the company. By 1986, he had run out of gas and the company needed new direction after two oil crises. **Donald Peterson** had become the first non-family CEO, and had already in 1979 started to orient the company toward quality. He developed the Taurus full-size car that was launched with a dramatic design and an emphasis on quality. It was a big success. **William Clay "Bill" Ford Junior** (1957-present) took over as CEO in 2001 to 2006, but failed to get the company going. **Allan Mulally** then came from Boeing to revive the company from 2006 to 2014, and Ford was the only American company to not require government bailouts in the Great Recession of 2008. In 2018, Ford announced that it was dropping most of its automotive sedans (other than the sporty Mustang) in favor of SUVs which after the fall in gas prices were ruling the roost in terms of sales. It had already dropped the Mercury division in 2011.The Ford 150 pick-up truck has been the leading vehicle sold in America for many years and in 2017 it was equipped with an aluminum frame to save gas. In 2017 Ford hired **James Patrick Hackett** as CEO who previously worked at PG&E and was the CEO of Steelcase, which is an office furniture manufacturer.

FMC often has a somewhat paternalistic presence among its workers and in Dearborn and Detroit Area compared to GM and Chrysler. This includes the Fairlane mansion, and The Henry composed of the Henry Ford Museum with the tools of industrialism, and Greenfield Village that is a nostalgic village with stores, homes, and other historical paraphernalia from the 1800s and early 1900s. Ford was the first to hire African-Americans in the 1920s, and provide them with a small city with decent homes—Inkster, Michigan. Ford also has a more cooperative relationship with the UAW, especially since 2005.

Deep Trouble Develops from 2001 to 2010: As Ford CEOs like Harold Poling, Alex Trotman, and Jacques Nassar became interested in taking over prestige auto companies or in cutting costs, interest in quality and teamwork waned. In 2001, the Ford Focus was launched in Europe and then in America. Vice President of North American Manufacturing, Anne Stevens said "within the first couple of months, there were 17 product recalls in what was supposed to be an easy launch. . . ". (Cutcher-Gerschenfeld et al. 2015:99). A number of factors contributed to this creation of a "burning platform:" half the suppliers were new, the product was under resourced and funded, and management fiefdoms were in conflict. Further, after Nassar brought in the Business Leadership Initiative (BLI) based on the GE model including cutting the bottom 10% of employees, there was a competition between Six Sigma, lean production and BLI. And clearly, the lean production focus on quality had been lost. To regain it, Stevens tried to improve quality through a strengthening the LIN with LMM being given greater authority. Plans were made to sell off the new acquisitions, but in large part, it was too late.

Ford reported losses of $12.7 billion in 2006 right before the Great Recession of 2008. FMC was again pressured to change its organizational culture in order to survive and compete in the hard-pressed auto industry. Different companies took different strategies as the overall auto market shrunk with higher unemployment and an endangered stock market. Ford's relations with the UAW are friendlier than other American company's relations with their unions.

In 2008, the UAW and Ford negotiated a contract to give them a two-tier wage structure. This transformation had a collaborative labor relations approach. Bringing the union into discussions on key issues such as product quality and mutual respect in the workplace had a transformative effect on the efficiency of assembly plants and the culture of Ford as an organization. These changes in workplace practices were initiated at the local plant level, and then spread to the larger firm. Changes included a lower entry-level pay for workers, the introduction of a team leader role with a higher wage. Ford was able to reorganize itself and return to profitability in 2010 without government loans, but the company supported these loans for GM and Chrysler. GM also received loans from the Canadian and Ontario governments. Ford earned the reputation of being the only self-sufficient American auto company, but it had mortgaged all its assets, the company name, and the Ford emblem to raise $23.4 billion in loans to finance new product development during another restructuring, which was more or less similar to taking government loans (but without the stigma). In other words, it pumped itself up with private rather than government debt, while the other two companies waited and ended up with government debt. And a transformative agreement with the UAW was signed in 2011 that further brought the UAW into cooperation with the company.[5]

Out of the Woods? 2011 to the Present

Ford, GM and Chrysler's efforts to become lean are so extensive that they are starting to be called the "lean three" instead of the "big three". This move was meant to change the public perception that the American auto companies are lagging behind their Japanese competitors in their efficiency, but was also meant to increases productivity and sales. Ron Atkinson, a former GM quality manager and past President of ASQ, asserted that lean had permeated the entire culture at GM and its Detroit rivals "to the point where the North American companies are as lean as any other organization in the world" (Cable 2009). The Lean Three are investing in global manufacturing strategies and flexible operations as a "Survival strategy" in a very competitive market. In many ways, the Big Three were at a disadvantage when the fierce competition with the Japanese lean companies started since the American factories were older and had union contracts in place that were quire rigid in protecting workers rights, whereas lean factories were flexible with job rotation and having newly built plants.

For instance, the 2018 Assembly Plant of the Year Award went to the Ford Van Dyke Transmission Plant in Sterling Heights, Michigan. According to the description of *Assembly Magazine*, the plant supplies more than 70 different transmission models to various Ford plants, fits all the criteria of a completely lean factory. This includes flexibility as the key to success, the use of lean methodologies—continuous improvement, standardized work, visual management, minimization of waste, and strong teams (Weber 2018a, b). However, one should be cautious since the Romeo Engine Plant got the Shingo Prize in 2002, and then the Ford Focus came out with a terrible quality rating (Cutcher-Gerschenfeld et al. 2015:132). In general, US auto companies are subject to "the red queen effect" (Barnett 2008). The American companies have gotten much better at producing quality automobiles; but Toyota and Honda are not standing still, and are constantly upping their quality game as automobiles become more and more complex. This is shown in *Consumer Reports* and J. D. Power quality ratings.

CEO Alan Mulally is given considerable credit for both the change in corporate culture and the debt restructuring, but UAW leaders who accepted the two-tier wage structure less so. Gerschenfeld et al. (2017) show that the Ford-UAW relationship was transformative in going forward including at the global level. However, cuts in employees at the blue-collar and management level continued to transpire. As a result, the rank-and-file UAW members at Ford found the restructuring processes to be difficult, especially because the two-tier wage structure, like the temporary employee situation in the Japanese transplants, caused considerable conflict within the UAW membership. The higher wages and protections of older workers were not available to the newly hired workers who were increasing over time. The UAW leadership accepted measures that clearly helped Ford (and the other two American automakers) stay in business, but they did so at great cost to their new and future members (Cutcher-Gerschenfeld et al. 2015). Further, the Taurus was starved for remodeling and upgrading funds, and it soon became oriented for the low-priced residual fleet sales for rental car and police cruiser market (USA Today 2006; MSNBC 2007). Then in 2019, Ford announced that it would no longer produce passenger cars (except the Mustang) and concentrate on the SUV and truck market.

However, the 2011 to 2016 Ford Focus had a major quality lapse that was a headline article in the *Detroit Free Press* including five full pages of text (Howard 2019). Ford was investigated for major transmission flaws that imperiled drivers with lurching starts and unexplained disconnects. Fortunately, no one was killed. The transmissions were made by Getrag, a German contractor that would not give Ford the design plans, which is highly unusual for a JIT supplier.[6] Under lean production, OEM manufacturers and suppliers share large amounts of design and production information. Further, Ford rushed the Focus into production apparently knowing that its transmission problems were not fixed. Now Ford faces over 2,500 complaints and many lawsuits even though the NHTSA declined to formally criticize Ford. Ford seems to have quality problems with small cars since this is the second major quality breakdown with the Focus since 2000. Also, this is a major quality issue since it occurred under Mullaly's time as CEO that emphasized quality, and it happened well beyond the time when Deming advised Ford on quality in the mid-1980s.[7]

We conclude that Ford did a lot to improve lean production, but it certainly had its ups and downs with competing issues like BLI, Six Sigma, buying prestige auto companies and then selling them at a loss, buying back stock, and paying out too much in dividends (Lombardo 2017). Its major quality lapses with the Focus in 2001 and 2011 to 2016 are disconcerting. However, concerning lean production, there are two main issues: teamwork and managers. Ford made many attempts at implementing teams, but for the most part, its lean teams were either only in a few plants, or otherwise too big. The "typical team" in 1988 consisted of a team manager, team leader, 3 engineers, a financial analyst, 3 team coordinators, 3 skilled trades persons, and 15 production workers—27 team members (Cutcher-Gerschenfeld et al. 2015:131). In the 2011 contract the production work group consisted of 20 to 30 members, and the Manufacturing Work Group had 10 to 20 members. In our interview with one HR manager, she said that teams were 10 to 17 workers, much too large compared to Toyota.[8]

Second, individualistic managers called "Cowboys" or "Marlboro Men" resisted teamwork and had a highly expedient attitude toward production. Decisions such as putting the Focus into production with known quality defects are not made by workers on the line. They are management decisions. One Ford manager we interviewed also said that when they were at a new plant in HR, they had three managers and only the last one had a positive attitude toward FPS or lean production. And this also affected the union. Dan Brooks, a top union team advocate, was taken aside by a union local leader into his office. The leader pointed to a date on next year's calendar, and said "this is when FPS will be launched here"—"It's the date I retire!" (Cutcher-Gerschenfeld et al. 2015:139, 177, 244). And a top level UAW official said he regretted opposing teams in the late 1970s, and that it was now hard to push teamwork given his past opposition. In general, managerial resistance was not overt, but it consisted of a lot of foot dragging or "We're getting to that, but first we have to do other things".

We concentrated on Ford in this discussion, but were need to say a somewhat different approach that GM took to lean production. CEO Roger Smith initiated a bold step in creating the Saturn Division that would become an American version of a Japanese transplant. The project adopted many aspects of lean production and had an innovative low pressure sales approach. The new division created an excellent automobile that sold relatively well. However, it failed to produce a profit and sustained losses for its whole existence. The former workers at Saturn, whom we interviewed, loved working there, but they were devastated that when the division was closed. The more powerful and historically significant Chevrolet division basically quashed it. Subsequently, the revived and more generously funded the Chevrolet division, especially the Impala, and some SUVs and smaller cars began to sell for GM (Keller 1993; Sherman 1994; Rubinstein and Kochan 2001; O'Toole 1996). However, the lean production methods did not survive except for a more extensive and well-managed supply chain with JIT principles. Currently, Ford and others are cutting sedans in favor of SUVs (Martinez 2016).

In conclusion, the Ford Production System is oriented to "increase its global flexible manufacturing to produce on average four different models at each plant around the world to allow for greater adaptability based on varying customer demand" (Ford Motor Manufacturing 2019). Boasting that its auto production has not stopped for over 100 years, Fords' current vision is "to design smart vehicles for a smart world". However, lean production, Six Sigma, and the business leadership initiative (BLI) program are competing models within Ford. The last two are inherently weak on teams, and the BLI can be perilously close to stakeholder value theory, which is the antithesis of lean production.

Lean Production and the Nissan Corporation

Nissan has been making cars under the Datsun brand from 1914 based on the first letters of the last names of their investors: Kenjiro Den, Rokuro Ayama and Meitaro Takeuchi plus the "sun" on the Japanese flag. It has gone through many name changes and mergers over the years with Nissan first being use in the 1930s. It was allied with Austin Motors from 1937 to 1960, merged with Prince Motor Company in 1966, and an American engineer William Gorham provided strong engineering technology to the company (Halberstam 1986:393). Nissan has been the most Fordist of the Japanese companies and reportedly is not a strong team-oriented culture. There is no evidence for the Nissan Production System as there is for the TPS. However, it does have a strong design and engineering, and a weak sales and marketing reputation. In some ways, it does not have the strong single family or entrepreneurial orientation of the Toyoda family or Soichiro Honda. Nonetheless, when the Japanese started exporting to the US, Nissan played a strong role in gaining sales through the Datsun name plate (later renamed to Nissan), and especially the Datsun 240Z sports car.

Box 7.2 The Nissan Corporation: Hashimoto and Ghosn

Nissan began in 1911 and became a Japanese multinational automobile manufacturer in Nishi-ku, Yokohama. The company sells Nissan, Infiniti, and Datsun cars as well as after-market performance spoilers and racing items under the name of Nismo. **Masujiro Hashimoto** (1875–1944) founded the Kaishinsha Motor Car Works as an automobile company (unlike Toyota, who only started making cars after World War II) and called its first car the DAT. It merged a number of times before World War II and started using the Nissan name, which is their stock exchange abbreviation. The company originated in the Nissan *Zaibatsu*, now a *Keiretsu* called the Nissan Group. After World War II, it began making the British Austin 7. After the 100-day strike in 1953, Nissan produced vehicles for the US Army and then merged with the Prince Motor company in 1966. It expanded in the 1970s and exported a large number of cars to the US in the 1970s. In 1970, it established its Canton, Mississippi plant, and in 1981 established a Deckard, TN and a larger Smyrna assembly plant in Tennessee. The upscale Infiniti brand was introduced in 1989 for sale in the United States (Skyline model in Japan).

With some major financial troubles in the 1990s, Nissan merged with the French Company Renault and the Japanese company Mitsubishi to form the Renault-Nissan-Mitsubishi Alliance. In 2013, Renault held a 43.4% voting share in Nissan, while Nissan only had a 15% non-voting share in Renault. From 2009 to 2017, the Brazilian born Frenchman **Carlos Ghosn** (1954-present) served as CEO of both companies. Under pressure, Ghosn stated that he would resign as CEO of Nissan on April 1, 2017, and the police then placed him under arrest for illegally using company assets, underreporting earnings, and engaging in tax evasion (Chozick and Rich 2018). In November 2018, he was fired by Nissan. It is important to note that Japanese executives receive much less pay and benefits than US CEOs, often by a factor of 10. **Hiroto Saikawa** is now the chairman of the board, and Thierry Bollore, the CEO of Renault, is the temporary CEO of Nissan.

In 2017, Nissan was the 10th largest automotive manufacturer, which is less than half the sales of Toyota and VW, but by adding in Renault and Mitsubishi, it is the sixth largest firm. Nissan is the leading Japanese brand in China, Russia and Mexico (Statista 2019). The Nissan Leaf is the top-selling all-electric vehicle to date. In January 2018, Nissan announced that after 2021, all Infiniti vehicles would be hybrid or all-electric vehicles.

Nissan has been prone to crises as the automobile industry is cyclical, but it has more trouble negotiating cycles than Toyota or Honda. In 2001, during a relatively mild recession but a major stagnation in Japan, Nissan was struggling with $20 billion of debt and the company was almost bankrupt. Carlos Ghosn, a French executive of Lebanese extraction, had extensive leadership experience at Michelin and Renault, and he was hired as part of a Renault-Nissan merger to straighten the company out (Ghosn and Riès 2005; Magee 2003; Wickens and Lopez 1988). The problems that he found were first of all that no executives or others understood why they were failing (Ghosn and Riès 2005:97). Ghosn found that the company had dropped in global market share from 6.6% to 4.9% from 1991 to 2001. He found weaknesses in concentrating on profit, neglecting customers, cross functional work; a long-term vision, and an "absence of a sense of urgency" (117). Ghosn developed a NRP composed of five parts (Ghosn and Riès 2005):

(1) Establish a clear focus on making profits rather than the volume of sales. This would involve clear attention to financial management (139)
(2) Cut factories from 24 manufacturing platforms in seven factories to 12 platforms in four factories
(3) Reduce purchasing costs by 20% over three years. (118–9). This would involve the Nissan 3–3–3 plan (107) that is Cut the number of

employees14% or 21,000 jobs evenly distributed between manufacturing, sales, and administration

(4) Get rid of extensive stock holdings in the Nissan *Keiretsu* where Nissan held stock in 1,394 companies, most of which had little to do with the auto industry
(5) Create a stronger brand image based on design and engineering centered around bringing back the sporty and muscular Z sports car which had been discontinued (resulting in the Nissan 350 Z (138–39))
(6) Improve transparency to the external world and communications within the company.
(7) Promote and pay employees based on performance and not seniority, and especially dismiss the practice of senior advisors composed of former executives that continue to receive pay and supervise an extensive but non-essential staff (150–1)
(8) And he would implement this by building his own cross-functional teams, and in fact promote cross-functional teams throughout the Nissan organization (Ghosn and Riès 2005:116–27, 128–35, 138–9)

Note that the emphasis on "profits" rather than market share in point one is largely the opposite of Toyota.

What is missing from this is a concept of teamwork that reaches the production shop floor? There is no mention of team leaders other than for executive teams. This is no discussion of continuous improvement (*kaizen*) or eliminating waste (*muda*). This approach was confirmed in our interviews with Nissan line-workers and managers at the Nissan assembly plant in Smyrna, TN. When we asked Nissan managers about teams, they sometimes referred to the whole plant or division as a team. And, so does Carlos Ghosn in his book with Philippe Riès (2005). When we asked Nissan workers and managers about how quality control was handled by teams, they looked puzzled. Then they pointed to their quality control department that had many employees engaged in the quality control process. Clearly, quality control was divorced from the day-to-day operations of stamping, assembly, and painting. There is a strong sense of flexibility, but this is not built into teamwork.

In Niels-Erik Weirgin's assessment of the Nissan Motors UK (NMUK) plant in Sunderland in the United Kingdom, workers are organized into teams from eight to 17 workers with an average size of 10 (2003:166–8). However, Weirgin says that worker autonomy is "very restricted" and according to the Amalgamated Engineering and Electrical Union (AEEU) regional office, work is dictated by management. Also, the NMUK company council said that "team members can make 'no decisions at all' " (2003:167). The team leader is "merely responsible for training and development in his team". Instead, the supervisor is responsible for nearly everything in his zone (2003:168).

Thomas Muramaki did a study of nine task areas concerning the autonomy of teams in Europe and Japan (1997). He used the Gulowsen scale in looking at teams: (1) selecting their leaders, (2) screening new team members, (3) distributing work within the team, (4) adjusting their work times, (5) accepting additional work and overtime, (6) representing the team to the larger organization, (7) deciding about methods of production, (8) deciding production output goals, and (9) deciding about production quality (1979:750; 2000). The Gulowsen scale is quite comprehensive and one would not expect particularly high scores on this measure. Muramaki concludes that autonomy of a work group is essential for teams, and he studied 19 different auto plants in Europe and Japan. The highest scores were 2.7 for the GM Eisenach plant and the Saab plant in Finland at 2.6. Germany could certainly be higher since it does give works councils some control over production methods. Although he does not report the scale scores for each plant, he found that German plants averaged 2.2 on the scale, British plants 1.6, the one Swedish plant 2.6, the Toyota plant in Japan 1.9 to 2.0 and the Nissan plant in Japan 1.9 to 2.0. But significantly, Muramaki comments that Nissan at Kyushu and Vauxhall at Luton were the only two plants out of nineteen where "'quality" remain(s) completely in the hands of management" (1997:754).[9] On the basis of this, one must conclude that QCCs at this major Nissan plant in Japan are not operational (see also Garrahan and Stewart 1992; Pardi 2010, 2005).

While Nissan does have job rotation and claims to have a strong *kaizen* program and effective training programs, it is not strong on teamwork. Thus, on the basis of this evidence of teamwork at Nissan not being oriented toward team-based quality control in the United States, Japan, and the United Kingdom, Nissan's implementation of lean production is quite partial, and one might also ask whether this is part of the reason why Nissan tends to get into financial trouble. Their plants will have continuing issues with teamwork, and their NMUK plant will face the Brexit situation and its attendant trade difficulties.

McDonaldization, Emotion Work, and Co-Production

McDonaldization was developed by George Ritzer in 1993 (2004, 2019), and it models itself on the rationalization and standardization of the McDonald's restaurant corporation. Relying on Max Weber's theory of bureaucracy and rationalization, it consists of five parts. First, efficiency experts discover the "best possible means" to produce services, which makes it resemble scientific management in the food industry. Second, calculability is part of all aspects of the restaurant business, especially the timing of cooking, the length of serving, and the designing of take-out windows. Third, efficiency and calculability are somewhat uniquely applied to the rituals of greeting and satisfying customers. Fourth, technology exerts great control over the food production process including the standardization of production (i.e., there are no "custom hamburger orders" making it antithetical to post-Fordism). And fifth, irrationality results from this process because it leads to ends that people do not want (e.g., obesity of customers, dehumanization of workers, mass-produced food, and a mind-numbing sameness of the product).

The originality of George Ritzer's concept of McDonaldization is that he takes Weber's principles of efficiency, predictability, control, and calculability from the organizational (meso) to the societal (macro) level, showing the unintended consequences of bureaucratic labor processes on the overall society. Today, all aspects of advanced industrialized societies from media to banking, housing, entertainment, education, health care, travel, and family are permeated of the rational, and ultimately irrational, principles of McDonaldization. Whereas McDonaldization highlights the homogeneity and predictability of production, a rising countertrend in post-industrial societies is eBayization, which provides variety, adventure, freedom and risk (Ahuvia and Izberk-Bilgin 2011). At the same time, a complementary notion to McDonaldization emerged called the Disneyization of society. Though not quite as prevalent, it is based on theming, dedifferentiation of consumption, merchandising and emotional labor (Bryman 1999; 2003).

Box 7.3 The McDonald's Corporation: The McDonald's and Ray Kroc

Richard (1909–1998) and **Maurice** (1902–1971) **McDonald** started the first McDonald's restaurant in Los Angeles with hot dogs and then hamburgers for ten cents in 1937. In 1952, they modernized and improved their processes into a fast food restaurant. In 1954, **Ray Kroc** (1902–1984), a milkshake machine salesman, helped modernize and expand the restaurant as a franchise system. Their business model is that the main company owns all the land, which it rents out to franchises. These rent payments are about 25% of each McDonald's restaurant's income. McDonald's went public in 1965, and in 1968, they opened their 1,000th restaurant. Ray Kroc became chair of the board for the next five years. In 1974, the first Ronald McDonald House to give back to community for seriously ill children. In 1975, they opened their first drive-thru window following Wendy's lead. In 1961, McDonald's opened its Hamburger University offering a "bachelor in Hamburgerology". The restaurants expanded throughout the world and the first restaurant opened in China in 1990. By then, sales started to lag, so the company varied the menu and cut back expansion (Ritzer 2004).

By 2000, sales were falling and McDonald's eliminated 700 restaurants. There were also scandals about using lard in their French fry mixtures and it became a target of anti-globalization campaign. In addition, criticism came from nutritionists and the movie *Supersize Me* showed the impact of eating at McDonald's with the protagonist becoming overweight on such a diet. In the late 2010s, McDonald's continued to make changes and built their largest restaurant for the 2012 London Olympic games with 1,500 seats. McDonald's also began a local approach with specialty menus for different countries with rice in Indonesia, prawns in Singapore and Japan, beer in Europe and New Zealand, meat pies, and a *Hallal* menu in Muslim countries. It also introduced McCafe with upscale coffee products (Ritzer 2019).

Like other fast-food chains, McDonald's has been criticized for paying employees a low wage with some employees being on welfare. There were also a number of strikes. Now, the number of controversies that involve McDonald's is enormous due to the size of the corporation (McDonald's Corporation 2018:3–9).

While Ritzer is a critic of McDonaldization, we have to be quite critical of his view of McDonalidization as the new model of the economy, not because it does not exist, but rather because McDonaldization is simply "food Fordism". All of the first four principles are present in Fordism, except perhaps something equivalent to the "five-dollar day" (i.e., pay in McDonaldization is relatively low). Ritzer directly applies Weberian rationality to food preparation, and this flies in the face of the flexibility, limited production runs, and unique and high quality products of the previous two models. While we contend that McDonalidzation as a whole is not new, two aspects of this business model can be considered novel.

The first new aspect is the rationalized emotion work involved with friendly greetings and attention to satisfaction concerning customer service. This is new in that good times and "joy" ("I'm lovin it") are engineered into the teachings of the McDonald's University with locations in Chicago, Sydney, Munich, London, Tokyo, Brussels, and Beijing. But again, even this "emotion" is standardized at every location. Since emotional labor is regarded as a key feature of McDonald's, its global expansion brings into question whether a standardized emotional display can be perceived positively across diverse cultures. Bryman (2003) gives examples from the restaurants opened in Moscow and Hong Kong where the happiness and friendliness of McDonald's workers with customers are treated with suspicion. In fact, workers who are having too much fun on the job are perceived as not working hard enough, thus enjoying themselves too much at the expense of consumers and management (Watson 1997). Even in the United States, the company is struggling to live up to its ideals of happiness and friendly service. As of April 2013, McDonald's received mounting criticism from customers that its employees are rude or unprofessional, which made the company appeal to its employees to fix "their broken service" (Jargon 2013). Further, crew members make only an average of $7.68 per hour or

$15,974 annually, with cashiers making five cents an hour more. Promotions may lead to be a trainer ($16,869 a year) and a shift manager ($20,113 a year) (Glassdoor 2013 with averages based on 1,112 workers reporting). This is among the lowest pay in the industry. Current workers are not saying "I'm lovin' it". Instead they are saying "I'm hatin' it!" And in 2013, this has resulted in demonstrations with many asking for a "living wage".

The second innovative aspect of McDonaldization involves the term "consumptive labor" that describes the rising category of consumers who perform work tasks from filling their drinks to cleaning their trash in the manner of quasi-employees (Koeber et al. 2012). The expansion of self-service through fast food dining, online and ATM banking, distance education, online shopping, walk-in/walk-out medical centers, and self-check-out at grocery stores brings the consumers as co-producing agents in the new division of labor. Thus, organizations outsource their service to customers in exchange for keeping the costs down.

However, McDonald's experience with lean production was limited. It does have an impressive JIT system in its supply chain. However, a pull system is more difficult. In 1997 McDonald's developed a new JIT system called "Made for You" in order to deliver fresher and hotter food. This entailed an expensive changeover costing $25,000 per restaurant with the corporation paying for about half the cost. However, by 2000, the "Made for you" system lost traction because service times lagged and it was too labor-intensive. It was abandoned as McDonald's closed a number of stores in the early 2000s (Canedy 1998), but brought back on a limited scale for one item in 2018 (Kline 2018). However, McDonald's has been able to avoid wastage in mass producing food items through a computer system and a version of continuous improvement (Dixon 2010).

Similar to Fordism, McDonaldization has two opposite forces in its labor model. On one hand, its bureaucratic principles of efficiency, calculability, control, and predictability ensure that both customers and employees are treated similarly regardless of their gender, class, age, or social class. On the other hand, this excessive organizational rationality creates dehumanizing and alienating workplaces where employees feel like robots working on the assembly line of fast food (Ritzer 2004). For some, McJobs have become the symbol of low skill, low pay, dead-end jobs (Gould 2010) with a record turnover rate of 60%, according to a 2010 report of National Restaurant Association (Jargon 2013). For others, particularly managers, McJobs offer great training opportunities for junior employees, tuition money for workers going to college, employment opportunities for people with disabilities, and a good record of hiring and promoting minorities and senior citizens (Ritzer 2004).

While Ritzer portrays McDonald's as a paragon of Fordism, the company has tried to implement some aspects of lean production. McDonald's has implemented a more flexible form of production avoiding stockpiling burgers before the rush hour and trying to provide a "pull system" whereby orders are produced just after they are made. From the greatest restaurant chain in the world to a symbol of American imperialism, McDonald's is a fascinating example on how the lowest-paid industry at one time became the fastest growing industry in the United States. While incorporating a small amount of ethnic food and showing appreciation for local cultures at its worldwide stores, it is also portrayed at the same time as the evil side of globalization. But getting back to the basics, McDonaldization is simply scientific management aimed at restaurants with a two original features—emotional labor and self-service.

Conclusion: Semi-Lean Production

This chapter has discussed three corporations—FMC, Nissan Corporation, and McDonald's Corporation—that have only a partial approach to lean production. Many firms are partial adopters of lean production, and to a large degree most firms have strongly embraced JIT inventory. But the part of lean production that is most downplayed has been strong teams. Teams and committees exist at the middle- and upper-levels of most firms. But the question is whether they exist on the shop floor. Some are truly open and creative teams (e.g., the Taurus team at Ford), but others are what we call "management teams" that are largely directed by the managers in focusing on tasks that workers have to figure out or take directions. Real lean has entirely different forms of teams.

The future of semi-lean production is hazardous since its environment is changing. For the two automotive companies, automobiles are entering into a new technological revolution. As second in terms of production and quality to Toyota and Honda, Ford and Nissan are facing the "internet of things" in auto production with smart devices being inserted in automobiles. This has resulted in the implanting of Silicon Valley firms and technologies into automobile production. Cameras on the outside of the automobile body are now *de rigueur* for automobile safety. Electric cars and hybrids have been in production for at least 10 years. Nissan has a bit of an edge with the all-electric Leaf. Self-driving cars are the technological challenge of the future, which then leads to renting out cars rather than actually owning them. Ford faces the consequences of a unionized work force with major divisions due to a dual wage structure, but in 10– 20 years, this will disappear as the lower-tier workers become dominant. These technological changes have driven some mergers or sales to China and India, while new entrants like Tesla push the technological envelope. Can Ford and Nissan survive without more effective teamwork throughout the organization rather than just at the top?

The future of McDonald's faces two different foci. On the one hand, the fast food market is highly saturated, especially in the United States. Craft and more artisanal production and conceptions of food are a challenge to a mass-produced food preparation chain—Arbys, Wendy's, Carls Jr., Sonic, etc. McDonald's has branched out into Mexican food with a Quidoba look alike in Chipotle. To counter or copy Starbucks, McDonald's is offering coffee specialties that are often quite good. But two forces are on the horizon, one technological and the other demographic. Food delivery is clearly a force in the future as groceries are being delivered to one's home now, and many restaurants have followed pizza delivery with similarly delivered meals. Will McDonald's be able to do this? Second, there will be a labor supply crunch as the baby boom cohort leaves the labor market and succeeding cohorts are smaller. The young adult labor supply is dwindling and food service has already moved into hiring senior citizens who often have poor pensions or none at all. This has already created high pressure on wages and movements for increases in the minimum wage, especially in cities, some of whom have enacted higher minimum wages. With high turnover in fast food service, especially due to low wages, will McDonald's adapt again to a changing environment? However, lean and quality control teams with specialized production operating according to a pull system are unlikely—can one envision a McDonald's version of a *konditorei* or *charcuterie*?[10]

In sum, the Achilles' heel of semi-lean production is their use of teamwork in the quality control process. Ford and Nissan have taken steps in this direction, but they cannot seem to fully implement teamwork-based quality control and are often distracted by other issues from BLI to mergers and financial crises. In the fast food industry, turnover is so great often because of low wages, that team-based quality control does not seem to be a realistic possibility.

Notes

1. The second author owned a new Ford Fairmont in 1980, and it had difficulty making it over the Bay Bridge from Oakland to San Francisco. We avoided the larger hills in San Francisco.
2. Statistical quality control was not a foreign concept at Ford. In 1947, at the beginning of Henry Ford II's revamping of Ford Motor Company after his father's decline at the helm, Hank the Deuce hired 10 men from the Office of Statistical Control in the War Department (Taub 1991:34–35). Although they knew nothing about automobiles, they used "statistical quality control" in their operations and became known as the "whiz kids". The most prominent among them was Robert McNamara who became the first non-family President of Ford and then went on to head the Defense Department under Presidents Kennedy and Johnson. However, statistical quality control fell back into simple inspection on most automotive assembly lines.
3. There is evidence that the quality and safety of the Taurus was not as high as claimed. Clarence Ditlow of the Center for Auto Safety (and co-author of the *Lemon Book* (2007) with Ralph Nader) found numerous problems with stalling, steering, piston scuffing, rough downshifting, and electrical systems (Taub 1991:252). Ditlow sent a letter to CEO Peterson saying that the car should be recalled. Also, the initial Taurus did poorly in the DOT's crash-worthiness tests in 1986. The car was not recalled, and by 1990, quality had improved (Taub 1991:253).
4. A number of other cars were introduced and discontinued during this time. Ford of Europe launched the third generation Escort, which became the European Car of the year for 1981. In the United States, the Lincoln Town Car and Ford Escort

were introduced. In 1982, Ford of Europe introduced the Sierra and ended the Cortina/Taunus. A new Thunderbird came in 1983 and the Grenada was discontinued. The Ford Tempo and Mercury Topaz were introduced in 1984, replacing the Ford Fairmont and Mercury Zephyr. A very exciting Ford Scorpio was launched in Europe, becoming the European Car of the Year for 1986, but it was too expensive in the United States. The Transit van was launched in 1986 and this model is still a maintain for its market today. The Ford Explorer in 1991 and the Ford Aerostar was *Motor Trend* Truck of the Year, and the Lincoln Town car was the *Motor Trend* Car of the Year.

5. This agreement is involved with many innovations, which are discussed by Cutcher-Gerschenfeld et al. (2015), but they take us a bit off track from our focus on lean production.
6. Getrag is a well-respected transmission maker in Germany that had a good reputation for supplying transmissions to high-end automobiles including Porsche. The DPS6 dual clutch transmission was considered to be innovative because it would improve fuel economy. Getrag went bankrupt and was purchased by Magna International, a Canadian supplier to Ford (Howard 2019:17A).
7. The American auto industry has repeated claimed that no profits can be made from small cars. It may be that management is culturally imprinted with the idea that small cars are a step-child model. As a result, designing and producing them is a low status position. However, other than the statement about the difficulty making profits, we do not have further evidence for the imprinting hypothesis.
8. The Ford quality initiatives include the QOS with QOSCs from both management and the union. The QOS meetings, however, tend to be more in the area of cross-functional teams at a higher level than quality control circles. The represent a laudable improvement in company and union cooperation, but they are not a shop-floor initiated process (Cutcher-Gerschenfeld 2015:279–98).
9. One might ask: why we do not have the same assessment for Toyota since its score is the same as Nissan? The first reason is that we consider the issue of "deciding about production quality" to be crucial. The second reason is that the Gluowsen scale is quite comprehensive since it is derived from socio-technical theory (1979, 2003). As a result, team autonomy at Toyota while higher for quality than Nissan, may be lower for other issues such as deciding about production methods and output goals. Also, as per Ghosn, Nissan does not maintain strong control of its different plants, and variations may occur (Ghosn and Riès 2005).
10. A *konditorei* is a shop that offers high-end pastries and coffees, similar to but with more offerings than Starbucks. *charcuterie* is usually a small shop in France that prepares gourmet foods from pork and other meats. Bacon, ham and sausages are offered along with galantines, ballotines, pates, and confit. Both are a major step above a common bakery or delicatessen.

8 Creative Teams at Home and Fordism Abroad

Design and Production at Nike, Apple, and Google

Silicon Valley in California has become famous for their technological innovation concerning computers and other electronic products, and with these epoch defining products, it has also become a symbol for a new informal way of working that is seen as especially encouraging creative work among highly talented people. Gatesism is mentioned by Gaëtan Tremblay as a movement toward the information or interactive society (Tremblay 1995, 2008a, b; LaCroix and Tremblay 1997; Stross 1997). Gary Hytrek (2008) mentions siliconism as how globalization is transforming stratification. Gatesism and Siliconism are less often discussed and somewhat hard to delineate. But they represent what is distinctive about how work is organized among the high technology firms of Silicon Valley in California.

In terms of technical and social aspects, there are some mixed features of this approach. Siliconism includes five features. First, the product is often not physical and is more often computer code written into programs such as operating systems, word processing programs, and various applications. But there are some physical products such as Intel's chips or Apple's i-Pads, but much of the manufacturing of these products may be off-shored to China, Taiwan, and elsewhere, whereas the knowledge and design work are done in its high-tech hub in Silicon Valley.

Second, most work in the central offices of large corporations in advanced industrialized countries is often creative so workers are organized into semi-autonomous teams solving problems in small groups sometimes having started much earlier in a garage or dorm room. This model generally follows many aspects of sociotechnical theory, where teams have considerable freedom to take advantage of their creativity. This has led to a tradition of informality on the Silicon Valley campus-style workplaces with various amenities such as gyms, high-end restaurants, coffee shops, and other activities. Dress is often extremely informal, and the workplace is pervaded with amenities. The 2013 movie *The Internship* with two middle-aged and unemployed watch salesmen trying to make it at Google is a good (if slightly exaggerated) description of Google's culture with intern beanies, free food and drink, relaxation pods, and group work (Twentieth Century Fox 2013).

Third, the organization is directed by professional managers with some residual founders in specialized niche and authority roles. This mix of entrepreneurs and managers is intended to maximize creativity and business acumen. Fourth, employees often work very long hours as a norm, and employees are encouraged to stay late into the night and sometimes even have relationships with each other (e.g., Google encourages its employees to date and perhaps marry each other). And fifth, pay and benefits are often quite high though job security is not always guaranteed. Workers may move fluidly between companies on different sorts of projects. In its more organizational form, it consists of project teams of highly skilled programmers and scientists working together on new and innovative projects. Some have referred to this as the home of the "protean worker" where job switching and learning new skills are the norm rather than the exception (Hall and Mirvis 1995).

The sociotechnical aspects of siliconism generally follow the work started by Kurt Lewin and furthered by the Tavistock Institute in London along with consultants. This approach emphasizes "semi-autonomous teams" that have a great deal of freedom to pursue their work, which is especially good in high tech and design environments. They are involved in larger operations that involve cross-functional teams with people from many different units. Lewin's approach consisted of four points: (1) democratic values in that participation is a basic value because "we are likely to modify our own behavior when we participate in problem analysis and solution" (Weisbord 2012:89), (2) force field analysis helps us to get out of what appear to be frozen conflicts, especially when we see the driving and restraining forces to any problem; (3) the nature of change is based on learning in organizations by viewing any structure as frozen, unfreezing, moving to a new situations, and then refreezing; and (4) the primacy of group dynamics in that people respond to tasks in groups (Sherif et al. 1988). This approach is sometimes known as the human relations approach to work. It assumed that work is organized in systems, technology and social concerns are optimized together, intrinsic properties of jobs (variety, challenge, autonomy, etc.) are more important than pay; and teams and groups can be rather autonomous with multiple skills.

The implementation of this approach in a changing environment consists of four parts. First, top managers establish the need to redesign the organization, often by visiting innovative workshops, gather ideas, and convene the steering group (Weisbord 2012:314). When they find a window of opportunity they assemble employees and convene a "steering committee". Second, this committee does an environmental scan is set up to examine the past, present and future using fair and open communication. Third, from this scan, the steering committee composed of all levels of the organizations (i.e., stakeholders) manages the goals and values, and appoints a "design team". Fourth, the design team does a technical and social analysis to come up with the solutions to implementing the changes. The design team is a semi-autonomous group that is responsible for new solutions.

They do the sociotechnical analysis and provide the solutions to problems that the steering committee reviews. However, the socio-technical theory with its semi-autonomous teams was only employed by a small number of firms, and otherwise ignored until Silicon Valley embraced it (Sachs and Rühli 2011; Kenney 2000).

Along with the more organizational aspects of the work, there are several socio-political aspects that should be considered. First, the Silicon Valley companies thrive on deregulated labor and product markets. They avoid bureaucracy in their organizations but also tend to overlook formal aspects of employee protection, which means that they are often quite anti-union or simply do not see the need for unions since they provide their employees so well with wages, perks, and benefits.

Secondly, these Silicon firms tend to avoid paying very much in taxes by generally using tax havens and loopholes. For instance, Apple, Google, and Facebook are known to use "double-Dutch" or "Double-Irish" tax dodges by putting offices in countries that allow putting money in Bermuda or the Cayman Islands. Their taxes often amount to 5–12% compared to the corporate tax rate of 35% in the United States. A recent Congressional inquiry finds that Apple even claims that its subsidiaries are stateless and "beyond any tax authorities reach" (Schwartz and Duhigg 2013). Similarly, web and internet firms have long enjoyed tax exempt status even though they are profit making (e.g., Amazon), and recently legislation trying to remove this tax exemption is facing heavy opposition from the Silicon Valley firms. And the CEOs are the stars of these corporations making very high salaries with stock options, while employees make much less (Fitzgerald 2011; Rao and Scariffi 2011; Kenney 2000; English-Leuck 2010).

Third, more directly connected to politics, Marc Andreessen, a Silicon Valley venture capitalist, sees valley politics as going through three stages: (a) 1970s to 1980s—"just leave us alone", which tends to embrace a form of libertarianism, (b) 1990s and 2011—"focused narrowly on pet issues", which involve repatriating profits or getting more HB-1 high tech visas; and (c) most recently from 2012—"equip more Americans for the digital age" perhaps through "Citizenville" and represented by the Zuckerberg inspired *FWD.us* lobbying organization (Packer 2013; Newsom 2013; Johnson 2012). Silicon Valley, with the creation of major fortunes, has produced inequality and some embarrassments (e.g., David Sacks estimated 1.4 million dollar "Let Him Eat Cake" birthday party based on a Louis XVI theme (Bindley 2012), and numerous high-end art, mansion, and automobile collectors) (Thomas 2012). But on the whole, the Silicon Valley influence has become liberal in order to support open communications.

This model is the least cited in the literature because few people have pushed it and it is so specific to Silicon Valley itself (even though Microsoft and Amazon are in Washington state). Also, it is restricted to programming activities and as such mainly refers to Silicon Valley with its culture of entrepreneurial start-up and more established firms. It could also refer to scientific discovery processes in drug and chemical, and new model or engineering developments in some auto companies. In some ways, it is more of a product design or scientific discovery model rather than a production process model (Morgan and Liker 2006; Kenney and Florida 1993).

But since Apple strongly resembles Nikeification, we see it as being an even more high-tech part of that model, which has led to major technological breakthroughs that have been effectively marketed to mass markets. This chapter examines Nike, Apple and Google, and as can be seen in Table 8.1, each one of them has been very successful. We start with Nike because they first established the high tech model with shoes, and more importantly were the first to engage in off-shoring and the creation of the "donut" corporation.

Nikeification and Off-shoring

Nike emerged out of the initial low cost production advantage of Japanese production of shoes in the early 1960s to become the leading worldwide producer athletic of shoes and apparel. After a brief initial attempt to produce in the United States, Nike moved to the lower wage area of Japan

Table 8.1 Revenue (in $US) and Employee Statistics at Nike, Apple and Google

Organization Outcomes	*Nike*	*Apple*	*Google*
Revenues in 2018	$ 149.6 billion	$ 265.6 billion	$ 136.2 billion
Net income 2018/2017	$ 1.93 billion	$ 3.13 billion	$ 1,074.24 billion
All Employees in 2018	73,100 (few mfg ees)[2]	132,000 (few mfg ees)[2]	$ 98,771 (no mfg gees)
2017	70,000	139,000	$ 94,000
Revenue per employee (ee)[1]	$ 2,038,739/ee	$ 2,012,121/ee	$ 1,379,150/ee
Entry level pay in 2018	$ 64,000/year	$ 133,000/year, Engineer	$ 100–126 thousand/ year, Engineer
	$ 31,000/year	$ 30 thousand, Apple Store	$ 84 thousand/year, intern
			$ 64 thousand, Adm Asst.
CEO pay per year	$ 13.9 million/2017[3]	$ 15.7 million/2018[4]	$ 470.0 million/2018[5]
	$ 47.6 million/2016	$ 101.9 million/2017	$ 556.0 million/2017

Notes

1 Revenue per employee is revenues divided by the number of employees, which can be inflated by the number of part-time and temporary employees.

2 The vast majority of Nike and Apple manufacturing takes place off-shore where employees are from subcontractors who pay quite poorly.

3 Nike had a bad year in 2017, so pay was cut 70% for the CEO.

4 CEO Tim Cook's pay in 2018 does not include stock options, which for Apple, might add as much as a $100 million to his pay.

5 The Google CEO took a pay cut in 2018.

and then to other factories in Asia. Nike came out of Oregon with Bill Bowerman, a coach for the University of Oregon track team, and Philip Knight, a runner turned entrepreneur after getting an MBA from Stanford University. Knight wrote the marketing plan to sell Tiger running shoes manufactured in Japan with low-wage labor to compete with Adidas in the United States. In 1962, after visiting the Onitsuka distribution center in Kobe, Japan, Knight became their distributor in the United States. Bowerman and Knight became partners as they placed more orders for shoes.

As the heads of Blue Ribbon Sports, Knight and Bowerman started producing shoes on their own in Japan and then moving in 1974 to Korea, Taiwan, and Maine (for only two years) and later to Indonesia. Adidas did not start off-shoring until 1989. Knight subcontracted numerous factories to keep up with the rapid change in technology with each factory producing different products. The production wage difference between the US and Asia at that time was 8 to 11 dollars an hour versus a mere one dollar. As a result, the New Hampshire and Maine assembly plants were closed. In 1979, they changed their company name to Nike and went public (though Knight keeps the company under his control through special stock arrangements).

Box 8.1 The Nike Corporation: Bowerman and Knight

In 1964, **Bill Bowerman** (1911–1999) and **Phil Knight** (1938-present) founded Blue Ribbon Sports (BRS) by University of Oregon runner Phil Knight and his coach, Bill Bowerman. The company started as a distributor of Onitsuka Tiger footware. Bowerman then made his own shoes, making most sales at track meets out of Knight's trunk. In its first year in business, BRS sold 1,300 pairs of Japanese running shoes in 1965 and did most of its advertising in running magazines. BRS opened its first retail store in 1966 in Santa Monica, California. A year later, BRS expanded retail and distribution operations on the East Coast. As the contract with Onitsuka neared its end, BRS started making its own shoes in 1971 using the Swoosh emblem of Nike, the Greek goddess of speed and victory. By 1980, Nike had attained a 50% market share in the US athletic shoe market, and the company went public. Nike became an aggressive advertiser of their shoes in print and in the media. The "Just Do It" ad campaign in 1988 was one of the top five slogans of the 20th century. Throughout the 1980s, Nike expanded its product line to encompass many sports and regions throughout the world. In 1990, Nike moved into its eight-building World Headquarters campus in Beaverton. Niketown, a dramatic retail store, was opened in Portland, Oregon, and it then spread to major cities in the United States. In 2016, Phil Knight stepped down as chair of Nike.

Nike markets its products under its own brand, as well many others (e.g., Nike+, Air Max, etc.), and it has subsidiaries Brand Jordan, Hurly International, and Converse. In addition to manufacturing sportswear and equipment, the company operates retail stores under the Niketown name. Nike sponsors many high-profile athletes and sports teams around the world, with the highly recognized trademarks of "Just Do It" and the Swoosh logo. Nike has operated an international supply chain for its manufacturing. Most of these factories are in third world countries where workers are paid miniscule wages. Nike has been extensively criticized for this (Janoski et al. 2014). Nike is the world's largest supplier of athletic shoes and sports clothing.

Nike, like Apple following in its footsteps, is a "donut corporation" with no production facilities actually existing in the United States, where most of the sales and advertising take place. This is not flexible production with teams. Nike corporation itself produces different-sized shoes but not different widths (i.e., the promise of flexibility is limited). However, it is constantly producing a wide variety of shoe designs, colors, and new materials that make the shoe constant news in advertising. The system uses an elaborate system of supply chain management, so its supply system is as sophisticated as its marketing. But the production methods used overseas are rather primitive compared to teamwork standards in lean production.

In Nike's first decades of off-shoring, the company largely ignored complaints of poor working conditions in its supply chain, as reports of exploitation remained local and Western news outlets ignored them. There have been several investigations into Nike production and exploitation, and monitoring groups have been set up to improve working conditions (Locke et al. 2006). Their subcontracted sweatshops were found to be in violation of ILO conventions 29 (forced and compulsory labor), 98 (rights to organize and bargain), 100 (equal pay), 105 (forced labor), 111 (discrimination) and 138 (employing minors) (Merk 2015:126). The working model of manufacturing employment does not at all appear to be of the lean production model used by Toyota. Instead, Nike factories resemble Fordist sweatshops (Anner 2012; Chang 2008; Lee 2007; Silver 2008).

Nike was the first brand name to disclose the names and addresses of its suppliers, and it slowly started to embrace corporate social responsibility. So, the company went from denying to taking responsibility of these independent suppliers in the early 1990s to taking responsibility for their treatment of employees in the 2000s (Merk 2015:127). Eventually, Nike and about 30 other companies signed on to the FLA monitoring program, whose strengths and weaknesses are discussed at the end of this chapter. Nonetheless, the system of production is still largely Fordist, while its design and marketing facilities in the United States use teamwork (close to self-managed teams in socio-technical theory) to maximize creativity. Thus, Nike remains a donut corporation with a division between sweatshops, and enlightened and creative teams.

Their comparative advantage is two-fold: (1) create cutting edge technology with the marketing of the shoes around famous athletes, and (2) produce the shoes at a very low cost in Asia. A third competitive advantage was that their main competitor, the German firm Adidas, went into a tail spin after the founder Adolf Dassler (d. 1978) gave the firm over to his son Horst, who then died at the early age of 51 in 1987. Bernard Tapie, a French industrialist bought the firm, but he went bankrupt in 1992. Nike then accelerated its market share (Smit 2008).[1]

According to the *Fortune 100* (2006), Nike employs more jobs within the United States than outside of the states. Nike's headquarters in Beaverton, Oregon has nearly 8,000 workers most of whom are quite well paid (average 100,000 dollars), but the overwhelming majority of Nike's employees in the United States are part-time. In East-Asia, production factories have many more workers and a number of controversial issues—physical and verbal abuses, working conditions and hourly wages (Todd et al. 2008). These subcontractors who make the shoes are not included in the official Nike employment statistics, but the company now has a staff of 90 employees to monitor them.

Marketing and advertising along with research and development comprise the largest portion of Nike operations in the United States. Using star athletes like Michael Jordan, Tiger Woods, and LeBron James, Nike spent more than $2.7 billion on marketing in 2012, which includes sponsorship contracts, especially on hard edge "Just Do It" television spots and extensive promotions (*Sports Business Daily* 2012). Nike also moved from independent marketing agents in the field to corporate sales within the company. By the late 1980s, Nike created a specific product for each sport and outfit for each team with their individualized production. Nike provided these employees with higher salaries and health benefits, but cut their commissions drastically. Each territory in the United States is equipped with at least two footwear and two apparel-representatives (Geisinger 2012).

Nike's worth is largely due to their ability to off-shore production. They create a product that is manufactured by low-wage labor in overseas plants that are not officially part of Nike. In doing so, Nike subcontracts with three different types of factories (Donaghu and Barff 1990; McCann 2009). First, it has quality partners that were first set up in Japan, but are now in Taiwan and South Korea. They make the top quality shoes in smaller batches (25,000 or less per day), and are more likely to collaborate in innovations with Nike. Quality partners may also use second- and third-tier subcontractors. Second, it has volume partners that are large factories producing large batches of standardized, lower-priced footwear (70,000–85,000 pairs a day). This production may serve companies other than Nike. They tend to own their own leather tanneries and rubber factories, but they are not particularly innovative.[2] And third, it has new partners that are located mostly in Thailand, Indonesia, and China because of their low labor costs. These factories are often joint ventures with Nike, Taiwanese, or South Korean firms. Each of these partners has an ability to meet the needs of Nike on an individual basis. Numerous partners allow for flexibility within the market, especially with innovations in technology and design. While wages in these Nike factories are extremely low, they have increased by 18% in the ten year period from 2001 to 2011 (Bain, 2018; Lemon 2018; Holmes 2007; Walters 2007). Nike also reported in 2007 a new emphasis on sustainability and corporate responsibility (Nike, Inc. 2007).

To keep innovation going, research and development is a key to competitive advantage. Knight's *Futures Plan* requires that new designs have to be finished a year in advance to let retailers review new product lines before committing to the 6-month lead time orders. R&D also invents new technologies to maintain their high price premium lines (air cushions and spring pumps, Dri-FIT, and AirFree). Part of this technological shift to new materials is to avoid US taxes because rates on these new products are often based on the percentage of leather in each pair of shoes. The end result is that Nike is almost in the public relations business with the Oregon football team being in the lead of the current revolution changing football uniforms.

Nike became a model for other firms who have decided to off-shore their production. For instance, computer and electronics firms such as Dell, Apple, and nearly all TV manufacturers adopted off-shore production by the mid-1990s. As a result, Foxconn or its parent company Hua Hai Precision Industries became one of the largest non-state employers in the world with over a million employees. They produce nearly all the i-pad, i-phone, and i-pod products for Apple, not to mention items for other corporations. As such, even though it was not particularly technologically complex, Nikefication became the model for off-shoring production, especially in China.[3]

Nike has off-shored its physical production facilities for shoes, and Apple has done so for its physical production of i-products. But what explains the off-shoring of administrative and technical services (ATS) and innovative research? Andy Grove (2010) warned that production expertise would soon lead to research expertise such that comparative advantage would shift to the off-shored production facilities. But Lewin and associates (2009) say that innovative work is now off-shored as an attempt to gather the world's talent into one organization. This seems to go a bit beyond the Nike model, but it may be hard to stop given Grove's comments. In any event, the Nikeification model can have some rather far-reaching implications.

The Apple Corporation

From nearly bankrupt two decades ago to the most valuable company in the United States, with a market value of more than 1 trillion dollars, Apple Inc. has touched lives around the world through its products: iPhone, iPad, iPod or iMac. From beginning in a garage, three friends created Apple (Rawlinson 2017), Apple defied all odds to become one of the most popular and profitable tech companies of our times despite selling expensive hardware (Manjoo 2018). Along with the rise of other powerful tech mega-companies, such as Facebook, Amazon, and Google,

Apple contributed significantly to the digital revolution that transformed society, but also engages in extensive off-shoring.

Box 8.2 The Apple Corporation: Steve Jobs and Tim Cook

Apple Inc. is located in Cupertino, California. It designs, produces and sells consumer electronics, software, and online services. Speculation abounds about the origin of the company name and the bitten apple logo: Steve Jobs's visits at apple orchards while on a fructarian diet, a "fun, spirited, and not intimidating" symbol, the symbol of knowledge from Adam and Eve in the Bible, or the apple hitting Isaac Newton on the head leading to the invention of gravity (pictured on their early advertising) (Rawlinson 2017).

Apple makes the following hardware: the iPhone smart phone, iPad tablet computer, iMac personal computer, iPod portable media player, Apple smartwatch, Apple TV digital media player, and Homepod smart speaker. They also make software including the MacOS and iOS operating systems, Safari web browser, iTunes media player, iLife and iWork, and more professional programs—Final Cut Pro, Logic Pro and Xcode. Apple also has various online services: the iTunes store, iOS App Store, Mac App Store, Apple music, and iCloud.

Steve Jobs (1955-2011), **Steve Wozniak** (1950-present), and Ronald Wayne (1934-present) created Apple to sell Wozniak's Apple I personal computer. Wayne sold his interests early for $800. Incorporated as Apple Computer in 1977, it grew fast with Jobs and Wozniak hiring a staff of engineers, designers, coders, and an American production line. Apple went public in 1980 and over the next few years, Apple shipped Macintosh computers with many innovative and attractive graphical user interfaces. Its commercials billed the company as the upstart David against the IBM PC's goliath. However, its high prices and few software titles were a disadvantage, and power struggles ensued between Apple executives. In 1985, both Wozniak and Jobs resigned, but Jobs founded NeXT with some former Apple employees. As the market for personal computers increased, Apple's computers lost share to the PC that used the Microsoft Windows operating system. After more executive job shuffles in 1997, CEO Gil Amelio brought Jobs back. Jobs regained leadership within the company and became the new CEO shortly thereafter. He began to rebuild Apple's status, opened Apple retail stores, and acquired numerous small innovative companies, and changed some of the hardware. The company returned to profitability. In 2007, Jobs introduced the iPhone, which became immensely popular. In August 2011, Jobs resigned as CEO due to health complications, and **Timothy "Tim" Cook** (1960-present) became the new CEO. Two months later, Steve Jobs died.

Apple is valued at over a trillion dollars. Apple has more than100,000 full-time employees, but this does not count the millions of employees who manufacture most of its physical products in China.

Figure 8.1 Apple Corporations New Headquarters

Apple employs about 132,000 employees in and connected to the United States, which includes designers, programmers, and Apple store employees. A sophisticated simplicity permeates the entire Apple culture from its simple name and logo, to its clean retail stores. Its new corporate headquarters for more than 12,000 employees in Cupertino, CA was built in 2018, and resembles a circular spaceship with four stories tall glass walls, covered in solar panels, a garden with fruit trees, the Steve Jobs Theater with 1,000 seats, basketball courts, yoga rooms, and state of the art gyms and cafeterias (Leswing 2018). In some ways, the ring shape structure employs some principles of an *obeya*. It also has a natural ventilation system, which does not require heating or air conditioning nine months per year (Apple Press Release, Feb. 22, 2017). Oddly enough, the building appears to symbolize an off-shoring "donut corporation" (Janoski et al. 2014).

The Apple Way

Operations management is hailed as the key to Apple's success. Apple meticulously analyzes every phase of the supply-chain management (Satariano and Burrows 2011). Shipping by air freight had stunned many of the competitors; as for suppliers, working with Apple is lucrative because of the high volume of products. Apple engages in hard bargaining tactics and months-long negotiations with targeted suppliers (Satariano and Burrows 2011). It demands very detailed breakdowns of each estimate from of its 9,000 suppliers, including labor and material costs and projected profit. Its suppliers are required to keep two weeks' inventory within one mile from their plants in Asia, although it is a regular practice for Apple to pay them back with a delay of 90 days.

Various accounts reveal that workers at the asian suppliers clock in many hours of overtime before major product launches; others describe how secret surveillance cameras may be placed in boxes with parts to monitor behavior at home and abroad. In this way, secrecy is preserved before major product launches. Other accounts describe Apple products being stored and shipped in unmarked boxes to prevent pre-launching leaks (Satariano and Burrows 2011). In fact, one of the famous mottos attributed to Tim Cook is: "nobody wants to buy sour milk" pointing to the importance of JIT production in the Apple Way. Tim Cook has also reportedly offered George Stalk and Thomas Hout's *Competing Against Time* (2003), a book about supply chains as a strategic weapon in business, as gifts to his co-workers (Satariano and Burrows 2011). This concept of "time compression" is also developed by Robert Hall (2014), a management theorist of lean.

In a short essay published in the *New York Times*, Mihir Desai (2018) explains why Apple is the future of capitalism. The supply chain is largely sustained on the suppliers' back who are willing to wait 100 days for Apple to pay them back. Apple is described as the epitome of an "asset light company" (i.e., donut corporation) because it has very few hard assets but a large cash flow to run the company. Minimal inventory and keeping the suppliers waiting had become a dominant pattern in large corporations after it was adopted by Apple (Desai 2018).[4] Driven to its logical conclusion, a company should be a designer with no actual productive base—"we write code and design luxury products that demand a premium price".

The Apple culture inspired by Steve Jobs focused on several simple principles: accountability, attention to detail, perfectionism, simplicity, and secrecy (Elliot 2012). Apple is a flat organization, with a small team structure, open communication style down to each and every store. The lean culture is embedded in the Apple Way by embracing simplicity. Simple is a core value that sets Apple apart from the other technology companies (Segall 2012). The past and current Apple employees can attest that "the simpler way is not always the easiest", and that it takes trial and error, and lots of time and energy to achieve it. The obsession with simplicity is engrained in Apple's DNA; one simple button becomes the ultimate sophistication which opens up to a world of possibilities (Segall 2012). An interesting aspect of the Apple Way is the Apple ecosystem: products that are highly integrated with each other in such a way that if you start using one application, say the app store or the iCloud, it is less likely to move to an Apple competitor and more likely to adopt the entire Apple lifestyle (Allford 2015).

Steve Jobs with his "Think Different" approach—Apple's management principles—lists seven different principles of doing business in a new way (Segall 2012):

(1) **Communication**. Short, direct, straightforward, honest, even brutal communication is preferred. Steve Jobs promoted a blunt and direct communication style as if there is little time to waste when you're working in a competitive environment.)
(2) **Teamwork**. Small teams of smart people are preferred to large groups. Every single member of the team has a purpose in the meeting; unnecessary people are eliminated from the team. Small teams are considered better at producing quality thinking and maintaining the agility of a start-up culture and flat bureaucracy, and avoiding growing bureaucratic complexity. Valuing process and procedure over simplicity and creativity could trigger excommunication from the Apple team. "Steve had a rule that there could never be more than 100 people on the Mac team. So, if you wanted to add someone, you had to take someone out. And the thinking was a typical Steve Jobs observation: "I can't remember more than a hundred first names, so I only want to be around people that I know personally" (Segall 2012:32). However, as described, these are extremely large teams compared to Toyotism.
(3) **Minimalism**. It ranges from the minimalist design of Apple stores, the structure of corporate presentations where few but solid facts are requested to back-up a point, to the minimal number of products of the highest quality on the market.
(4) **Continuous Improvement and Forward-Thinking**. Be constantly on the move, looking forward. New versions of products do not

allow them to get old. Think ahead, create the future you envision. "Sometimes when you innovate, you make mistakes. It is best to admit them quickly, and get on with improving your other innovations" (Segall 2012:213). Jobs had quoted Henry Ford on several occasions: "If I asked people what they wanted, they'd say a faster horse". Therefore, he saw Apple's responsibility to be forward thinking and create revolutionary products.

(5) **Iconic Products.** Apple does not shy away from challenging its business rivals. Although Apple was not a pioneer in the field by creating the first computer, M3 player, tablet or smart phone, it has radically improved them and turned them into revolutionary products that looked stunningly different than everything before. Whereas, mass production generated an entire car culture a century ago, the Apple products are credited to have stimulated an entire digital revolution in the 21st century. [5]

(6) **Personalized Relationship with Technology**. The "i" of the Apple has become a quintessential element of the Apple brand, reflecting the same basic simplicity in branding and a personal relationship with your computer/gadget. In many ways, the Apple products have become extensions of the human body, customers not being able to imagine leaving the house without their iPhone in their hand or MacBook in their bag. The Apple vision is to connect technology to humanity by creating "the most personal computer" and providing easy access to photos, music, movies, browsing the Internet and the personal assistant Siri (Segall 2012).

(7) **Obsession with Quality**. Apple does not make profit its first priority, but is grounded on the assumption that a high quality product will generate profit. Therefore, Apple's priority is with the customer experience, especially inducing an emotional experience in the Apple users. Their focus is not excessively on the product, but embracing a broader vision of "enriching peoples' lives". Cool design and high quality are embedded not only in the products, but also in the packaging and the stores. It turns out that the Apple stores have more visitors than all the Disney's four biggest theme parks combined (Segall 2012).

These principles are expressed more succinctly in Box 8.3.

Box 8.3 The Apple Manifesto: The Think Different Campaign, 1997

"Here's to the crazy ones.
The misfits.
The rebels.

The troublemakers.
The round pegs in the square holes.
The ones who see things differently.
They're not fond of rules.
And they have no respect for the status quo.
You can quote them,
Disagree with them,
Glorify or vilify them.
About the only thing you can't do is ignore them.
Because they change things.
They push the human race forward.
And while some may see them as the crazy ones,
We see genius.
Because the people who are crazy enough
To think they can change the world . . .
Are the ones who do".

Source: www.youtube.com/watch?v=z4BQD8Uu14

According to specialists, the invention of the iPod, the iPhone, the iPad, iTunes, and the Apple Store started the greatest decade of innovation in business history (Galloway 2017). Apple used scarcity to market its products as luxury items. Apple's earth-shattering success, according to marketing specialists, comes from several attributes that make it a luxury brand (Galloway 2017).

One of the most important business principles were: "to follow the principles of human engineering to build "friendly" products whose simplicity and ease of use make them natural extensions of their owners" (Eliot 2012:16–17). According to former Apple VP for HR, Jay Eliott, some of Apple's most important business values were: (1) Empathy for users (solving customer problems while not compromising or integrity in the name of profit); (2) Setting aggressive goals; (3) Positive social contribution (making the world a better place to live by making products that expand human capability); (4) Innovation and Vision (making leadership products that are new and needed); (5) Individual Performance; (6) Team Spirit (Apple employees are encouraged to interact with each other and management to share ideas on how to improve the effectiveness of the company); (7) Quality; and (8) Individual Reward (Apple shares rewards with its employees; some may be financial or psychological); and (9) Great Management (that should make sure the Apple values are implemented).

A company must have a vision in order to transform society with its products. Steve Jobs' vision was to create products that are friendly, human, and appealing. He had also excelled at pulling all the team members to embrace this vision (Eliot 2012). Jobs wanted to avoid Apple turning into a traditional corporate structure, and in one of the Macintosh team meetings in the 1980s, he explained his vision for a team culture of Pirates, not the Navy, working together. This means that the non-conformists, the rebels, the risk-takers, the dissenters, the rule breakers, "the positive deviants" were highly valued to infuse the Pirates culture and to bring a competitive edge. Since this meant mostly 22-year-olds, the pirates must be balanced with people of wisdom with enough business experience and common sense. The "magical sense of exhilaration" of working for an innovative team was so high that sometimes software engineers had to sleep at work to meet deadlines, but that still did not dent their joy of working for Apple (Eliot 2012).

All official Apple employees, which do not include the off-shore employees who actually make Apple products, receive stock options, and are eligible for profit sharing and bonuses from the first day of employment (Elliot 2012). Apple invests strongly in a thank-you culture that supports the employees as stakeholders of the organization. For the past 30 years, Apple also had a mentoring program, called the iBuddy program, that allowed new employees to be connected with older employees, sometimes from different teams or parts of the organization, who would teach them the ropes (Elliot 2012).

The culture of secrecy was very important in teams since leaks about new tech innovations would find their ways to the press. Secrecy was enforced with confidentiality agreements and infused from day one in the new employee training. Lockdown rooms are available for teams experimenting with secret projects. The Research and Development division had to have a pirate culture in a Navy organization, or in Jobs' words, a team of A players in a larger organization of B and C players. The pirate culture implied also the dominating power of a strong leader. Steve Jobs understood the relationship between space and the work environment and did his best to design a new headquarters that fosters innovation and facilitates teamwork (Elliot 2012). Consider, Steve Job's view on innovation and teamwork:

> Process makes you more efficient. But innovation comes from people meeting in the hallways or calling each other at 10:30 at night with a new idea, or because they realized something that shoots holes in how we've been thinking about a problem. It's ad hoc meetings of six people called by someone who thinks he has figured out the coolest new thing ever and who wants to know that other people think of this idea. (Segal 2012:140)

Attracting the best innovators that push the company forward is central to Apple's HR strategies. Its hiring strategies include have three basic principles: (1) clear, but not rigid, hiring requirements; (2) involve the team in the hiring process, and (3) set a broad candidate pool, but hire only a handful of very talented people. For instance, only about 6% of job applicants are hired in the Apple retail stores, while Steve Jobs himself was credited for personally handpicking more than 5,000 of the Apple employees (Elliot 2012).[6]

Numerous examples from Apple's history highlight their obsession with quality. With his uncompromising attention to high standards, Steve Jobs specifically ordered that printers that were not of the best quality be destroyed instead of being pushed on the market for a considerable discount. A considerable loss for the company was preferred instead of associating the Apple name with inferior products, even if that meant destroying 200,000 Apple III computers that were not selling well (Elliot 2012). As an example of these innovative principles, Steve Jobs reaction to an early model of the iPod:

> he said it was too large. His developers explained that they have used the absolute minimum size of case that would hold the necessary components. Steve carried the device to a fish tank and dropped the iPod into the tank. As it sank, a stream of air bubbles gurgled to the surface.
> (Elliot 2012:98)

Jobs said look at the bubbles. There must be considerable space that would be eliminated. And his iPod team made the iPod much smaller.

Apple's Fordist Side and Off-shoring

The crisis of Fordist capitalism in the 1970s had generated a new international division of labor with global supply chains that outsource manual labor from the core countries to the periphery. As such, the rise of the neoliberal globalization forces and especially the new-found identity of China as the "workshop of the world" has undermined the successes of organized labor in old economies. Often times, Chinese contractors compete fiercely with each other to get the Apple business and the easiest way to reduce costs is by reducing the labor costs (Sandoval 2013). Long working hours, low wages and poor occupational health and safety have been cited often as unethical labor practices at the Apple subcontractors despite the company being awarded the most admired company by *Fortune* magazine for ten years in a row since 2008. This shows a clear disconnect between the Western public that is fascinated by Apple products and the actual workers who make them. Chan et al. (2015) argue that Apple concerns itself little with the working conditions of Foxconn workers as Apple continues to press workers to work overtime under difficult conditions. Apple reportedly asked for some last minute changes on a strict deadline, and Foxconn in Chengu (aka, IPad City) got workers up at 2:00 AM to work to get Apple products manufactured for critical deadlines (Chan et al. 2015:87–88). Twelve hour days are common, overtime is often not paid, and that overtime is compensated by less work in slow periods rather than paying overtime (Chan et al. 2015:89). A series of young workers leaping off Foxconn roofs to their deaths between 2007 and 2010 pushed Foxconn into the limelight and raised awareness on the abuses that happen in China and other countries with loose labor regulations.[7]

Most of these manufacturing workers are from rural areas and because the *Hukou* passport system are denied many rights when they enter urban areas where the main factories are located. The Hukou system is intended to prevent excessive migration to the cities so it penalizes workers who do so in a number of ways. One important part of it prevents the children of rural workers from going to the public schools in these urban areas and only Apple. As a result, staying in the urban areas to raise a family is not really possible. As a result, there is a circular migration of rural workers from the rural areas to the cities and back again (Chan et al. 2015:87–88).

Several watchdog organizations, such as the FLA, SACOM, China Labor Watch, and the European organization "makeITfair" conduct research and raise awareness on the labor conditions and lives of manufacturing workers, present their reports to the electronic companies, and demand answers and corrective measures from them.

Most consumer electronics, such as Macs, iPhones, iPads and iPods, require minerals that are usually extracted from mostly conflict zones, for instance, cobalt from the Democratic Republic of Congo and Zambia. Swedwatch and Finnwatch have previously reported the rampant use of child labor or female workers pushed into sexual exploitation by rebel groups that control the mines (Sandoval 2013). Environmental degradation of forests and waters occurs around the extraction mines threatening the native population's livelihoods whereas work in these mines have an elevated safety risk. Many of the Chinese workers are predominantly young, female, and migrant workers who lose protection benefits (unemployment, parental leave, etc.) when they migrate from their provinces to the manufacturing facilities. SACOM documented numerous cases of workers being intoxicated with chemical substances, such as n-hexane or aluminum dust, while polishing the screen and covers of the 100,000 iPhones produced per day.

However, Apple claims that respect for people and human rights are infused at the deepest levels of their supply chain network. Since 2013, Apple had made considerable progress to be in accordance with the OECD (2018) regulations regarding the resourcing of minerals from conflict-affected and high-risk areas. One hundred percent of its supply chain for minerals had been reviewed by third-party audits. Apple also supports efforts to provide more vocational educational training for his mineral suppliers in the Democratic Republic of Congo, so they can expand their job skills other than mining. Regarding environmental protection, Apple was ranked number one by the Business and Human Rights Resource Centre's Corporate Information Transparency Index for the fourth straight year (*Apple Supplier Responsibility 2018 Progress Report*). However, Apple refuses to join the FLA as Nike did, and SACOM still reports that Apple has failed "to end grueling conditions at Foxconn Factories" (Garside 2012; SACOM 2018a, especially their "zero inventory" policy "as required by Chinese law" (SACOM 2018b).

In Asia, the 2010 and 2011 SACOM report that Apple sub-contractors (mainly Foxconn) still have poor ventilation, insufficient protection equipment, and workers not being informed of the harmful chemicals they encounter, which led to two major factory explosions that killed three workers and injured 75 others in 2011. The long overtime hours seriously affect the workers' well-being and social lives. The migrant workers do not have a support system at their new locations, and in their very limited free time, they can only eat and sleep (Sandoval 2013; Chan et al. 2015).

Minimizing waste is one of the tenets of the lean philosophy. True to its lean culture, Apple launched the Zero Waste Program in 2015. This program helps suppliers to recycle and reuse materials in order to divert from landfills. By 2017, 100% of iPhone final assembly facilities have received the Zero Waste to Landfill Certification. As of 2018, the company reports that all its operations use clean and renewable energy to address climate change (of course, this does not necessarily include contractors). Apple is also using lighter packaging to reduce logistics costs and the MacBook Pros use 61% less power than the previous generations. However, Greenpeace accuses the company of planned obsolescence by making upgrading and repairing of Apple products difficult (Fingas 2018).

Over the past decade, Apple had made sustained efforts to improve its corporate responsibility ratings. Recent developments show that Apple is being recognized as a top company for responsible sourcing of conflict minerals and has paid more than $30 million in excessive recruitment fees for more than 35,000 employees in bonded labor. Although bonded labor, charging a fee for intermediating an employment opportunity, is a legal practice in China, Apple had announced that starting in 2015, there will be zero tolerance regarding bonded labor at its supplier facilities. For instance, some Chinese suppliers hire foreign workers through HR agencies used as intermediaries. The supplier employees in Asia receive training on their rights, receive vocational education or higher education support, and health education. According to corporate reports, Apple conducted 756 assessments at 95% of its suppliers in 2017 alone. Only two cases of underage labor were identified where the minors have used false IDs to obtain work at the Apple suppliers. In both instances, the children were returned home, enrolled back in school and were promised their jobs back upon graduation while the supplier continued to pay their wages while they were in school. Moral support teams at the manufacturing facilities are in charge with hearing employee grievances, organizing employee forums, and bringing these complaints forward to the Apple management (*Apple Supplier Responsibility 2018 Progress Report*).

In terms of Labor and Human Rights, Apple had scored 86 out of 100 in regard to wage and benefits, involuntary labor, and student intern management at its supplier facilities. Apple is currently paying 125% for overtime, even in countries that do not have legally established overtime rates, and the number of student interns was limited to no more than 10% of the entire workforce. Out of the 756 facilities that were inspected in 2017, the audits have identified 38 violations included working hours violations (e.g., more than 60 hours of work a week or no rest day after six consecutive days at work), three bonded labor violations and two underage labor (*Apple Supplier Responsibility 2018 Progress Report*). Apple had also received a 90 out of 100 score for Health and Safety Hazards, the company losing points mostly for not installing fire detectors in dorm areas. Apple publicly admitted that they are concerned and seeking solutions to children's iPhone addiction phenomenon, as well as protecting consumers' privacy and safety (Manjoo 2018).

President Trump has pressed Apple to reduce its off-shore production and have its products produced in the United States. To reinforce this, he has imposed tariffs on goods produced in China, which would consist of many of Apple's products (Mickle 2018). Apple CEO Tim Cook filed objections with the US Trade Representative that these tariffs on Chinese goods would hurt Apple, lead to higher prices, and disadvantage Apple products vis-a-vis foreign competitors. President Trump urged Apple to build its next largest factory in the United States (Mickle 2018). Apple has decided to invest in a major headquarters in Austin, TX, but it does not appear to replace its massive reliance on Foxconn in China. Foxconn announced an innovation center to be built in Wisconsin, but after receiving subsidies and building it, Foxconn announced that the center would open in 2020 with fewer employees than expected (Rushe 2018; Statt 2019). The growing social and political concerns against the rise of mega-companies and the trend of luxury tech products, such as the Apple products, point out to the increase of social inequality.[8]

Google and Alphabet

Google has some of the same innovative and teamwork principles as Nike and Apple, but Google produces few manufactured and physical products.[9] Instead they produce a wide variety of computer programs and systems. Its main product consists of creating, developing and servicing computer programs and their code (e.g., the Google search engine, G-mail, Google Drive, Google Maps, Google Assistant, and so forth). This makes Google a good example of how quality control procedures apply to the writing of computer code. While making most of its revenue from advertising, Google's major products are computer programs run by computer code that users desire and use. Similarly, Apple has a huge investment in writing code, but we did not cover the "quality" aspects of computer code in the discussion of apple because the products were important. For a brief history of the Google Corporation, see Box 8.4.

Box 8.4 The Google Corporation and Alphabet

Google is a multinational corporation in Silicon Valley that provides internet services and products, online advertising technologies, the Google search engine, cloud computing, hardware and other software. It is often considered to be the most innovative corporation in the world. Google was founded by Larry Page, the son of a Michigan State University computer science professor and Sergey Brin, the son of a Moscow University mathematics professor who immigrated from Russia when Sergey was 6 years old. As graduate students at Stanford University, Page and Brin created Google in 1998. They offered it to the public in an initial public offering (IPO) in 2004, and they currently own about 14% of its stock and control 56% of the "super voting shares", which gives them complete control of the corporation. In 2015, they formed the Alphabet Corporation, and Google became its main subsidiary. Google is now housed in the post-modernistic Googleplex in Mountain View, California. Sundar Pichai (1972-present), who immigrated from Tamil Nadu in southern India to the US after completing his BS, was appointed CEO of Google in 2015, replacing Larry Page who became the CEO of Alphabet. Pichai is noted for both his diplomatic and technological skills, and he became CEO of Alphabet in 2019.

The company has grown rapidly and is the darling of the internet for its Google search engine and expansion into many other products—Google Docs, Google Sheets, Google Slide, G-mail, Google Calendar, and Google Hangouts, Google Chrome, Google Maps, Google Translate, and so forth. Google developed the Android mobile operating system for smart phones that competes with Apple's I-phone systems. Google also has some hardware with a smart phone, headsets, wireless routers, and the Goggle Chrome computer. It has also ventured into being an Internet carrier with Google Fiber and Project Fi, which combines Wi-Fi and cellular networks from different providers. Google Station is a product to make public Wi-Fi available throughout the world.

Google services are among the world's top 100 most visited websites, and Google is among the top internet corporations in the world. The company is known for their irreverence for business decorum and has a seemingly laid back but intense campus with strong teams, company bicycles, excellent cafeterias, free snacks and lunch, informal dress, and diverse informal seating. While not the target that Facebook is, Google has been criticized for privacy issues, tax avoidance, antitrust, censorship, and search neutrality. In their mission statement, Google states that they exist "to organize the world's information and make it universally accessible and useful" with a slogan "Don't be evil." In 2015, this slogan was replaced in the Alphabet Code of Conduct with "Do the right thing" (Moyer 2015).

In this section, we cover specifically how Google transcends Apple in some ways as an open source corporation, how it uses STS principles perhaps updated by "burning man" (Gilder 2018:32–35), and a central question about how Goggle uses lean production in the quality control of writing code.

Google, Secrecy, and Boundaries

George Gilder (2018:29) says that "Apple is an old-style company, charging handsomely for everything it offers". Apple emphasizes secrecy and boundaries to protect company patents and new launches. Google is entirely different. Google makes most of what it produces open sourced to the public. This cannot be totally followed since it protects the privacy of its employees and has some secrets. But Google operates more on the principle of "we are the leader, and when others follow us, it makes us stronger". This model comes out of the development of Google at Stanford where the university and others supported the fledgling company. They are a big factor in the view held by many computer techies that "data want to be free", which has linkages to techno-utopian thought (Gilmore 2010). Much is made of their slogan "Don't do evil", and controversy has raged over this with Google removing it from their code of conduct (Moyer 2015; Turner 2017). On the issue of secrecy, Google quietly bought the property of a former "energy hog" aluminum smelter for its Dalles Data Center. In the state of Washington on the Columbia River, this site has access to tremendous electrical sources from the Dalles Dam and 1.8 gigawatts of power. According to Gilder (2018:51–54), it holds 75,000 interlinked computer servers. This facility uses massive amounts of electricity especially for the air-conditioning to keep it cool. "The Cloud", which is firmly on earth and quite hot (perhaps "The Sun" would be a more appropriate metaphor), is similarly housed by other companies such as Apple at diverse locations. While this is not exactly a dastardly deed, it does show that "no secrecy" has some important bounds and that the cloud is burning large amounts of energy.

Google as an Innovative Corporation Based on Socio-Technical Theory

Much of Google's approach to employees and organizational dynamics is based on socio-technical theory. However, they would never identify it as such. More of an update of STS are the basic principles of Burning Man (see Box 8.5), a 33-year ritualistic, artistic and countercultural event in the Nevada desert.

Box 8.5 The Culture of Burning Man: 10 Principles

Burning Man is a temporary city erected each year in in the middle of nowhere in northwest Nevada. It is a combination festival, artistic event, and "happening" based on new age values. In 2017, over 69,000 people attended. It has a highly participatory culture, and at the end, the "burning man" is ignited with everyone subsequently leaving without a trace. Its values include:

"Radical inclusion" for participating in the eight- to 10-day event
"Gifting" or offering valued items with no expectation of return
"Decommodification" with exchange avoiding monetary markets and exploitation
"Radical self-reliance" in using one's own inner resources
"Radical self-expression" in terms of art as a gift
"Communal effort" in getting things done
"Civic responsibility" that values civil society and obeys laws
"Leaving no trace" in the environment
"Participation" is radically supported
"Immediacy" in the sense that immediate experience is ultimately valued.

Source: https://burningman.org/culture/philosophical-center/10-principles/

Socio-technical theory is not so much event or artistically focused, but "radical participation and communal effort is certainly central to its principles" (Kim 2017). Much of this is new-age thought based on communitarian theories in many ways, but it accords very well with socio-technical thinking concerning radical participation and humanistic ideals. The idea of "do no evil" comes from these principles, but also from the second generation PhD environment they came from and were incubated within at Stanford University (Gilder 2018:25–35; St. John 2017; Gilmore 2010; Turner 2009; Gilmore and van Proyen 2005; Schmidt and Rosenberg 2017).

The Quality Control of Computer Code

More down to earth is the requirement that Google produce excellent and bug (mistake)-free code in the design of its programs, which again are the basis of their products. James Whittaker was the director of test engineering at Google (specifically with the creation of Chrome), and with the company's approval, he, Jason Arbon, and Jeff Carollo have written *How Google*

Tests Software (2012) on how the quality control process works at Google Inc. Historically, Google has had three well-defined for the writing of code. First, software engineers (SWE) develop the code for new applications and areoften considered the most creative workers in the company. Second, software engineers in test (SET) test the code developed by the SEs and interpret the results. Third, test engineers (TE) test code from the perspective of the user, provide suggestions for improving code, prepare programs for external checking, and prepare programs for external checking such as beta testing, crowd sourcing, and early adopters. They are slightly more in the management area. The testing process goes through eight states. In the internal phases, there are three phases with the SWE, SET and TE processes. In the external phases there are four tests: crowd sourcing tests, dog food tests (internal use of the programs by Google personnel themselves), beta users, and early adopters. Two steps less used are contract testers with outside firms, and service pack corrections, which result from quality control failures or improvements that fix released programs (Whittaker et al. 2012:6–8). Whittaker et al. (2012) also view these processes in four channels: canary channels, dev channels, test channels, and Beta or release channels (Whittaker et al. 2012:11–12).

Although Google does not mention lean production by name, these processes are closely related to lean production for seven different reasons. First, from 2007 on Google has loosened the division of labor on roles and has made all three roles (SWE, SET, and TE) co-responsible for quality. This lessens the "throw it over the wall" tendencies of certain roles focusing on "just their job" and taking responsibility for quality. Second, the status of testing has been raised to be coequal with the development of new products and management (i.e., lean flattening of organizations). Third, there is a strong focus on teams rather than individual job descriptions, and the teams often meet at the end of the week for beer. Fourth, Google is very wary of hiring too many testers and this prevents role crystallization (Microsoft appears to be less wary).[10] Fifth, team members are rotated at least every 18 months so that knowledge is diffused throughout the organization and people develop new skills and experiences. Sixth, Google still uses temporary workers like Toyota and other lean companies. And seventh, 20% of employee time can be used in developing your own projects. All seven factors make this process close to lean production; however, we do not necessarily review these processes as good ones especially the use of temporary workers (Levy 2011; Gilder 2018).

These seven processes reinforce creativity and concern for quality. All in all, these processes are a combination of lean and semi-autonomous teams that gives a bit more freedom to self-managed teams than QCCs. The adherence to the lean principle is quite strong, especially in the recent overlap of the three major roles on quality. Apple and Nike are quite similar in their programming and production design, but despite some use of temporary workers, Google is not a donut corporation because they produce rather than sub-contract their main products. Consequently, when it comes to quality and new products, Google, Microsoft, and the states-side units of Apple are firmly within the processes of continuous improvement (*kaizen*) and lean production (Whittaker et al. 2012; Humble and Farley 2011; Humble et al. 2015; Whittaker 2009, n.d.a, n.d.b; Page et al. 2008; Poppendieck and Poppendieck 2006).

More generally in terms of teams, Google has done numerous studies of teams. Project Aristotle gathered data on 180 teams studying their effectiveness. According to Julia Rozovsky (Google's people analytics manager) coming across research on "group norms"—the traditions, behavioral standards, and unwritten rules that govern how teams function. These norms can be unspoken or openly acknowledged. Five types of norms were important: (1) dependability of team members in getting things done on time and meeting expectations; (2) structure and clarity of goals and roles but "nudge" (don't shove); (3) meaningful work that has significance for each team member; (4) the group believes that it will have an organizational impact; and (5) psychological safety and trust (recall, Deming's "drive out fear") (Schneider 2017; Bock 2015:118–49, 182–317). These principles fit well with lean production and socio-technical autonomous teams.[11]

Movements Against International Sweatshops

The political model of Nikeification largely resembles what was discussed earlier with Siliconism—largely free market economics; however, the large dose of off-shoring has some strong political implications for the degradation of labor standards in these off-shore locations. Nikeification involves avoiding the politics of the long-standing ILO and the creation of their own weaker organizations, such as the FLA, which has been the most prominent player in human and labor rights issues.

Much of this movement has involved clothing manufacturers (e.g., Nike, Liz Claiborne, Reebok, Patagonia, etc., but not Walmart). But it also involves Apple and other electronics manufacturing firms. Started in 1997, the FLA often operates by what we call *regulation by scandal*, in which the corporations react to media reports of labor abuses by instituting new forms of labor rules or standards—but most often, these new rules are merely half-measures, ineffective, and usually unenforced. The FLA process—a contemporary form of "muckraking"—usually starts with a reporter visiting a factory or meeting with the workers of a multinational corporation and collecting their stories of abuse and exploitation. Naturally, these workers are taking quite a risk to talk to these reporters, but many cooperate. The reporters then write stories and supply pictures to document abuse and exploitation to American and European media audiences on television, radio, or the Internet. For instance, Apple did not enter into the agreements with the FLA until its scandal with Foxconn intensified in 2010.

The FLA is mostly financed by their corporate members, but some money also comes from universities because their apparel licensing contracts can be quite lucrative (e.g., Oregon, Michigan, etc.). The FLA's tripartite board is composed of corporations protecting their product image, university administrators protecting their school logos on T-shirts and sweatshirts, and the NGO itself which is actually interested in protecting workers. The ultimate power in the relationship is the corporate brand's leverage to discipline subcontractors. But the process requires generating media publicity, so that reporters can "shame" the parent corporation into working with the FLA in their investigation to correct the abuses of workers in the international production of their highly successful products.[12]

There are some incredible deficiencies in the FLA's "scandal" or more "shaming" approach to labor standards (Bartley 2014; 2018a,b; Bartley and Child 2011; Bartley and Zhang 2016; Zwolinski 2007). First, the FLA is supported by multinational employers and universities who sell athletic apparel. This means that the FLA is a "captured" agency (i.e., a company union). When news outlets episodically report FLA results, pressure is put on multi-national corporations to discipline their subcontractors. Without this publicity, enforcement can be ignored. Second, the firm voluntarily complies with the proposed resolution and in order to find out whether or not it is followed, the FLA or other organizations have to investigate periodically to see if any corrections have been made.

Third, China, Vietnam, and a few other countries have laws against collective bargaining or strikes by organized labor. Consequently, state enforcement of employee protection does not have much pressure from labor. In China, a few foreign companies (notably Japanese corporations) have been forced by the Chinese government to treat their employees better (e.g., the aftermath of Toyota and Honda strikes with government enforcing wage increases). But for most workers, there is little or no impact on internal labor standards. Distelhorst et al. (2015) show that Hewlett-Packard's poor performance on worker protection in China compared to wealthy Singapore and the Czech Republic but also Mexico and Malaysia. And after labor problems at Foxconn, Apple's own labor responsibility reported "Core Violations Found and Actions Taken—None" (Apple Inc. 2013:37). But Distelhorst et al. (2015) also suggests the possibility that when lean production is actually implemented, labor protections may improve.

Fourth, the pressures brought about by reporters are highly episodic and this means that most labor violations are ignored and unregulated. And fifth, many corporations do not belong to the FLA and place little emphasis on enforcing labor or human rights. In sum, labor standards exist mainly to protect corporate reputations especially when they are exposed to public scrutiny. Labor regulations at the national level have actually declined in recent years, and the UN, while it supports the ILO, has provided little guidance or leadership in this area (Anner 2012). So, it is clear that these failures of labor standards are endemic to the Nikeification model.

Conclusion

It is a widely known fact that Apple's late charismatic CEO, Steve Jobs, was a challenger of traditional corporate structures and a big fan of the Nike brand. Indeed, the two companies share the "simplicity as the ultimate sophistication" philosophy. For this reason, we call this model "Nikefication". This may seem odd since Nike simply makes shoes and Apple creates new products. Nonetheless, Nike began the process of becoming a donut organization with a big hole where the manufacturing process would exist and replaces it with off-shored manufacturing taking place mostly in East Asia where low wages prevail. The strong teamwork and extensive benefits trumpeted in the United States by these companies only minimally exists off-shore. Instead, these operations, though improved from their purely exploitative bases in the 1990s and early 2000s, are not the same as those that exist in the home country. Further, whatever improvements that have taken place in the off-shore factories have not come from the largesse of these corporations but instead from social movements that have shamed the practices of these large corporations. As for lean production, we can say that these companies are the masters of supply chain management according to JIT principles, but they are wholly inconsistent in their applications of teamwork, high wages and benefits, and firm loyalty (Rothstein 2016). Generally, these multinational corporations try to escape these charges by saying that they subcontract the manufacturing aspects of their products and are minimally responsible in what goes on in these plants and factors. Social movements have done much to counter these claims, but regulation by shame and publicity is episodic and not the most effective way to solve many of these problems. However, in the last analysis, the firms represented by Nikeification are partially in the lean production category and partially out of it.

Notes

1. Adidas then recovered and prospered under later CEOs Robert Louis-Dreyfus, Kasper Rorsted, and Herbert Hainer.
2. Being a tanner is one of the main occupations of the *Burakumin* people in Japan, who are a low-caste group discriminated against because Buddhism considers working with hides and leather to be unclean. Nike's production in Japan used leather products and may have employed *Burakumin* workers.
3. Later under additive production, we discuss how Nike has recently adapted 3D printing to the production of shoes (*WSJ* 2013). This process is quite technologically advanced.
4. Actually, the TV maker Vizio is one of the most asset-light electronics companies with very few employees compared to their massive sales throughout the world (Janoski et al. 2014).
5. Apple had clear intentions to design the iCar and started an electric self-driving car program in 2014. The car with luxurious interiors was designed to allow passengers to face each other in a lounge type setting. Under the leadership of Jonathan Ive, Apple's design chief, software programmers, automotive engineers, rocket scientists, and industrial designers experimented with various innovative concepts such as

augmented reality and holographic displays embedded in windshields and windows, a sunroof that reduces heat from sun, and windows with adjustable tints (Nicas 2018; Phillips 2015, 2018).

6. After the 2015 terrorist attack in San Bernadino that killed 14 people, Apple's values of secrecy were put to the test after the government asked Apple to open the terrorists phone contacts and messages. Apple refused on the basis of the First Amendment (Etzioni 2018) and stood their ground.
7. Another issue is that most consumer electronics, including smart phones, require minerals that are usually extracted from mostly conflict zones (e.g., cobalt from the Democratic Republic of Congo and Zambia). Swedwatch and Finnwatch have previously reported the rampant use of child labor or female workers pushed into sexual exploitation by rebel groups that control the mines (Sandoval 2013). Environmental degradation of forests and waters occurs around the extraction mines threatening the native population's livelihoods whereas work in these mines have an elevated safety risk. SACOM documented numerous cases of workers being intoxicated with chemical substances, such as n-hexane or aluminum dust, while polishing the screen and covers of the 100,000 iPhones produced per day.
8. There is also an issue with tax evasion and keeping large amounts of money offshore in foreign bank accounts. President Obama met with Steven Jobs to ask what it would take to bring that money back to the United States and Jobs replied that he would like a tax amnesty. Obama demurred. However, since then, President Trump signed a tax reform package that greatly reduced the taxes on repatriating these massive amounts of money.
9. Exceptions to this are Chromebooks and Pixel, which are "physical" computers and phones. However, this is not where Google's competitive advantage is lodged. As some claim about Google's bottom line, it is a software-oriented advertising company.
10. The "gig-economy" of temporary workers in Silicon Valley has also been criticized for employment instability and their resulting low yearly wages (Wood et al. 2019; Petriglieri et al. 2018).
11. There are problems at Google (Bock 2015:318–35), but many of them are in managing expectations and feelings of entitlement that can get out of hand (e.g., free lunches leading to employees taking food home for their families or for the weekend, and ending the Prius subsidy of $5,000).
12. There were stronger watchdog groups and unions that broke off from the more conservative FLA. The Worldwide Responsible Apparel Production (n.d.a, n.d., b) has a stronger monitoring system, but manufacturing corporations are less interested in them.

9 High-Powered Merchandizing as a Special Case of Lean

Walmart, Costco, and Amazon

The largest merchandizing firms in the United States and also in the world now play a leading role in the world economy. In 2015, the two largest—Walmart and Costco—were followed by Kroger, Walgreens, Tesco, Carrefour, Amazon, Metro Group, Home Depot, and Target (NRF 2019; Carpenter 2016). Since then, Amazon has moved up. In 2018, Wal-Mart Stores, Inc. had revenues of $514 billion and 2,200,000 employees, and Costco Wholesale Corporation had worldwide revenues of $129 billion with 245,000 employees (Statista 2019). In the same year, Amazon generated $178 billion in revenue with 647,000 employees, which is now larger than Costco but below Walmart (see Table 9.1). However, large portions of Amazon revenues come from media and computer services, which are not strictly retail. Nonetheless, merchandizing firms have never before held such power in the economy.

Not only are these retail corporations larger than any manufacturing or oil corporation but they have reversed the power relationship between manufacturers and retail such that merchandising is now in control. Thus, their importance to the world economy is unassailed as they have expanded from clothing, books, everyday accessories, and electronics into food, pharmacies, home improvement, landscaping, tires and batteries, and even automobile and vacation sales. More importantly, merchandising firms are now stronger than manufacturers in the supplier-seller relationship.

As merchandisers, these iconic and somewhat unique corporations represent three distinct models of conducting business, especially working conditions, wages, and benefits. First, Walmart's model can be generalized as Waltonism based in the same sense as Fordism was an important model (Vidal 2010; Janoski and Lepadatu 2014). This model is intensified since Walmart is now the third largest employer in the world after the US Department of Defense and the People's Liberation Army in China. Further, many merchandisers have followed its price cutting model. The second model is the Costco that presents a positive model for treating employees, while still making considerable revenues and profits. And the third approach is Amazon which is distinct since it is an entirely web-based form of merchandising that now shows incredible growth and constant

Table 9.1 Revenue (in $US) and Employee Statistics at Walmart, Costco and Amazon

Organization Outcomes	*Walmart*	*Costco*	*Amazon*
Revenues in 2018	$ 514 billion	$ 129 billion	$ 232 billion
Net income	$ 6.67 billion	$ 3.13 billion	$ 10.1 billion
All Employees in 2018 (% Temps)	2,200,000 (50%)[2]	245,000 (41.6%)	647,500 (>50-%)[2]
2017 (% Temps)	2,300,000 (50%)[2]	231,000 (42.4%)	566,000 (>50-%)[2]
Revenue per employee (ee)[1]	$ 233,646/ee	$ 526,530/ee	$ 358,301/ee
Entry level pay in 2018	$ 11/hour	$ 15–15.50/hour	$ 15/hour
CEO pay in 2018	$ 24 million/year	$ 6.6 million/year	$ 19.7 million/year[3]

Notes

1 Revenue per employee is revenues divided by the number of employees. They can be impacted by very low prices and Walmart's very low prices make revenue per employee (and profit margins) very small.

2 A number of sources put Walmart temporary employees at 50% but we do not have a number that is more precise. Amazon employees were only 17,000 in 2007 and 33,700 in 2010 but they have grown tremendously since then. We do not have a good figure on part-time employment at Amazon. Recent hiring has been strongly temporary, but recent acquisitions include many full time employees. More than 50% is an estimate.

3 Founder and CEO Jeff Bezos' pay was only a million and a half, but he is already the richest man in the world and needs little in terms of income. We used instead the pay of two sub-CEOs, who each had $19.7 million income in 2018.

expansion into new areas of commerce including movies, TV and product servicing.

After presenting these three very important models, we discuss some of their impacts and controversies, and then systematically compare each model in how they adhere to the principles of lean production. In general, their adoption of lean production principles is significant but also very incomplete. This chapter concludes by looking at the implications and problems of each model.

Waltonism and the Relentless Minimization of Costs

After working for J.C. Penney, Sam Walton purchased a Ben Franklin franchise store in 1950 modeled on Woolworth and Kresge that emerged a half century earlier. After some success in Arkansas, he transformed this operation into the Walmart Discount City stores that eventually spread throughout the rural South. The model was so successful at keeping prices low that by 1980, Walmart had bested their top competitor, K-Mart (K for Kresge), even in the urban areas where K-Mart had reigned supreme. Now, Walmart and the bulk store called Sam's Club became the largest corporate employer in the world with more than 2.2 million employees in 2010 (*Fortune* 2012).

Box 9.1 The Walmart Company and Sam Walton

Sam Walton (1918–1992) began with purchasing a Franklin 5 & 10 in 1945, and expanded it and renamed it as Walmart in 1962. After decades of explosive growth, his company along with Sam's Club is the largest private employer in the world.

Walton was born on a Farm in Kingfisher, OK, and moved to Columbia, MO during the Depression. He held a variety of jobs and attended the University of Missouri as a ROTC cadet, and he received a number of honors related to fraternities and the Scabbard and Blade. When he graduated in 1940 with a degree in economics, he was voted "permanent president" of the class. He joined J. C. Penny as a management trainee, and worked in US Army Intelligence in connection to aircraft plants and prisoner of war detention centers. He reached the rank of captain and married **Helen Robson** (1919–2007) while in his military uniform in 1943. She was the daughter of a banker/rancher Leland Stanford Robson, and after becoming valedictorian of her high school class, she graduated with a degree in finance from the University of Oklahoma (her graduation and finance major were unusual for a woman at that time). After the war, Sam and Helen started the Walton's five and dime that they moved to Bentonville, Arkansas with a large loan from her father. They eventually expanded that business into Walmart and Sam's Club—a monolithic merchandising business (Walton 1992; Bergdahl 2007).

They also established the Walton Family Foundation giving $300 million to the University of Arkansas, and a $50 million to the Walton School of Business. Sam Walton died in 1992 from bone cancer, and Helen lived until 2007 but she suffered from dementia beginning in 2000. Helen also established the Walton Scholar program that gives 150 students scholarships and also brought Central American students to the United States to learn about democracy and free enterprise. Both Sam and Helen were lifelong Republicans and they donated millions of dollars to conservative interests. However, in 1991, a year before his death, Sam Walton contributed to Bill Clinton's campaign, and the Clinton's ties to Walmart started with Hilary joining the company's board of directors. However, since then the largest contributions go to Republican causes, especially GOPAC. Walmart has participated in some philanthropic endeavors, but much of the family fortune has remained in Arkansas. The Walton heiress Alice Walton established the Crystal Bridges Museum of American Art, which may be the world's best collection and most expensively acquired museum of American art (Mejia 2018).

Walton was a folksy southerner, but he was clearly driven and did not let anything—from competitors to his own "living high on the hog"—get in his way. While he did give store managers considerable leeway in managing, he was very successful in imposing his values of low costs on his organization and on suppliers to this day.

Walton's business strategy has been strongly influenced by the Great Depression, which he experienced firsthand. He professed 10 rules (Walton 1992:312–17; Bergdahl 2007; Rice et al. 2016):

(1) Commit to your business with a passion by catching it "like a fever"
(2) Share your profits with all of your associates and treat them like partners
(3) Motivate your partners with high goals and give them a share in the business
(4) Communicate everything with your partners by empowering your associates
(5) Appreciate everything your associates do for the business and provide sincere praise
(6) Celebrate your successes with enthusiasm and find some humor in your failures
(7) Listen to everyone in the company and percolate good ideas up from the bottom
(8) Exceed your customers" expectations with "satisfaction guaranteed"
(9) Control your expenses better than your competition by cutting prices (and wages?)
(10) Ignore conventional wisdom and swim upstream

Sam Walton definitely followed rules 1 and 9, especially in forcing suppliers to cut prices and imposing low wage schedules on workers. He and his managers led a spartan existence in Bentonville, Arkansas often sitting in supplier donated sample lawn chairs. On rules 4 and 7, he may have listened inside the company for good ideas, but many of his best ideas came from his competitors like Costco, Target, and even the dreaded-but-conquered K-Mart. However, in terms of communicating, listening, sharing, and celebrating with associates and workers, (items 4, 5, 6, and 7), Walton's actions toward workers were more rhetorical than operational.

Walmart has five major characteristics. First, Walmart is singularly focused on lowering prices and costs. Its motto "Always Low Prices, Always" is faithfully and relentlessly followed throughout the corporate structure, even at the level of corporate offices that are spartan and often involve sample furniture (Fishman 2006).[1] This has allowed Walmart to generate high revenues with low margins. Second, Walmart generates high absolute revenues with very low profit margins, which requires a very high operating volume. This massive amount of products is controlled by a very efficient and innovative supply chain management system (i.e., a JIT inventory system). This often requires using an aggressive approach to expansion of new transportation and inventory control techniques. Third, these low prices then dictate their treatment of employees. While retail is noted for its low wages, Walmart especially keeps wages and benefits low, and in the process hires many temporary workers who often do not qualify for benefits because they are kept under 40 hours a week. Some studies have found a large number of Walmart workers receive welfare benefits because they are poor. This is the opposite of Fordism, which in its beginnings paid a high wage of $5 a day. Fourth, Walmart exerts tremendous pressure on its suppliers to offer the lowest prices on products. This hard bargaining offers them large volume sales with very low profit margins. This can put extreme pressure on suppliers, and in the last 20 years, this has led to Walmart pushing some suppliers close to bankruptcy and to then go off-shore, especially to China. Fifth, Walmart uses a heavy union avoidance strategy that has been mainly successful, although in the last five years, there have been worker demonstrations for higher wages (Sainato 2018; Reich and Bearman 2018). There are also indications of Walmart's political inclinations going in the conservative direction to protect their positions.

Matt Vidal (2010) and Nelson Lichtenstein (2006, 2010) call this model *Waltonism*, and its unique contribution to the division of labor is represented by the intense bargaining process between merchandisers and their suppliers. Consequently, the Waltonism model has a strong impact on global production because of its domestic and growing number of foreign stores, and also especially concerning its off-shore locations. Some studies have shown that Walmart even has an impact on reducing the overall inflation rate in the United States (Jantzen et al. 2009).

Walmart is a merchandising firm, which means that their main task is to purchase goods from other suppliers and manufacturers, and then resell them in a retail store. Since they do not produce any physical products (other than the big box buildings and some rebranded clothing), their manufacturing story is nil. Their supply chain network is perhaps the most effective JIT supply chain management system in the world (Vidal 2010; Kharif 2007). Walmart has a very advanced JIT inventory system using advanced lean production techniques. In 1987 it developed a 24 million dollar satellite communications system. It links all operating units of Walmart stores, warehouses, suppliers and headquarters with voice, data and some video communications (Wailgum 2007). It tracks products and containers using bar code labeling and also "radio frequency identification" (RFID) tags, which actively initiate communication throughout a "collaborative planning, forecasting, and replenishment" (CPFR) program (Nguyen 2017:104–9). In 1996 they started using the internet for most of these functions, and developed on-line platforms to counter the threats from Amazon movement from books into nearly all products available. With this new direct on-line presence, there is some evidence that Walmart has become effective in "beating back" Amazon's challenge to their retail business (Halzack 2019; Cheng 2018). Overall, Walmart's JIT or supply-chain management system is state of the art and it develops many new and copied technological innovations. To some degree, it has to be so advanced because it is tracking so many different products. In terms of logistics, a trucking fleet engages in "cross-docking techniques at distribution centers whereby items are moved from multiple delivering trucks to multiple distribution trucks so that inventory in warehouses may even be avoided. Cross-docking is an advanced inventory reduction technique that has a number of facets and variations (Krajewski et al. 2015; Bartholdi and Gue 2004; Napolitano 2000).

The most important point about Walmart's cost-cutting is that their bargaining process offers a big payoff—a national and even worldwide contract for producing millions of units—and a major downside—Walmart's purchase of these products is at a very low price. Walmart is relentless in pursuing this low price and this has caused them to largely abandon suppliers in advanced industrialized countries and pursue low wage production in less-developed countries, especially but not restricted to China (Walton 1992; Vidal 2010; Abernathy et al. 1999; Lichtenstein 2006; Fishman 2006; Gereffi 1994; Gereffi et al. 2005).

Walmart invites suppliers to come to Bentonville, Arkansas or negotiation centers in Asia to bargain about the prices of the goods that suppliers would like to sell to Walmart. The incentive to suppliers is nearly irresistible because it gives suppliers access to Walmart stores all over the world (e.g., Walmart sells more in one day than Lowes, a major hardware store and lumber supplier, sells in a year). However, the price-cutting pressure that Walmart exerts on these suppliers is massive. *Frontline* (2004) tells the story of how Rubbermaid, a major supplier of US kitchen and household ware refused the Walmart price pressure and how they were almost put out of business. Charles Fishman illustrates how Vlasic, an upscale producer of Kosher pickles, accepted a deal by which it would sell a gallon of Kosher Dills for $2.97 which is less than what they charge for a quart of the same pickles in other grocery stores (2006:79–83).

There are three consequences for suppliers in this system. First, they are often asked by Walmart to reduce the number of features or quality on products so that they can be offered at a lower price. Second, they are repeatedly asked to lower costs over and over again. And third, after making these requests, Walmart is rather cavalier about dropping suppliers. So, operating at a very high volume of supply to Walmart can be dangerous to suppliers, especially small ones. While larger firms can survive losing Walmart contracts, small firms who have become oriented toward Walmart's high volumes can then be in grave danger when Walmart drops them as a supplier, which often comes when price cuts put small firms into unprofitable positions.

Walmart's use of the JIT philosophy extends not only to inventory, but to labor as well. According to Kris Maher (2007), Walmart's computerized scheduling and labor optimization system makes employee work schedules much less regular and predictable for workers. The software improves efficiency for the store and improves short-term profits by determining how many employees are needed based on demand and store traffic. In this system, employees are "on-call" and have to come into work when business surges or sent home when there is a lull. For the low-paid hourly employees at Walmart, reductions in hours and sporadic work schedules increase stress and financial burdens on employees (Trottman and Maher 2013). This is a stark contrast to Toyotism, where employees are kept and trained for long-term and given competitive wages and benefits.[2] But for Walmart, the pressure to cut labor costs is due in part to Walmart's low-cost business model. The low margins that they receive on sales and the high volumes required to make a profit necessitate cutting costs elsewhere, so labor costs need to be kept as low as possible (Vidal 2010). With their large number of workers they employ, Walmart influences the minimum wage for the US (Vidal 2010; Hamilton 2013).

Walmart's treatment of its own employees has been a constant source of criticism. The *Wall Street Journal* grades the corporation's treatment of "employee engagement and development" as a cause for concern (Cutter 2019:R4). Walmart also lost a lawsuit concerning meal breaks in 2017 and has to pay 6 million dollars in back pay (*Hamilton v. Wal-Mart Stores* 2017). First, US employees are rarely organized into quality teams with effective input into the merchandizing process (Sainato 2018; Vidal 2010; Ingram et al. 2010; Harney 2008; Fishman 2003). However, they do use a form of pep rallies that praise the firm and worker dedication. Second, Walmart reaps record profits while offering employees" wages that start at $11 an hour which only comes to a salary of 22,880 dollars a year for full-time associates (Glassdoor 2013). These wages are so low that a report prepared by the Democratic Staff of the US House Committee on Education and Workforce in 2013 showed that full-time Walmart sales associates and cashiers needed an average of $5,815 a year in medical and other welfare benefits from the government (Worstall 2013). Assistant store managers made $44,634 a year on average so presumably they would not need government assistance. But the workers in other industries that have become unemployed when their employers are forced to go off-shore receive even more government assistance. All in all, there is a strong element of immiseration in Waltonism (Gereffi and Christian 2009; Fishman 2006).[3] This leads to strikes in Germany where Walmart eventually pulled all their stores and operations in 2006 (Christopherson 2007; Thomasson 2019; Thomasson and Inverardi 2018).

Foreign employees of the subcontractors that Walmart uses are subject to constant human rights violations that the company supposedly corrects, but which they constantly tolerate. The teamwork elements of production in foreign subcontractors are often advertised but not actually implemented. Their involvement with the Tarzeen Fashions Ltd. fire in 2012 when 112 garment workers died and the Rana Plaza fire that killed 1,129 workers in Bangladesh are horrific examples (Yardley 2013). In the aftermath, however, Walmart is not cooperating with other retailers in the FLA or other watchdog groups because they claim that they are establishing their own monitoring system that will expel suppliers from their network if they do not comply (Baja 2012; Yardley 2012). So much for an independent monitoring system (even one financed by employers like the FLA).

However, despite the maltreatment of employees at home and abroad, our main point here is that Waltonism is a unique supply-chain model that exerts tremendous pressure on suppliers to cut costs. As such, it does not employ much in the way of long-term philosophy in its employment system or closely knit and highly trained teams as we shall see in the lean production model. For instance, lower-level employees at Walmart are not empowered for problem-solving as they are in Toyotism. Their bargaining model with suppliers has led to greater reliance on off-shoring because this has proven to be a very effective way to cut costs, which is the *raison d'etre* of Walmart. However, teamwork, constant improvement, and organizational aspects lean are not present. Hence, it is a very partial model of lean.

Costco: Wholesaling and a Different Employment Model

Sol Price, the effective founder of Costco, had an entirely different background than Sam Walton. Price's father worked with David Dubinsky and helped organize the International Ladies Garment Workers Union. However, his father had TB and they moved to San Diego for health reasons. In this move across the country in 1928, the 12-year old Sol Price saw the depths of the depression. After working as a lawyer doing extensive *pro bono* work, he worked for Convair on B-24 airplanes while maintaining his legal practice. This led to starting Fed-Mart in San Diego where there were a great number of federal employees. As his business developed, he maintained that while keeping prices low he would not take it out of workers' wages and benefits. James Sinegal and Jeffrey Brotman followed the same principles as they declined an offer from Sam Walton to merge with Walmart, and then merged Price Club and Costco after Sol Price moved on.

Box 9.2 The Costco Wholesale Corporation and the Son of a Labor Organizer

Costco (founded 1983) began with **Sol Price** (1916–2009) and his son, **Robert Price**. Sol Price had an entirely different background than Sam Walton. Price's father worked with David Dubinsky and helped organize the International Ladies Garment Workers Union.

However, his father had Tuberculosis so they moved to San Diego for health reasons. In this move across the country in 1928, the 12-year-old Sol Price saw the depths of the depression. After working as a lawyer doing extensive *pro bono* work, he worked for Convair on B-24 airplanes while maintaining his legal practice. Eventually, he saw a major opportunity, and started Fed-Mart in San Diego, where there were a great number of military personnel and other federal employees. As his business developed, he said he would keep prices low but not reduce workers' wages and benefits.

Sol Price established Fed-mart with Hugo Mann, but was forced out by his partner, and Fed-mart later went out of business. Sol Price and his son Robert opened their first Price Club warehouse in1976. In 1982, Fed-mart was liquidated and many of its stores were taken over by Target. But the Price Club was a new concept in retailing—a warehouse club that sold to the public when they bought a membership (Fedmart had been only for federal employees). The Price family placed Price Club Warehouse #1 inside a series of old airplane hangars previously owned by Howard Hughes, and Costco Warehouse #401 is still in operation today. Walmart also followed the Price Club example, and Sam Walton said "I guess I've stolen—Actually I prefer borrowed—as many ideas from Sol Price as from anybody else in the business" (Sellers 2009; Price 2012).

James Sinegal worked his way up from grocery bagger to executive vice president at Fed-mart. He was a protégé of Sol Price in following his principles of fair work and long-term growth. After a stint with Builders Emporium from 1977 to 1978, he became executive vice president for the Price Company from 1978 to 1979. With Jeff Brotman, he co-founded Costco in 1983 and served as CEO and President until retiring in 2011. Unlike other CEOs, Sinegal traveled to all the Costco stores each year. He made Costco the first warehouse club to include fresh food, optometrists, pharmacies, and gas stations. Like Price, Sinegal is known for treating his employees well because he says they in turn will treat customers well. Sinegal provided his workers with higher wages and compensation than other merchandizers, including health insurance. This in part causes Costco to have the lowest turnover rate in the industry. In 1993, Price Club and Costco Wholesale merged into PriceCostco, opening new stores in the United States and abroad. Robert Price then went his own way, as Sinegal remained the CEO of the newly named Costco in 1997 and stayed until 2011. Sinegal specifically says that he does not care about Wall Street analysts who criticize the firm for paying employees well. He discounts short-term pursuits of profits and share prices, and takes a long view of growth.

With its headquarters in Seattle, James Sinegal and Jeffrey Brotman opened the first Costco warehouse in 1983, and it soon became a top-ten retailer in the world. James Sinegal began in wholesale distribution industry by working for Sol Prince at Fedmart, and Jeffrey Brotman was a lawyer from a family who owned a Seattle retailing firm. Costco's business model and size were similar to those of Price Club, which made the merger more natural for both companies. The combined company became PriceCostco, and memberships became universal, meaning that a Price Club member could use their membership to shop at Costco and vice versa (Eisner 2009; Berman 2011). PriceCostco was initially led by executives from both companies, but the Price brothers left the company in 1994 to form Price Enterprises that would operate in Central America and the Caribbean current Costco. The company changed its name in 1997 to Costco Wholesale Corporation and all remaining warehouses were renamed Costco. As of 2015, Costco is an American corporation that operates 759 warehouses in the United States and ten other countries. As of 2016, Costco was the world's largest merchandizer of rotisserie chicken, prime beef, organic food, and both low- and high-priced wine. Although Costco is well behind Walmart and recently Amazon, it is ranked fifteenth American Corporation by revenue in *Fortune* magazine's top 500 corporation rankings and it continues to grow (Petterson 2019).

Two points are important about Costco. First, Costco is a decent employer. Walmart often pays its employees low wages and actively promotes workers being slightly less than full time so they do not have to pay benefits. In comparison to the industry as a whole and especially Walmart and Amazon, Costco pays moderately high wages and provides good benefits, which earned the company the "best employer" award for a number of years. Costco sometimes even deals well with unions where they are involved (Calfas 2018; McCarthy 2017; Ton 2014). As a result, they have the lowest turnover in the industry a 7% a year compared to 50 to 70% at other merchandising firms (Relihan 2018). This is an unprecedented turnover figure and actually close to Toyota and Honda turnover numbers. Costco still tries to find the lowest prices, but does not cut employee pay or pressure retailers as Walmart does. Thus, it provides a major counter example in the merchandising field. Second, Costco has actively participated in Lean Six Sigma programs. They have sent workers and managers to the Lean Six Sigma Training Certifications (Lean Six Sigma 2017; Miller 2013; Calfas 2018; McCarthy 2017; Berman 2011; David et al. 2015; Jakobson 1988).

However, Costco is not a perfect company. From out interviews in the metro-Atlanta area, we found that the many workers who give out food samples are subcontracted from Club Demonstration Services, whose only client is Costco. These contractor associates are closely supervised by the subcontractor and are paid less than Costco associates. As a result, they have higher dissatisfaction with their jobs. In many ways, they have similar experiences with temporary workers in the auto industry. They are doing similar jobs, but are paid much less and are actually jealous of the high morale, camaraderie and family spirit among the Costco employees. But in an indirect way, this points out the positive work environment for regular Costco employees.

Since Costco has a strong JIT inventory and SCM system, the question arises about how it treats suppliers? Evidence shows that Costco can be very tough on suppliers who do not give them the prices that they expect. Nonetheless, Costco follows similar cross-docking techniques and has an elaborate system of training supplier's goods. Costco emphasizes supplier diversity (i.e., minority-owned and women-owned businesses) as well as making sure that suppliers conform to their Ethic Code. They have 8 regional offices that operate directly with suppliers from their region. To compete with Amazon and increasingly Walmart, Costco is quietly pursuing an online presence. However, Costco does not hold the worldwide heft that Walmart does, so suppliers are confronted by a smaller presence. To some degree not as much has been written about Costco's treatment of suppliers (Soni 2016) because in many ways Walmart attracts so much attention.

Given that Costco seemingly has higher costs than Walmart due to lower volume and higher wages, how can it survive in this highly competitive industry? Two factors are important. First, Costco is a membership store that requires that those who want to shop there must pay a substantial fee ranging from \$60 to \$120 dollars to obtain a membership card that has to be shown to get into the store and again when one pays for goods (also they restrict payment options preferring debit cards though they recently allow VISA card payment). Upon leaving, the clerk checks customer's receipts against the goods in their cart. This started at their roots as FEDCO in the 1970s was originally a store only for federal employees. Being a membership organization becomes a guarantee that customers are able to pay for the goods they purchase. It is also an unseen way to screen customers and actually generate money (David et al. 2015). Second, Costco has found a very effective way of controlling shoplifting. With controlled entry and exit with inspection, the store broadcasts that customers will need to show proof of every item they have purchased. So, the initial screening by paying the membership fee is then followed by checking "legal purchases" at the exit to create a competitive advantage in shaping and narrowing its customer base.

The marketing strategies of the two companies also differ. First, Walmart tends to focus on lower income customers. Although the two stores have a considerable overlap, Costco focuses more on middle- and upper-middle-class shoppers. Costco carries high-priced goods with more advanced features. For instance, when we exit the store, we pass Costco's auto purchase vehicle display, which in Lexington, are Cadillacs and Volvo SUVs. While Costco's program does not stock cars and one has to go to the participating dealer, one would not find Volvos and Cadillac's at Walmart. Numerous other examples include jewelry, electronics, and other products. Second, Costco sells more products in bulk than Walmart. Again, while there is some overlap, Walmart generally sells single-item products, while Costco may package two to four of them together. For instance, Walmart sells a pound of coffee, while Costco has two- or three-pound bags. Or Walmart sells a single bottle of mouthwash, while Costco sells a package of three bottles. Or one can buy a half pound piece of Salmon at Walmart, while one has to buy a five-pack of trout or a two-pound piece of salmon at Costco. As a result of these two factors, Costco generates a higher rate of profit, but its total revenues are nowhere near as large as Walmart and there are some communities in which Costco avoids building a store. Third, Costco carries only a fraction of the products that Walmart carries, and they pursue a "skimming the cream" strategy to obtain higher value-added products that can attract customers (Wulfraat 2014; Soni 2016). This is referred to as a "treasure hunting strategy" that leads customers to products that they might not consider to be staples. By contrast, Walmart is clearly a staple store. A fourth factor that has a lesser impact is that Costco pays its CEO and top executives much less than Walmart and inconceivably less than Amazon. Jim Sinegal, the CEO, received $350,000 in 2005 (Cascio 2006). In 2019, Costco CEO Craig Jelinek received $6.6 million, Walmart CEO Doug McMillon earned $24 million, and two CEOs at Amazon got $19,700 million (Salary.com 2019).[4] Thus, these other CEOs make three to four times as much as the Costco CEO.

Walmart has copied Costco to some degree with the bulk sales at Sam's Club, but the upscale selling is not prominent. Again, there is overlap since both stores carry a gallon of milk and Walmart does have its gallon of pickles. Finally, Costco has extensive sampling of food products in their stores with many of them cooked. Walmart avoids this practice.

In sum, Costco is somewhat of an the anti-Walmart in many respects.

Amazon, Warehouses, and the Internet

Amazon is similar in some ways to Walmart except for the fact that all of their operations are online. Amazon started out as a book seller and received considerable public subsidies in that internet companies did not have to pay state taxes. This enabled the Amazon CEO, Jeff Bezos, to become the wealthiest man in the world. Amazon also sought further subsidies and tax breaks through city and state competitions for the location of their offices.

Box 9.3 The Amazon Corporation and the Voracious Appetite of Jeff Bezos

Jeffrey Jorgensen (1964 to present) was born in New Mexico but changed his name when he was four to Bezos after his mother Jacklyn divorced his father and married Miguel Bezos. The family moved to Houston, TX, where his father worked as an engineer for Exxon after getting a degree from the University of Mexico. After being the valedictorian at his high school, Jeff Bezos graduated from Princeton with a BS in electrical engineering and computer science. Since he had a GPA of 4.2, he received many offers but chose to work for Fitel, a telecommunications start-up. From there, he moved to Bankers Trust as a product manager, and then joined D. E. Shaw and Co. rising at the age of 30 to the fourth senior vice-president of this lucrative hedge fund. With a loan of $300,000 from his parents, Bezos started an online bookstore named Amazon in his garage in 1994. He honestly warned his initial investors that there was a 70% chance that Amazon would go under. Bezos says that he had always planned to expand to other products beyond books. Three years later, Bezos took Amazon public in an IPO. Although it took at least five years for Amazon to make a profit, Bezos predicted that the growth of the Internet would overtake competition from large book retailers and smaller bookstores. From there, Bezos has added a bewildering number of other products making it a major merchandising firm, and has expanded into media (movies and the *Washington Post*), programming assistance, used products, and even space exploration.

Bezos' management style involves a "regret-minimization framework". He describes this life philosophy by stating: "When I'm 80, am I going to regret leaving Wall Street? Will I regret missing the beginning of the Internet? Yes". During his first 20 years at Amazon, Bezos tried to quantify all aspects of the company, using spreadsheets and basing all his decisions on data. His mantra seemed to be "Get Big Fast" by bringing the company to scale in order to achieve market dominance. He often plowed profits back into the company in lieu of paying taxes and allocating dividends among Amazon's shareholders. Bezos uses the term "work—life harmony" because balance implies a zero-sum relationship while harmony implies that you can have both. Bezos believes that work and home life are interconnected, informing and helping each other. The Bezos principle is that those who cannot tolerate criticism or critique cannot do anything new or interesting (Mossberg 2016). Bezos enforces a two-pizza rule that keeps meetings small enough so that two pizzas will feed everyone.

Located in a new headquarters building in Seattle, Amazon has become the largest e-commerce marketplace in the world and the second largest employer in the United States. Amazon's supply-chain is rather complicated but incredibly efficient. It operates as a large warehousing system all over the world and these warehouses are their most prominent physical locations.

But it has also moved into secondary sellers' markets, whereby suppliers use the Amazon platform but maintain their own inventory and deliver the products to the consumer. The company has employed a "non-mandatory" semi-membership system reminiscent of Costco through its Amazon Prime feature. This requires a membership fee of $119 or $11 or $13 a month (students pay $59). Prime membership entails free shipping of products, unlimited movie and TV streaming, unlimited music streaming, photo storage, a lending library, and many other benefits. Like the Costco membership card, it creates a strong incentive to buy products and use services on the Amazon platform. And this may then lead to Amazon being criticized for monopolistic competitive practices. Low prices and incredibly fast delivery are major advantages that Amazon presents. In fact, in some locations Amazon guarantees two-hour delivery (some urban centers), and it now promises one-day delivery. Future deliveries may be made by drones, which may be controversial.

Bezos published fourteen leadership principles of the Amazon website building on the five principles he put forward in 1997. They center on a long-term orientation toward leadership, and many of them related to lean production: (1) Leaders stay focused on customers ("customer obsession"), not competitors; (2) Leaders think in the long-term in terms of ownership; (3) Leaders invent and simplify; (4) Leaders are right a lot; (5) Leaders are never done learning and are always curious; (6) Leaders hire and develop the best; (7) Leaders insist on the highest standards; (8) Leaders think big; (9) Leaders are biased toward action; (10) Leaders are frugal and accomplish more with less using invention; (11) Leaders earn trust; (12) Leaders dive deep to stay connected at all levels of the organization; (13) Leaders have backbone and are not afraid of conflict when they commit; and (14) Leaders deliver results of high quality and on time.

This list differs in employee orientation from his previous list in 1997. The list obviously focuses on leadership, but item 11 does refer to staff morale and might have to do with teams and empowering people.

Amazon has been heavily criticized as a harsh and low-paying employer (Geissler 2018; Stone 2014; Bloodworth 2018a, b; Cain 2018; Galloway 2017; Onetto 2014; Sainato 2018). Its main workforce is in very large warehouses, and the pressures of very fast delivery are then put upon the workers in the warehouses. A number of exposés have highly criticized the treatment of Amazon warehouse workers who face high pressure to deliver goods, work in overheated buildings, and struggle to find time for bathroom breaks (Geissler 2018; Scheiber 2019; Bloodworth 2018a, b; Bair and Bernstein 2006). The complaints are also present in highly skilled occupations. In one of our interviews, an ex-Amazon software manager described a harsh regime of work intensification amounting to "insane" working hours that were more than he had experienced at any other high-tech company in his software career. At his exit interview, his managers wanted to genuinely find out why no software team manager stays on the job for more than a year. He cited work-life balance as one of the main reasons. Also, Amazon uses aggressive anti-union tactics, and an internal company video details how to avoid union organizing campaigns (Menegus 2018; Shaban 2019). Nonetheless, the company has dealt with unions in New York and especially in Germany facing some strikes (Streitfeld 2013; Thomasson 2019). Currently, a strike is planned at the Amazon warehouse in Minnesota concerning Prime Day that sees a tremendous increase in work hours (Shaban 2019). However, since 2016, Walmart has raised its entry wage to $15 an hour due to strong pressures on low paid workers in the United States.

In many ways, employment at Amazon is replacing Walmart as the chief villain of employment critiques. And as Amazon relies on JIT, they are also completely Fordist in their production methods in their warehouses. Their delivery is largely outsourced to the Post Office, UPS, Fedex and so forth. These companies are a mix of unionized and non-union features, and they are not generally individual with teamwork though their JIT principles are major.

New York's rejection of one of Amazon's two new headquarters shows a turning point in the public relationship and assessment of Amazon. Creating a major competition between states to create 25,000 mainly low wage jobs in exchange for billions of dollars of subsidies can be viewed as a tragedy of the commons where communities gain short-term outcomes in terms of job but lose long-term revenues. The result of creating these state versus state and city versus city competitions yields no overall gains for the country as a whole. While the discourse does not often use the term, these exchanges are a clear form of "corporate welfare" in addition to tax avoidance at start up (Leonhardt 2019).

Comparisons of Three Corporations

Are any of these three companies—Walmart, Costco, and Amazon—lean organizations? We can look at five criteria: long-term orientation, JIT inventory, continuous improvement, emphasis on teamwork, and loyalty to employees (Cascio 2006). We make a rough estimate of these criteria in Table 9.2.

First, concerning long-term goals, Walmart and Amazon are pursuing aggressive growth as a long-term goal. While Walmart has branched out to many new services for retailers like pharmacies and auto repair, Amazon has gone even further from its book and electronics base with entertainment provision (Amazon Prime) and even making movies and TV shows. It currently challenges Netflix for providing online movies.

Costco is much more prudent and careful about its expansion. Since it appeals to a somewhat higher level market niche, store expansions remain more conservative. In terms of employee's long-term goals, Costco is clearly considered to be the best place to work of the three. It provides more training and promotion opportunities. However, at the higher echelons of each

Table 9.2 A Comparison of Lean Production at Walmart, Costco, and Amazon

Components of Lean Production	*Walmart*	*Costco*	*Amazon*
1. Long-term goals:			
a. For firm	Yes, growth	Mixed=1/2	Yes, wide ranging growth
b. For employees	No	Yes	Maybe=1/2
2. Just-in-time Inventory:			
a. Strong supply chain	Yes, strong	Yes, strong	Yes, mixed
b. Relations with Suppliers:	No, Exploitative	Mixed=1/2	Mixed=1/2
3. Continuous Improvement:			
a. Lower costs and prices	Yes, for price	Yes, for price	Yes, for price
b. High quality	Varies=.25	Yes, higher quality	Varies=.25
4. Emphasis on Teamwork:			
a. Teams are emphasized	No, weak	No, weak	No, weak
b. Employee training	No for employees	Yes for employees	No for employees
5. Loyalty to Employees			
a. Treatment of Employees	No, low wages	Yes, moderate wages	No, low wages
b. Temporary Employees	Yes, but overused	Yes, seasonal	Yes, temporary ee's
Total	42.5%	80%	52.5%

corporation, these opportunities may be more equal. In Table 9.2 we give Walmart a one, and Costco and Amazon a one and a half.

Second, all three corporations have very strong supply chain processes involving JIT inventory. The only exception to this might be Amazon's use of book suppliers that more or less provide delivery on their own (i.e., the product does not go to an Amazon warehouse) even though the financial processing goes through Amazon. In this sense, the supply chain is more regulated by supplier evaluations provided by customers. After that, the process is more of a "buyer beware" decision on whether to buy something from a supplier with a lower rating. But for most products that are warehoused, the Amazon supply chain is quite strong.

In lean production, suppliers are brought into the corporation's decision-making and although the OEM wants lower prices, the OEM works with the supplier on improving their services and after they qualify as an approved supplier, they have a strong and loyal connection to the OEM. On this point, Walmart is the worst. They treat suppliers in a total market orientation, and may even pressure suppliers to provide products at a loss. They also will ask manufacturers to provide a product with fewer bells and whistles to reduce costs but also reduce the quality of the product. Walmart's treatment of suppliers is nearly universally criticized and cannot be considered to be in the lean production tradition. Amazon's treatment of suppliers is also suspect. They have forced book publishers to reduce their prices much lower than what they want and have even engaged in self-publishing without editors (Packer 2014), and Amazon carried on a feud with Hachette recently, maintaining its control over the prices of its e-books (Streitfeld 2014). Little is known about Costco's relations with its suppliers other than the company's emphasis on buying at wholesale and in large lots. While "no news is good news', there is some recent evidence that Costco has been squeezing suppliers in its competition with Amazon (Boyle 2018). Hence we give Costco and Amazon a mixed rating in Table 9.2.[5]

Third, the meaning of continuous improvement is somewhat different for a merchandizing firm. For the most part, we can say that each of these firms emphasizing lower costs and prices. Perhaps Walmart is the best in this regard, but Costco and Amazon are certainly price cutters. In terms of the constant improvement of quality, both Walmart and Amazon have engaged in practices that reduce the quality or features of products that they sell. This is not exactly creating low quality or mistakes, but it is not the higher quality that one might find at Costco. Hence we give a one and a quarter score for Walmart and Amazon and a two score for Costco.

Fourth, teamwork is an intrinsic part of lean production. It is here that both Walmart and Amazon clearly do not have a strong teamwork focus. Management may talk about teamwork in a generic sense of everyone in the firm working toward a common goal, but actual teams with interdependent tasks are largely absent. Costco adheres more to Six Sigma, but even this process itself is not strong on teamwork. Kate Mulholland (2011) describes teamwork in supermarkets as "a fig-leaf for flexibility" in the sense that there was little interaction between workers other than relief workers sent to new locations in times of need. She calls this "the single company-wide team" (2011; 219). Only two areas of the supermarket had four to five people working in bakery and code checking, but this does not resemble any kind of *kaizen* or quality circle processes. Related to teamwork is job training and to some extent job rotation. Training and possibilities of promotion seem to be a bit stronger at Costco, as this is a weak point for Walmart and Amazon. As a result, we give zeros for Walmart and Amazon, and a one out of two for Costco.[6]

Fifth, loyalty and treatment of employees is a critical part of lean production. Although the retail or merchandizing sector is known for its low wages, there are major differences between these corporations. Currently in 2018, Walmart pays entry-level employees $11.00 an hour, while Costco pays $15.00 to $15.50 an hour and Amazon pays $15.00 an hour (Taylor

2019). Costco has opted for the high road in their "good job strategy". They are boasting that they offer the highest wages in retail and that 70% of every dollar spent by Costco goes to employee wages (Relihan 2018). As a result, Costco is ranked by *Money Magazine* as one of the best places to work in the United States, while Walmart is not (Calfas 2018, McCarthy 2017). Amazon is ranked as the best place to work in 2018, but that rating is done by LinkedIn, which emphasizes higher level employees and definitely not warehouse employees (Roth 2018).[7] Thus, Costco gets a two and the others a one.

In sum, Walmart and Amazon only score 42.5% and 52.5% respectively on lean production. Costco comes closer at 80%. However, the weaknesses in terms of lean production for all their companies are the treatment of suppliers and the presence of teams. Based on this comparison, Costco comes much closer to lean production principles than either Walmart or Amazon, who nonetheless have a lean production supply chain with considerable JIT inventory.

Critiques of the Big Box Store Approach

While not directly connected to lean production, one cannot talk about big merchandising operations without mentioning many of the critiques that they have garnered. We have already discussed: (1) low wages and benefits, which are endemic to merchandising, but Walmart has shown the low road and Costco demonstrates a higher path; and (2) how large monopolist organizations can exert tremendous power over suppliers once they become dedicated to the big box store and their demands sometimes end up with supplier bankruptcies. Three other issues are of importance.

First, big box or internet platform merchandizers tend to put local stores out of business because of their economies of scale and sometimes predatory pricing. As demonstrated in the 1883 novel *Au Bonheur des Dames* (or *The Paradise* in a recent BBC production) by Emil Zola (1995) describes how custom made dress, belt, shoe, and other specialty stores cope with an early version of the department store in 1864 complete with loss leaders and offering exclusive products bought at much lower prices than the specialty stores could ever provide. A number of studies have shown that Walmart Supercenters put local shops out of business within five years. On the other hand, large chains such as the once overwhelmingly dominant Sears Corporation and its catalogue have faced decimation in the modern period. But do these big box stores, which have dropped the services of the large department stores of the 1940s and 1950s, decrease overall employment in the communities in which they locate? Studies largely show that they do decrease employment within five years (Gereffi and Christian 2009; Basker 2007; Basker and Noel 2009; Basker et al. 2012; Bloom and Perry 2001; Stone 1997).

Second: do big box stores increase or decrease employment and general well-being in the communities in which they are located? Neumark et al.

(2007) show that the impact of Walmart stores on the local labor market reduces county-level retail employment by 150 workers, and county-level retail earnings are reduced by $1.4 million or 1.5 percent. However, Hatamiya (2014) shows a positive effect in his study that was published by Walmart (Bonanno and Goetz 2012) show that big box stores have a small negative impact on local development. Hicks (2015) argues that big box stores increase the community's dependence on government welfare, and specifically increase Medicaid expenditures by $898 per worker but did not increase AFDC/TANF expenditures and may even reduce them. And Wolfe and Pyrooz (2014) show that big box stores actually increase the crime rate after controlling for numerous factors. Many of these questions were not asked about Toyota or Honda, but manufacturing facilities attract suppliers. Big box stores negotiate with suppliers but do not actually bring suppliers to the community where the big box stores are located. However, Courtemanche and Carden (2014) show that Walmart clearly decreases grocery prices in communities (Mulholland and Stewart 2013). There are also issues about lower housing prices (Pope and Pope 2015).

Third, are big box stores and their headquarters worth it in terms of tax evasion and subsidies? Based on the mix of negative and very small positive outcomes from attracting big box stores, it does not really seem worth it for communities to forego taxes or pay subsidies for these stores. Further, one must really question the tax free status of internet companies like Amazon given that many of these firms have enriched their owners to immense degrees (e.g., Jeff Bezos of Amazon being the richest human being in the world) (Vandergrift and Loyer 2014).

Fourth, human rights abuses in off-shore production are an issue for Walmart since they promote off-shore production and sell so many products made in China. The rise of the neoliberal globalization forces and especially the new-found identity of China as the "workshop of the world" has undermined the successes of organized labor in old economies. Often times, Chinese contractors compete fiercely with each other to get the Western business and the easiest way to reduce costs is by reducing the labor costs (Sandoval 2013). Long working hours, low wages and poor occupational health and safety have been cited often as unethical labor practices at many subcontractors. This shows a clear disconnect between the Western public and the workers who make the actual products (Bartley 2018a, b; Bartley and Child 2014; Bartley et al. 2015). Chan et al. (2015) argue that Walmart concern themselves little with the working conditions of off-shore workers who experience overtime under difficult conditions. Most of these workers are from rural areas and because of the *Hukou* passport system are denied many rights when they enter the more urban areas where factories are located. Several watchdog organizations, such as the FLA, SACOM, China Labor Watch, and the European organization "makeITfair" conduct research and raise awareness on the labor conditions and lives of manufacturing workers, present their reports to the companies, and demand answers and corrective measures from them.

Walmart states on their corporate responsibility website that the "primary responsibility for compliance with Walmart's Standards for Suppliers rests with the supplier" (Walmart 2019). Walmart says that they can "leverage" its size and influence to "assist" suppliers in having an impact on human rights and reducing exploitation. They specifically mention the Bangladesh garment industry, Mexican farm workers, and the Thai seafood industry. Walmart claims that respect for people and human rights are an important value in their supply chain network. However, Walmart, Costco and Amazon refuse to join the FLA as Nike and other major corporations have.[8] And the Asia Floor Wage Alliance states that Walmart supply chain workers in Bangladesh, Cambodia, India and Indonesia have had "persistent rights violations" (AFWA 2016; Smith 2016). SACOM still reports that electronic manufacturers have failed "to end grueling conditions at Foxconn Factories" (Garside 2012; SACOM 2018a, SACOM 2018b). And Walmart sells many of these products. SACOM reports that electronics sub-contractors (mainly Foxconn) still have poor ventilation, insufficient protection equipment, and workers not being informed of the harmful chemicals they encounter, which led to two major factory explosions that killed 3 workers and injured 75 others in 2011. The long overtime hours seriously affect the workers' well-being and social lives. The migrant workers do not have a support system at their new locations, and in their very limited free time, they can only eat and sleep (Sandoval 2013; Chan et al. 2015).

Fifth, environmental and sustainability issues are also important. Quiñones et al. (2019) give the example of salmon fishing in pens in Chile as an example of an unsustainable industry that spoils large tracts of undersea lands with water. Similar issues concern the use of heavy carbon footprint items. The breakneck competition, heavy manufacturing, and loose environmental regulations have led to high levels of air, water, and land pollution, and even to the rise of "cancer villages" in China. Intense and persistent smog often requires people to wear masks in industrial China, especially in Beijing (BBC 2013). While Walmart paid the second largest environmental fine in history for water pollution generated by pesticides (Kaufman 2017), Amazon is being criticized for the pollution due to air and road transportation (i.e., trucks and aircraft) due to the demands of JIT delivery (Nguyen 2017; Natto 2014).

Over the past decade, Walmart and Amazon have put corporate responsibility for off-shore production on their agenda (little is known about Costco). However, their argument is that they buy from manufacturers who then subcontract to off-shore companies. Hence, merchandizers are twice removed and highly indirect parties to the exploitation that occurs. On the other hand, when they purchase goods directly from the third world, they are as responsible as Apple or other manufacturers.

While many might not consider these issues to be related to lean production, they do concern externalities or byproducts of production that involve waste in the larger scheme of global society.

Conclusion

Walmart, Costco, and Amazon provide a strong contrast in their implementation of lean production. All of them have adopted very strong JIT inventory practices to their SCM system. However, Walmart and Amazon's treatment of employees and their questionable use of teams makes their lean regimes highly partial. Costco is clearly superior to Walmart in terms of wages and benefits; however, its use of lean principles is still quite partial compared to the leader in lean production—Toyota. Walmart has faced continued criticism for low wages, despite recent wage increases, and its discriminatory treatment of workers. Similarly, Amazon has faced a wall of criticism about the working conditions in their warehouses (Geissler 2018; Bloodworth 2018a, b; Cain 2018, Sainato 2018). Further, Walmart and Amazon's treatment of suppliers does not at all accord with the more nurturing nature of supplier relations promoted by lean principles. Meanwhile, Costco is ranked as the best company to work for two years in a row (Calfas 2018). The contrast between these two approaches to lean systems represented by Walmart and Costco in comparison to Costco could not be greater. Clearly, large merchandisers have a choice in how to organize and treat their employees, and there are many fewer who take the high road concerning employees and suppliers. Concerning off-shore production, less is known concerning how these firms deal with sweatshops. But this is a problematic area for most merchandisers who nominally claim that they simply purchase products and do not produce them. Under pressure each company has increased their social awareness of "externalities" of their supplier purchases, but progress has been slow in improving the lot of off-shore production.

In sum, Costco has demonstrated much higher wages, benefits, and employee satisfaction, and a stronger focus on quality and customer service than Walmart, and to some extent than Amazon. It has consequently reduced turnover in its employees, thus saving training costs and improving the morale of its organization. In terms of efficiency, this has resulted in much higher revenue per employee at Costco—double Amazon and over a thousand times Walmart. However, we have not been able to ascertain how different these three organizations are concerning their treatment of suppliers. Nonetheless, Costco's model toward employment and organization was profiled in President Barack Obama's 2014 State of the Union speech (Carré and Tilly 2017:84; Ton 2014). This does not make Costco perfect and to some degree, retail is retail with comparatively lower wages. Nonetheless, Costco demonstrates an effective implementation of lean principles though it and the other two companies could be much better in the implementation of teamwork and monitoring supplier's abuses of workers in third world countries.

Notes

1. In 2007 after 45 years, Walmart introduced a new slogan: "Save Money Live Better".
2. Toyota uses temporary employees but they do so mainly to buffer permanent employees, not specifically to reduce costs in the short-term. Also, hiring temporary workers

performs a preview function for hiring new workers (Lepadatu and Janoski 2011, 2018).

3. The German grocer and merchandiser ALDI has also been criticized for its treatment of employees (Winkler 2016).
4. Nominal CEO Jeff Bezos's salary is only a million and a half, but this may be due to the fact that he is already the richest man in the world with a net worth of almost 150 billion dollars.
5. Massive changes are occurring in JIT inventory and supply chains. According to an *Economist* special report, "the golden age of globalization" is over (2019) for five reasons: (1) Wages are rising in the BRICS, while the costs of manufacturing are declining in the West, (2) Trade wars are affecting "hyper-globalized supply chains" (3) Technology is changing at an accelerating pace—3D printing, drone delivery, robots in warehouses and driver-less vehicles in warehouses, block chain technology, and artificial intelligence accurately predicting supply rather than actual feedback from production—causing delivery times to drop precipitously; (4) Multinational corporations are backing away from China for many reasons and moving towards regional rather than global supply chains; and (5) Security concerns are growing (e.g., from competitor, independent and political hackers), and they are increasingly coming from suppliers (e.g., the Capital One hack by an Amazon web employee in 2019) (see also, Lund et al. 2019). Walmart and Amazon are leading these efforts that shorten delivery times through new technology in JIT inventory, and in the regionalization and revolutionizing of supply chains. However, we are in the midst of these changes at the moment and do not really know the final results.
6. One generally searches in vain for references to teamwork in books on Walmart and Amazon (Galloway 2017; Lichtenstein 2006; Fishman 2006) and Amazon (Stone 2013). One must differentiate a lean production team (i.e., a problem solving quality control circle) from a functional group of people who do a particular task. In the latter case, "cooperating to get a job done" is not the same as regular team meetings focused on continuous improvement with statistical charts, fishbone diagrams, and Pareto charts. However, at higher levels in these corporations there often are problem solving teams (not production teams) that work together. Jeff Bezos has a "two pizza rule" that limits to about 6 to 8 people (Schmidt and Rosenberg 2014:46–47). But again, the "two pizza rule" is more for professional problem solving groups than the more numerous locators on the warehouse floor.
7. Wages in retail are higher with more benefits in much of Western Europe. Carré and Tilly (2017:114–48) show that this is due to a number of factors: collective bargaining, national regulations, more advanced vocational education, and US corporate strategy. This chapter on lean production notes these differences but does not explore them further.
8. Corporations that belong to the FLA are Nike, Adidas, Puma, Under Armor, Patagonia, Prana, Hanes, Fruit of the Loom, Barnes and Noble, Cutter and Buck, Gore Inc., Hugo Boss, and Nestle. But also note that Apple does not belong as they go it alone on their monitoring.

Part III

Syntheses and Conclusions

10 Comparing and Synthesizing Lean Models and Exploring Lean Politics

We presented two approaches to lean production in framing and managing lean production. In Part I we looked at how six disciplines—management, industrial engineering, sociology, labor process theory, labor and employment relations, HR—and three country models—diversified quality production from Germany, productive models approach from France, and State-led capitalism from China—frame lean production. In Part II, we examined how 11 corporations—Toyota, Honda, Ford, Nissan, McDonald's, Nike, Apple, Google, Walmart, Costco and Amazon—manage lean production. In this chapter, we will reverse the order of presentation and look at the corporations first and then examine the disciplines. Using the principles of lean production for the corporations managing organizations, we synthesize the 11 organizations into four models—Toyotism, Nikeification, Waltonism, and diversified quality production. Each applies lean production to various percentages of the full model. Second, we compare the disciplines but rather than synthesizing, we find each one of them partial and recommend that each one can take a broader view and include some of what the other disciplines do in their own work. And third, we look at the political implications of lean production in terms of national and global politics, which is something associated with some of these models and not so much with others.

Synthesizing the Eleven Corporations

It is helpful to systematically map out these 11 organizations. In Table 10.1 we use five variables of lean production to summarize a number of aspects of each model, and then provide an approximate score to compare them.

The first variable—a long-term and inclusive view of employees that ultimately builds a stronger society and community—is often essential for the long-term benefit of the country involved. Second, elaborate JIT inventory systems consist of complex and well-organized supply-chain supplier systems. The third variable is continuous improvement and strong flexibility in the production labor force. This is the opposite of the Fordist principle of detailed job descriptions, and most workers engage in cross-training and job

Table 10.1 Comparing the Eight Models

Division of Labor Model	*(1) Long-term Goals*	*(2) Elaborate JIT*	*(3) Continuous improvement/ Flexibility*	*(4) Strong Teams*	*(5) Loyalty to employees/ Off-shoring*	*Total Score*
1-Post-Fordism	No	Yes		No	No (not specified)	40%
2-Flexible Accumulation	No	Yes	Yes	Yes/No	Yes/No	60%
3-McDonald-ization	No	Yes	No	No	No	20%
4-Siliconism	No	Yes	Yes	Yes/No	Yes/No	60%
5-Shareholder Value	No	No (not specified)	(not specified)	No	Yes/No	15%
6-Lean Production One or Toyotism	Yes		Yes	Yes	Yes	100%
7-Nikeification	No	Yes	Yes	Yes/No	Yes/No	60%
8-Waltonism	No	Yes	Yes	No	Yes/No	50%
9-Additive Production	Maybe	Yes	Yes	Maybe	No/No	60%

rotation. Fourth, we look at the existence of strong team work systems that are actually productive of better products and loyal workforces. If the model describes teamwork but sees it as counter-productive we give a half point total. The most complicated variable is the fifth one. The complete absence of outsourcing is the Fordist model much like Henry Ford's conception of the Ford Rouge Plant that actually made its own steel. In other words, it was vertically integrated and avoided a supply chain as much as possible (albeit, they did not own iron ore mines though they did own ships to transport the ore). But we are not in that Fordist world. In column 1, outsourcing within a country is a positive feature of this model, and so is off-shoring for production to a foreign market. However, if off-shoring is fully intended for a home market (e.g., as with Apple or Nike), it is a negative factor because it contradicts the inclusive and long-term nature of the previous variable.

In the total score column in Table 10.1, each of the eight models of the division of labor could reach a score of 100% (see Table 10.1).

If strong flexibility is present (item 1), the model gets 20 points, and so on for elaborate JIT (item 2), inclusive and long-term employment (item 4). However, two variables are more complicated. With strong teams (item 3), a model could have strong teams in its domestic market but no teamwork at all in its foreign production facilities. This is the case with Apple, Nike, GE and other companies. As a result, it would only get half the points. The same applies to item 5. A firm outsources, which gives it 10 points, and if it off-shores for consumers in the country in which it is located, it gets another 10 points. However, if it off-shores for all of its production it loses 10 points. Thus, Toyotism receives a full score since it outsources within the United States and Japan, but it off-shores production to China and the US mainly to serve a local market. This is opposed to Nike and Apple, who off-shore nearly all of their production. Some aspects of each model are not specified in the theory, but they are often clearly implied, and we put "implied" in parentheses to indicate this. If they are totally absent, we indicate "not specified".

The results in the total score show that McDonaldization is really Fordism, it receives a low score of 20%. Similarly, shareholder value theory scores only 10% but for a different reason. It is not really a theory of production, so it contains no mention of a number of variables. Nonetheless, it often has strong implications for production firms. Post-Fordism is somewhat low at 40% but that is because some aspects of its theory were not totally filled out (especially outsourcing/off-shoring). Four theories are in the range of 50–60%, and they are flexible accumulation at 60%, Nikeification at 60%, Siliconism at 60%, and Waltonism at 50%. These four models tend to lack strong teams and long-term philosophy that includes job security. Lean production based on Toyotism comes in at a 100%. This is not a precise measurement model, but does give us an idea of the strengths and weaknesses of each one of these eight models.

In our choice of which models to synthesize, lean production one or Toyotism was included because of its high score. Next, we eliminated siliconism because although it is a very good place to work if you are in the home country (often the US), it relies on extensive off-shoring where people most often have a very bad place to work. Also, siliconism is simply too small in its coverage of the labor force since it mainly applies to Silicon Valley and a few other high tech locations. Next we eliminated McDonaldization and Shareholder value theories for perhaps obvious reasons of their very low scores. This leaves us with the choice of flexible accumulation (a somewhat better choice than post-Fordism), Nikefication, and Waltonism.

Flexible accumulation directly contradicted Toyotism on teams, so it was like a failed true lean model. Or to put it another way, it is like Nikeification but with production onshore rather than off. As a result, we did not include flexible accumulation because the other models covered most of the specific and important variations of it.

While Toyotism is the strongest model, the Nikeification and Waltonism models are weaker forms of lean production. But the later models represent two industries—high technology production and merchandising—that are of tremendous importance to the retail economy. The 100% score for lean production does not indicate that it is perfect. As previously discussed, there are some strong criticisms of Toyotism with their use of temporary employees who work in their US factories (i.e., workers who are paid less and receive fewer benefits), and in their intense work processes and extensive use of overtime (Lepadatu and Janoski 2011). And to some degree we had to wince a bit in giving ten points for outsourcing. Nonetheless, we view their approach to teamwork as highly beneficial and not a charade as flexible accumulation indicates. Thus, their long-term view toward employment (with of course the exception of temporary workers) and their approach to teamwork makes the Toyotism version of lean production a strong and desirable aspect of the present-day division of labor. Nikeification and Waltonism embody the hard aspects of lean production (JIT and flexibility) but not the soft aspects (teamwork and long-term employment).

Comparing the Five Disciplines and Corporate Models

Next, we look at the disciplines. In Table 10.2, we look at the various disciplines that address lean production concerning the same five points as the corporations about lean production.

Concerning management and HR in column 1, we find that strategic management approaches are somewhat involved with lean production, but many times it is secondary to other concerns like mergers. We score both at strategic management at 80% lean and HR at 90%. Teamwork and long-term goals tend to give way to larger strategic issues. However, shareholder value is largely agnostic about lean production, as it largely ignores lean issues and focuses on short-term goals of profitability for the next stock market reports. It is scored at 40%, which is less than half. Industrial engineering is one of the strongest disciplines connected to lean production, and lean has an honored place in its curriculum. However, it is not particular focused on teamwork (i.e., engineering is not focused on groups of people interacting with each other), and not that strong on loyalty to employees and off-shoring. This leads to a score of 80%. But again, industrial engineering is central to the academic promulgation of lean production. Sociology is rather mixed in its view of lean production (Row 3). The flexible accumulation theory scores about 50% and generally has a dim view of team work as worker cooptation or simply the failure of teams. The Cole and Lepadatu approaches are more favorable toward lean production with 100% and 90% implementation (Row 3, b, and c). Similar to this is labor process theory that largely rejects most of lean production other than JIT inventory and long-term goals. In Row 4, it receives a score of 45%. Labor and employment relations is a strong advocate of lean production, with only a small issue concerning off-shoring and a persistent interest in unionization. It scores 90% in Row 5. The next two schools are mixed. The Diversified Quality Production approach (DQP) is clearly a German model that has expanded to the European Union (Row 5).

Table 10.2 Comparing Disciplinary Perspectives with Lean Theory

Division of Labor Model	*(1) Long-term Goals*	*(2) Elaborate JIT*	*(3) Continuous improvement/ Flexibility*	*(4) Strong Teams*	*(5) Loyalty to employees/ Off-shoring*	*Total Score*
1-Management						
a-Strategic Management	Maybe	Yes	Yes	Maybe	Yes	80%
b-Shareholder value	No	Yes	Yes	No	No	40%
c-HR	Yes	Yes	Yes	Yes	Maybe	90%
2-Industrial Engineering	Yes	Yes	Yes	Maybe	Yes/No	80%
3-Sociology						
a-Vallas/Vidal	Yes	Yes	Maybe	No	No	45%
b-Cole	Yes	Yes	Yes	Yes	Yes	100%
c-Lepadatu/ Janoski	Yes/No (temps)	Yes	Yes	Yes	Yes	90%
4-Labor Process Theory	No	Yes	Yes	No	No	20%
5-Labor and Employment Relations	Yes	Yes	Yes	Yes	Maybe	90%
6-Diversified Quality Production (DQP)						
a-VW	Yes	Yes	Yes	Yes	Maybe	90%
b-BMW	Yes	Yes	Yes	Yes	Maybe	90%
7-Productive Models						
a-Toyota	Yes	Yes	Yes	Yes	Yes	100%
b-Honda	Yes	Yes	Yes	Yes/ Maybe	Yes	95%

It tends to have many of the lean characteristics, how its codetermination and works councils are somewhat weak on loyalty to employees. Finally, the French productive model approach sees lean production (Row 6) as particularly strong, and while it does not advocate lean production, its description of the Toyota and Honduran systems is a near perfect example of lean production. In sum, the Management (minus shareholder value), Industrial engineering, the sociology views of Cole and Janoski and Lepadatu, labor and employment relations, and Productive models have strong views of lean production (80 to 100%). Labor process theory and shareholder value theories are much lower due to vary being critical in the first instance, and ignoring lean production in the second.

There are also differences in these disciplines concerning the amount of determinism in their theories concerning lean production, and the amount of exploitation of the workforce that may be involved. No theoretical approach considers these factors equally. In Table 10.3 we do a simple two by two by three table that illustrates some of these differences.

Labor process theory sees lean production as involving a large amount of worker exploitation though it may differ between companies, industries, and regions (cell 1 in Table 10.3). Coupled with this exploitation is a view that the lean production version of the division of labor is coupled with an economic developmental process that is highly deterministic. Flexible accumulation is close to labor process theory but not quite as wedded to Marxist theory. On the other hand, industrial engineering and management theories see much of lean production to be the full force of the modern development of industry, but they tend to overlook or interpret its effects on workers as being beneficial for the most part (cell 3). In fact, they do not concern themselves too much with the impact of lean production on the workforce and see lean as leading to higher paying and more stable jobs because firms will be more profitable. They do not actually study the impact of lean production on workers very much. Social science approaches see the adoption of lean production to be less deterministic and much more contingent based on industry, national culture, and organizational approaches (cell 4). On this last point the regulation school or "productive models" approach sees significant differences even between Honda and Toyota. But like labor process theory, they view lean production as being rather exploitative of workers and employees and largely benefiting top level managers and stockholders. Deming and another approach to industrial engineering and sometimes management has a focus on "driving out fear", and fear has a strong connection to exploitation. This idea is not entirely developed by Deming, but it largely involves employees feeling secure enough (i.e., not feeling exploited) that they will open up to the new ideas of others, and also develop their own new ideas so that the *kaizen* process can develop fully. A management approach to open or semi-autonomous teams needs this buy-in by workers and employees. Management theories will focus on

Table 10.3 The Relationship of Determinism and Exploitation in Lean Models

	Worker Exploitation	*Mixture exploitation and not*	*No Exploitation*
Deterministic theory	**(1)** Labor Process theory, Flexible accumulation	**(2)** Cole	**(3)** Industrial engineering, Management$_1$ (Womack et al.)
Contingent, or non-deterministic theory	**(4)** Regulation Theory, Productive Models	**(5)** LERA, DQP, Lepadatu and Janoski	**(6)** Deming, Juran, Management$_2$

how to make teamwork much better and how the failure to implement teams may create problems, however, this comes close to being concerned about employee welfare and ideas, but it does not quite take the viewpoint of the worker through a process of empathy or *verstehen*. This is where the emphasis on Toyota principles of "long-term focus, loyal employees, and lean" must come together with a clear focus on what environments that employees will thrive within.

While classification systems try to derive clear principles and relationships, they inevitably leave some theories in a mixed approach. Robert Cole is one that could fit in different categories but really belongs in his own. Using a sociological approach that is even more like anthropology at times, he sees the structural encasement of workers in lean production, but at the same time, sees the freedoms that workers have to come up with new ideas. He praises lean production and is critical at the same time. For instance, there is life-time employment but management can make life difficult for employees so that they quite (mainly in third tier suppliers) or they get a window seat where they do not interfere with everyone else (mainly OEM firms). This allows a view that often sees Japanese firms as having too many employees and not at all lean (a point picked up by management theorists Osono et al. 2008). In another view of industrial relations (LERA), German approaches (DQP), and Lepadatu and Janoski (2011), lean production is generally quite successful but its use of temporary employees is quite exploitative. Other factors in lean production are not, especially teamwork processes, but the widespread use of temporary employees as buffers to a more permanent workforce at lower pay, no job security, and general uncertainty is not what Lean-Loyal-Long-term (L-L-L) approach promises. They are clearly a contradiction. Further, the DQP approach from the German angle provides a model that is more participative and less exploitative but their quality ratings for automobiles, despite their reputations, do not reach the high level of Japanese production techniques—a point often ignored by their theoretical enthusiasts.[1]

With regard to the disciplines, we do not synthesize a grand discipline since that would be somewhat impossible. Instead, we recommend that each one of the disciplines recognize the broad range of scholarship on lean production that exists in the world, and include some of it in their work. Much of this scholarship has been like the parable of the "Elephant and the Blind Men" only recognizing the small sections of the animal that they actually touch. Also, each discipline needs to include lean production as a more central part of their disciplines. Industrial engineering is the broadest of the disciplines with regards to lean production. Management needs to make lean production a more central part of its theories of strategic management. Sociology has to look much more at teamwork and how it is organized throughout organizations using lean production. HR needs to look beyond the scope of HR jobs in the corporate leadership structure, and Labor and Employment Relations has to develop larger theory for a non-union environment. The DQP model presents a lot to think about concerning the design of lean production organizations, especially concerning employee participation. And then there are two models antithetical to lean production—shareholder value theory, and state-led capitalism. We think these opposite approaches represent the least desirable aspects of capitalism and socialism.

Consequently, we present four organizational varieties of managing lean production because they have different involvement with teams, off-shoring and merchandising. In fact, teamwork is usually the first item that drops out of the lean model (e.g., Nissan has most of lean production but not strong teams). Further, the off-shore production facilities of Apple and others in China do not use teams. And as lean production moves into the service industries (e.g., the VA hospital in Pittsburgh in Grunden 2008), teamwork is difficult to implement among hierarchical professions (doctors, nurses, and aides). Nonetheless, many aspects of lean production included some teams are being used in hospitals with great strides in quality control (Toussaint forthcoming; Graban 2009; Chalice 2008). Thus, the new global division of labor consists of lean production with Toyotism being the full model with all features listed above, Nikeification being a partial production model that drops teamwork out in its off-shored but not on-shored plants, and Waltonism being the merchandising model that exemplifies the supply-chain aspects of JIT, but also not the teamwork aspects of the model.

The Political Component of the Division of Labor

We further extend lean production into political-economy, which involves five relatively different political forces. First, the Toyotism model favors employment security and long-term relationships. Nikeification has largely severed most of its production facilities from US territories, so they are in favor of continuing the off-shoring process and are largely against tariffs and any impediments to the free trade across the Pacific Ocean that benefits them. Waltonism has only a minor interest in job security and long-term employment of its workers. Since it pays such low wages, it is unlikely that they would want any legislation on this issue. But unlike the eventual Fordist political model with the welfare state, none of these lean production models wants effective government intervention.

Second, on the issue of trade unions, all three models of lean production favor the weakening of American and other country's trade unions. At present Nikeification industries (e.g., Nike, Apple, Visio, and other electronics firms) have little interest in unions since their employees are not subject to the NLRB and unionization rules. Toyota is not interested in unionization and strongly repels unionization attempts, but in actually, these have been few and far between. For them, this is largely a non-issue. However, Waltonism (e.g., Walmart, Target, K-Mart) has a large number of blue collar employees who could be unionized. In large part, they strenuously fight against unions and both Walmart and Target have had unionization drives that they have so far largely repelled. K-Mart has four distribution centers that are unionized (two UAW, one Unite, and one Teamster unions). In large part, Waltonism is the most vulnerable to unionization and offers the strongest resistance to it. The decline of unionization among autoworkers had contributed to a large extent to decline in membership of one of the most powerful unions in the country. Meanwhile the unionization rate has declined from a high of 35% after World War to only 10.5% of employees, and they are mostly in government employ (BLS 2019). The emphasis of lean production with subcontracting, outsourcing, off-shoring, and contingent employment makes organizing unions quite difficult.[2]

Third, concerning corporate taxes, Toyota, Honda, Walmart, and Target generally pay their corporate taxes. However, Nike, Apple and Amazon are long-term corporate tax avoiders holding large amounts of profits overseas. Steven Jobs in a meeting with President Obama stated that he would only bring Apple's manufacturing jobs back to the US if the president guaranteed the education of 30,000 more engineers and that Apple could bring in its billions of dollars in overseas profits into the country tax free. Amazon and other internet businesses have long avoided state sales taxes, which could pay for education, but some states have gotten them to pay sales taxes based on Amazon having warehouses within state boundaries. Since 2017, Amazon and other companies will finally pay sales tax on their transactions in all the states of America (Statt 2017).

Fourth, none of these three models has much interest in supporting the welfare state since taxes have declined for government services, especially for the welfare benefits. Lean has been clearly applied to government under conservative terms. Generally, business supports the view that a centralized welfare state can no longer meet the needs of the country, and each corporation makes some donations to voluntary associations and engages in some philanthropy (Clarke 1990:73–74). Since some Walmart, McDonald's, and other merchandizing employees rely on government benefits, it would seem that these corporations might support the welfare of their own workforce. However, this is a sore point that they would rather avoid. There is declining support for higher education from individual states" tax cuts. And each of these types of corporations makes some contributions to higher education though the amounts are small compared to the need.

Fifth, in terms of creating inequality, there is a strange dialectic between these three models. Nikefication (e.g., Apple, Google, Intel, etc.) promotes high wage earnings in the United States, and Waltonism creates a massive amount of very low wage earners. As both Nikefication and Waltonism lead to off-shoring, many former manufacturing and other goods producing workers have been unemployed and then have taken jobs in the lower wage sector, which is largely where Walmart, K-Mart, and McDonald's operate. The result is greater and greater inequality in American society, which has seen its Gini coefficient of income inequality go from the middle of the pack (the mid-.300s) to the most unequal society in the industrialized world (near .420) (US Census Bureau 2013). In essence, the middle class is disappearing. On the other hand, the Toyota model leads to many more middle-class jobs with fairly good salaries. Thus, it mitigates against the move toward inequality. But nonetheless, the forces of off-shoring and low wage lean production models have led to the largest amount of inequality in the United States since figures were first collected.

These points about political influence leave an unsettled picture (Lewin and Kim 2004). Is the larger picture of political Toyotism or lean production inherently conservative, anti-union, and prone to inequality? The Fordist model led to social democracy and strong unions, but the lean production model looks like it promotes neo-liberalism, inequality, and disciplined global workers. Taylor thought that his system would increase wages, but employers found a way to redo the piecework system to create a work speed up without wage increase. In a similar way, Toyotism may be reworked to leave out teamwork and decent wages *al la* Walmart or ship production overseas to Fordist plants *al la* Nike and Apple. It appears to be like a globalized corporatist welfare system for a tiny portion of the workforce and an unprotected privatized system reliant on charity and voluntary associations for the largest number of citizens. The large secondary labor market is then left relatively unprotected. Nonetheless, one must recognize that Toyotism in the form of Japanese, Korean and even German auto transplants are investing in American jobs with new manufacturing plants with good wages (Lepadatu and Janoski 2011; Perruci 1994). Thus, in the end, job security appears to exist for some, but not for most. This is a macro-politics much more friendly to conservative than more liberal political forces. And while Fordist firms did not campaign for greater equality, their emphasis on higher wages and the unionization that they finally accepted brought a more equitable society. Lean production does not seem to do that.

Politics, Economics, and the Japanese Economy

George Ritzer and a few others argue the Japanese production methods have been discredited due to the lost decade from 2001 to 2010. After the dot.com bubble and 9/11, the Japanese economy has not returned to its pre-crash level. Much of this has to do with the strongest Japanese companies competing with a rising Korean economy in automobiles and phones (e.g., Hyundai, Kia, and Samsung). It took 12 years for the Japanese economy to regain ground to its 1995 standing, and then the Great Recession hit. The country experienced chronic deflation and low growth and has one of the highest national debt loads most of which is held domestically. The argument is that this discredits lean production.

However, the consensus among economic analysts is that the lost-decade and subsequent problems is due to three main issues that have little to do with lean production. First, Japanese monetary policy could not lower interest rates because they were nearly zero to begin with. Firms largely relied on bank loans to accumulate capital and many firms were able to gain loans that would not be made in other countries. This was referred to as crony capitalism and it is largely connected to the *Kieretsu* system of strong networks of somewhat unrelated firms. These firms could not sustain growth and in fact faced many losses leading to the failure of many loans. With monetary policy ineffective firms looked to the government for economic stimulus. But monetary policy remained tight, and neo-liberal cuts led to more malaise. There are a number of competing explanations for the lost decade; however, nearly all of them are macro-economic explanations not connected to the firm's use of lean production techniques (Aramaki 2018; Hamada et al. 2010, Auslin 2017; Wakatabe 2015).

Second, the Japanese population is one of the oldest in the world because the birth rate is very low. At the same time, Japan is one of the most restricted countries in the advanced industrialized world concerning immigration, naturalization, and accepting asylum-seekers (Janoski 2010). This has a two-fold effect. On the one hand, the increasing supply of population begets its own economic demands, which lead to greater economic growth. As the population of Japan relative to other countries shrinks, demand is lessened. On the other hand, there is a labor shortage for Japanese companies. While Japan has mentioned that this will be handled by increasing robotics, there has been a tendency for companies to move where labor may be more available. So, immigration policy is not helping the Japanese economy; however, these issues are also not directly related to lean production.

Third, major Japanese companies have off-shored a large number of plants for quite different reasons. The auto quotas imposed on Japanese imports starting in the 1980s in the United States led Japanese auto firms to invest in new plants in America. This has led to an underinvestment in plants in Japan. Also, the opening up of China has led to new Japanese auto plants being built there due to Chinese restrictions on trade. The Japanese have largely adopted a "build it near the market where it is sold" policy, which has a number of positive features. In either case, production in Japan is not as high as it could be. These three factors have led to Japanese production techniques spreading to other countries in the world, but they have lessened production that could have occurred in Japan itself.

Evaluating the Updated Japanese Model

Much of what is discussed in lean production in Japan parallels the two German models. Japan has not stood still for 50 to 60 years since the TPS system emerged. In fact, one could argue that Japan has regressed since then (Ritzer 2004, 2019). In a sense, the early model can be called Japanese production methods or JPM-1. After 50 years, JPM-2 has emerged with similar but not the same weaknesses as the German model (Haak and Pudelko 2005; Herstaat et al. 2006). First, there has been considerable weakening of permanent employment and job security in general. Even Toyota has increased its hiring of temporary workers in Japan and is paying them less (Adams 2010). Second, Japan's birth rate has declined precipitously and its population is aging. However, the Japanese nation-state has not opened the doors to immigration. To some degree a great amount of off-shoring has occurred to the Western countries and China. Some of this was due to import quotas and domestic content laws imposed on Japanese cars by the US government. However, this does not alleviate the low supply of labor in Japan. Third, the *Keiretsu* system has declined extensively so the tight knit horizontal networks of different firms helping each other in time of need has clearly gone down. Consequently, trust within the firm with less job security and trust between firms in *keiretsu's* is no longer part of the Japanese model. Fourth, and perhaps most importantly, the Japanese model has been in a long-term period of reduced economic growth. One might ask, if the lean production system is so powerful, why isn't it working better in Japan?

The Global and World Systems Dimensions

The global structure of the economy is closely related to the political power of the leading nations. The division of labor we have discussed so far in this book has been focused on the production process and how things are made or services delivered. World systems theory has powerful (core) nations that largely determine the politics and the economies of semi-peripheral (developed but often smaller countries) and less-developed or much weaker (peripheral) countries, which can be very large in terms of land areas and population. World systems theory focuses specifically on the world capitalist system and indicates that the core countries largely control most of the wealth and economic development (Wallerstein 1974).

World economic development is unduly influenced by the core countries and this is specifically done through controls over investment, but more specifically through commodity or value chains (Arrighi 2007; Arrighi and Moore 2001, Arrighi and Silver 1999; Gereffi 1994; 2005; Gereffi et al. 2005). Value chains largely involve outsourcing and off-shoring arrangements which might be looked at as an inverted telescope. The core countries control the nature of the distribution of income from the product being produced, so the largest amounts of money go to the large barrel of the telescope, and then each extension of the smaller segments of the telescope get smaller and smaller portions of the profits. As a result, the core country corporations pay their employees and stockholders well, and the pay and benefits get smaller and smaller the closer the value chain is in the less-developed country. Wallerstein (1974) developed world systems theory to discover why poor countries tended to stay poor. The process is not unlike the supply-chain concept of lean production. The original equipment producer (e.g., Toyota) gains the higher portion of the profits and wages, the first tier supplier gets a bit less, and the second and third tier suppliers get even less. When this is done within a country, the wage differentials are large but not too great. When this is done between countries, the wage differentials become astronomic and off-shore production reveals incredibly low wages.

For instance, the average age of Apple product designers and Apple store sales persons is thirty-nine and they typically receive five times greater pay than the 15- to 30-year-old (average) village girls who assemble iPods and iPhones in the industrial zones of Shenzen, China (Nicas 2019; Glassdoor 2019). Before the 2019 increase, the lowest Apple store employees were 8 times the level of Foxconn assemblers (*Economist* Editors 2012). This is the basic logic of Nikeification but a smaller version of it is already in lean production. Waltonism takes Nikeification and spreads the last elements of the supply chain or value chain as far as they can. In other words, the making of iron skillets is not particularly technologically advanced and it is really just the last cylinder of the inverted telescope.[3]

China attempts to control the value chain process by insisting that technology transfer be part of the production deal within its borders. Large corporations from the core countries generally have no incentive to engage in technology transfer; in fact, it is largely diametrically opposed to their interests in a capitalist world system economy. It does not contravene value chains directly, but it represents a gradual leakage in maintaining strong leverage over wages and profits. Over time, workers gain the skills to do better jobs, college trained engineers become more and more adept in doing what it takes to produce these products, and then Chinese scientists and engineers can develop their own competing products. This is largely the last thing a core country corporation wants to see—serious competition from a less-developed country. So why did they agree to these joint ventures with technology transfer? There are three reasons. First, in my opinion the Chinese market will likely be the dominant market of the next century. If you are not in that market, you will be passed by those who are in the market. Second, I believe the Chinese government has enough military might and political power to enforce this deal on core corporations. This factor does not need a great deal of saber rattling, but it is there. And third, these corporations will get considerable profit in the short run from the incredible wage gap with Chinese production workers. In addition, China is a society with a strong work and education ethic. As a result, the probability of being successful is rather high. So, in some ways, this is a tradeoff of short-term (one or two decades) versus long-term (three to four decades). So why would Boeing give the Chinese air transport manufacturing industry the plans for producing their latest jet? First, they are in the Chinese market selling jets already. Second, they recognize the importance of China economically and politically. Third, they will get profits now and they are banking on their technology being obsolete by the time the Chinese are able to use it.[4]

Other countries in the periphery will not have the advantages that China has. For instance, I don't think that Vietnam or Indonesia will become the dominant markets of the next century. Neither one has the political or economic power to be a major player on the world economic stage. And corporations may make short-term profits in either one of these countries but if too much is demanded, corporations will be more than willing to move to a new location. The Middle Eastern countries are in a similar situation although they were able to use their collective political power to achieve a major redistribution of capital in their direction (i.e., via OPEC). However, they do not have a large internal market and their most important resource (i.e., oil) is being diluted by new discoveries in the United States and Canada, and the threat of wind and solar power. Further, the work and educational ethic in the Middle East and Indonesia is not up to the level of China. To finish this argument that the other countries do not have the economic and political power of China, consider whether any of these countries could ever make the major foray into Africa that China has recently engineered. It is largely inconceivable that any other less-developed countries other than China could do this.

As a result, the core powers and the major ascending power of China will likely control the direction of foreign direct investment in the next century. In my opinion they will manage the levers of the World Bank, the International Monetary Fund, and various United Nations peace-keeping and economic development efforts. But as one can see, this discussion of world systems theory (sometimes called world hegemony theory by political scientists) ventures into the area of international politics and global investment, and gets a bit away from how production is actually organized on the shop floor. It presents a big picture, but this view is so big that it knocks the workers and managers off the shop floor in how they are actually organizing work. In fact, it can operate just as well with a Fordist or Taylorist division of labor as it can with the newer approaches that we have discussed here.

The Possibly Emerging Localization of Labor

The emergence of additive technology has the potential to create a new and almost anarchistic structure of production based on local technological parks with 3D printers producing small batches of incredibly diverse products. For instance, stores could have 3D printers to produce items on demand for customers. That seems to be the immediate vision of this new technology. And it is highly persuasive. However, it is not easy to project a division of labor on an only emerging development. There are other scenarios.

So, if we are forecasting on only a few initial points, there are other scenarios that might develop. First, the development of the best 3D printers might need a very involved and detailed technology that could require hiring highly paid engineers and scientists. Hence an organization not unlike Apple could emerge that would be highly secretive about its technology and thus raise the price of high quality 3D printers to a large sum. Then corporations with large resources would be the only ones who could afford the technology, and indeed over time, might specialize the technology for their particular products. The shape of JIT production might change in that the 3D printer suppliers might be owned by the OEM, or the suppliers would need to be very close to the final product assembler. In either case, the vision of many local producers would be replaced by large corporations.

Second, what if the localization of production could be countered by the centralization of the raw materials? In this case it would be the ultra-violet curing resins that are the raw material from which additive production works. For instance, in the previous two centuries, back yards or communities were not peppered by basic oxygen process or open hearth furnaces to make steel. Instead, steel was produced by very large corporations. The only known example of community steel production was tried by China in the 1960s "great leap forward" and this was such a failure that it helped create a major famine that killed 25 to 40 million people. Currently, these resins are polyester, epoxy, and vinyl resins with glass components. But future resins or molecular raw materials might be of such a technical grade that they can only be produced by large companies with massive resources. For instance, communities having the access to rare earths to produce their own smart phones is rather inconceivable. One might need a molecular decompiler to use for input into a recompiling 3-D printer. The main point at this initial beginning of additive technology, is that it is extremely hazardous to predict the global implications of this new technology. But it promises ample speculation for the next decade on its development.

So, we end this chapter with the convergence of three models of lean production—Toyotism, Nikeification and Waltonism—within a global context of the world capitalist system, and the speculative divergences of the additive model of possible future production. And the fourth model of DQP still exists and has great influence in Europe and has been spread internationally by VW and Audi (and secondarily by Daimler Benz, BMW and other German firms). Thus, lean production in its three forms is the most accurate model of the current division of labor. It is specific enough to describe the basic operational principles of the system (i.e., not just flexibility), it has enough flexibility to show variations from Nike to Toyota to Walmart, and it describes the current conundrum that labor faces.

Notes

1. Sometimes this is puzzling. German automobiles have a reputation for quality despite the fluctuations from high to average in their quality levels in *Consumer Reports* and J. D. Power assessments. This may be due to two factors: (1) German automobiles may be built more solidly, especially since they often cost much more and sometimes have the reputation for being over-engineered, and (2) German cars are often luxury vehicles that have a high status, which then affects buyers' perceptions of quality.
2. Recent events show that the NLRB has ruled against Target for unfair labor practices, and that Walmart is fighting unionization drives (Jamieson 2012). Meanwhile, Republican filibusters prevented the appointment of replacement NLRB commissioners until recently. McDonald's is facing demonstrations about low pay and not providing a living wage in Chicago, Detroit, Milwaukee, and Baltimore. And Washington, DC may rule that Walmart cannot build a store there unless they pay a living wage (DeBonis 2013). In a comparison to 250 top corporations, Cutter shows that McDonald's has a cause for concern involving its "employee engagement and development" (2019: R6). This does not mean that we are on the verge of organizing this massive low wage sector, but it clearly indicates that the Waltonism model faces labor problems and is totally antiunion.
3. These low-tech products are produced by township and village enterprises (TVEs) rather than transnational joint ventures or state-owned enterprises.
4. President Donald Trump has recently told US corporations to leave China, citing the International Emergency Economic and Powers Act of 1977 (Baker and Bradshear 2019). While many doubt his ability to enforce this, many corporations are leaving due to increasing wages, but others will have great difficulty exiting China.

11 Conclusion

What the New Divisions of Labor Mean

Clearly, lean production with its more flexible brand of social organization has assumed the mantle of the dominant form of the division of labor from Fordism. But in our view, it has a variety of forms in the workplace with strong teams declining to none at all, and in politics it has reinforced neo-liberalism and state budget-cutting. Whether one agrees with it all, half of it, or a quarter or so, it is a force in our present economy and society. We have made the case that lean production should be considered the newer model of the division of labor. Donna Samuel, Pauline Found and Sharon Williams confirm our conclusion in assessing the relevance of *The Machine that Changed the World* that in the last 25 years:

> [L]ean has touched many aspects of our everyday lives beyond how companies structured, operate, and organize themselves. Lean has influenced the way we operate our education and healthcare systems. It is clear Lean has come a long way from its shop floor origins in the best car-making companies. It continues to evolve today and to infiltrate our strategic and operational management thinking into the twenty-first century.
>
> (2015:16)

As with Fordism, we could wait until the model goes into decline and a new model takes off. But why not analyze it and pose some alternatives while the model appears to be at the peak of its power? Osono et al. (2008) show that this model is more complex than currently portrayed, with contradictions and certain dialectical character, but that is what we should expect of a successful model—being opportunistic to develop its own potentiality. Only the consultants believe that there is only "one right way" to get things done.

Summary

In Part I of this book we discussed the disciplinary imagining of lean production. We largely found that the management and industrial engineering disciplines largely focused on JIT inventory, but they were not always interested in teams, especially semi-autonomous teams. The LERA approach of industrial relations and the HR approach were especially interested in employee and worker representation and protections. The social science approach, though it was quite diverse in its evaluation of lean production generally was more interested in teamwork and small groups. The Productive models approach stressed how different societies may implement lean in different way.

In Part II of the book, we formulated our models of Toyotism, Nikefication, DQP production based on the German model, and Waltonism. The Toyotism model is the strongest model in the world, but the German DQP model has assimilated some lean production but exists quite differently with strong employee representation. We compared and contrasted these models in Chapter 10.

The Consequences of the New Division of Labor

Richard Conney asked "Is "Lean" a Universal Production System?" (2002). In sum, we answer "yes", but emphasize four varieties of lean production that are part of this new and dominant division of labor:

(a) ***Toyotism*** with the full model of lean, long-term and loyal
(b) ***Nikeification*** with off-shored production that jettisons teams but may use teams on-shore for design and marketing
(c) ***Waltonism*** with its low-wage merchandizing and hard-bargaining off-shoring supply chain mode
(d) ***DQP*** with its German model exemplified by VW and Audi

One must recognize that the Japanese, Korean, and even German auto transplants are investing in American jobs with new manufacturing plants going up each year. Much of this involves Toyotism. Thus, in the end, the labor process and job security appear by comparison to be the welcomed part of the model, but the stress and overtime are not. Further, the macro-politics of lean production seems to be friendlier to conservative and business circles. In sum, as we synthesize these three models, lean production "is" the new and largely dominant form of the division of labor. However, we note that "shareholder value capitalism", though not being a production model in-and-of-itself, is antithetical to Toyotism but more ambiguously related to Nikeification and Waltonism.

To extend this model into political-economy, we push the lean production model just as Fordism was extended into the more social and political aspects of society. Lean production and Toyotism can be seen as entering into larger contexts to make a social model of the division of labor. The following social and political points are added to six technical points discussed earlier:

- Lean production fragments trade unions by dividing them with a differentiated labor force of subcontracted, out-sourced, contingent, and permanent workers. Its Japanese transplants also favor non-union locations.

- It provides uneasy support for the welfare state with declining tax revenues for the welfare state as more conservative (i.e., "lean") governments gain office (Clarke 1990:73–74).
- Lean production requires local and regional governments to provide heavy incentives for corporations to locate in their area (i.e., exemptions from 10 years of taxes and subsidies for land, education, or other matters); but supports more permanent employment and tends not to off-shore (though it outsources) production.

The completion of these tenets may leave advocates of lean production somewhat unsettled. We are sure that they would prefer to leave these social and political points out of the lean model, but just as the Fordist model led to a strong unions and a social welfare state in the United States and social democracy in Europe, the "lean" or Toyotism model tends to link up with neo-liberalism and disciplining workers throughout society. And in fact, the metaphor of "lean" has been extensively used by neo-liberal politicians who are intent on cutting worker benefits and state regulations (but not employer subsidies).[1] Again, our innovative extension of "lean" into the world of politics may be controversial with lean advocates, but we would be remiss as social scientists if we did not mention it.

In evaluating the specifics of lean production, there are positive and negative aspects for workers and society. Ritzer, in a quick summary, says that lean production was "tarnished by the precipitous decline of Japanese industry in the 1990s" and "there are great problems with these systems and they may even serve to heighten the level of exploitation of workers" (2011:304–5). One could add the safety issues that emerged in 2009 to 2010. But we take a more balanced view than Ritzer.

Thus, we can see that the consequences of lean production in all three of its versions are rather diverse. Not all firms embrace Toyotism, not all firms are like Apple, and not all firms are merchandisers like Walmart. However, the combination of these three types of lean production cover a large swath of the production process in the world today. Nonetheless, Fordism still exists as McDonalidzation shows, and most of production in China is largely operated under Fordist principles. Apple's production through Foxconn involves no teamwork or long-term employment prospects for their young village girls. and the consequences for the partially lean models of Nikeification and Waltonism consist of a large amount of off-shored production that have led to higher structural unemployment and the shrinkage of the middle class in the United States (Janoski et al. 2014).

Contradictions of the New Division of Labor

One major problem of the new division of labor generally represented by lean production (Toyotism, Nikeification, and Waltonism) is that it is largely exploitative of labor. Toyotism is the least exploitative in that it largely promotes the skill training of its workers and pays them well, but temporary workers are much less well treated and the major complaints about Toyotism go back to these temporary workers with lower pay, fewer benefits, and little or no employment security Lepadatu and Janoski 2011, 2018; Bernier 2009). It is one thing for Toyotism to use temporary employees as a buffer to protect the firm's permanent labor force and to use as a HR department for previewing and hiring new employees. It is yet another thing for firms to declare a large majority of their employees to be temporary employees. Further, it is one thing to push your suppliers for cost reductions and quality improvements. It is a quite different thing to push your suppliers for low prices that will bankrupt them and put them out of business. Toyota itself does not exhibit these two negative outcomes, but the lean production or Toyotism approach opens the door to both processes.

Overseas with off-shoring, Nikeification and Waltonism promote large scale exploitation with little or no worries about injuries or even deaths. The fact that Foxconn had to install suicide nets around their factories in Shenzhen, China shows that they recognized a problem but Apple, when confronted by the media, resorted to the weakest response possible by funding their own NGO to investigate and make changes (Anner 2012). Further, Apple and others should recognize that they benefit from the internal *Hukou* system that imposes internal passport restrictions on these workers. This means that the workers who produce Apple's products are second-class citizens in their own country and even resemble illegal aliens in the United States due to their vulnerability. These non-resident workers do not have basic citizenship rights in the factories where they work because they cannot send their children to the local schools, they are highly restricted in the health care services they receive, and so forth. Their rights reside in the villages from which most of them come. As a result, they tend to be circular migrants limiting their work in world class factories by eventually going back to their villages, and then returning again. But in my opinion most American purchasers of Apple iPhones or Nike shoes could care less about the conditions under which these products are made. The most egregious situation involves the fires at the Tazreen Fashion factory in April 2012 that killed 112 workers who made clothes for Walmart. This was followed by the collapse of the Rana Plaza factory building that killed 1,129 people. I don't believe that these tragedies are unlikely to produce any convictions of the people directly responsible (i.e., the subcontractors) and the big corporations that sell these products may try to avoid taking responsibility (Gross 2013; Yardley 2013; Claeson 2012). Subcontracting seems to be a near total inoculation for large corporations against the clear exploitation of workers in third-world countries.

Limitations or Weaknesses of Lean Production

There are six limitations or weaknesses of this new division of labor based on lean production. First, although many Japanese firms and a few American and European firms in China use lean production methods, for the most part, Chinese firms rely on Fordism in their production process with the exception of JIT inventory. The test case for lean production is implementing strong and participative teamwork. The same criticism may apply to other Asian countries outside of Japan and Korea, and to most of Africa and Latin America. And of course, in the advanced industrialized countries like the US, Japan and Germany, there are numerous examples of Fordist, Taylorist and even craft production methods being used in various industries. And in the big electronics factories of Foxconn who supply Apple with nearly all their products, the process is clearly Fordist.

Second, lean production leads to demands on workers' skills and time that are hard to take for extended periods of time. The requirements of mandatory overtime, or socially pressured overtime lead to workers being totally encompassed by "their employer's demands". We have referred to this as "team intensification" and others have called it "management by stress" or "lean and mean production" (Juravich 1985; Babson 1995). If you consider yourself as number one in your industry, as many Toyota workers do, then you may have the motivation to continue. However, as lean production diffuses throughout the economy, lesser companies may not be able to support these demands as they face increased worker turnover. Related to this is an argument that Jerry Useem makes about the implementation of downsizing, reengineering, transformational technologies, and "minimal manning" brought to the US Navy by Secretary of Defense, Donald Rumsfeld (Useem 2019). While he does not mention lean production, the emphasis on workers/sailors learning many different skills then leads to "the end of expertise", which was the strength of bureaucratic organization. It raises the question whether a future dominated by lean production will lead to emphasizing the "generalist" over the "specialist", and whether we can afford to lose these specialists" knowledge.

Third, lean production contains a critical component of temporary workers, and some may claim that it has led to the increase in outsourcing and off-shoring. This creates an internal conflict within organizations based on the job security of permanent workers, and the lower pay, lower benefits, and tenuous job security of temporary workers. This internal conflict has a negative impact on teamwork, especially concerning problem solving and *kaizen*. But temporary workers are an important part of establishing loyalty to permanent employers and creating trust. This internal contradiction appears to be a permanent part of lean production.

Fourth, lean production makes certain demands on managers, which are difficult for many Western managers to accept. Japanese senior management accepts relatively low wages compared to their Western counterparts. Based on the information in Tables 6.1 and 7.1, American and German automobile company CEO salaries are 6 to 7 times greater than Toyota CEOs, and 23 to 28 times larger than Honda salaries (although Toyota did hire a French executive at a very high salary). The problem here is not so much the money as it is the sense of equity within the corporation, and the effect on managerial motivation concerning social mobility. The managers seeking promotions and high pay leads to having a short-term orientation that are not in lean productions stress on long-term goals. So inflating your numbers in the short-term gets you the promotion, but it comes back to haunt the firm's overall performance. It is difficult to see how Western CEOs and upwardly mobile managers can avoid these incentives and very high salaries that are contrary to the values of lean production.

Fifth, JIT inventory and SCM have been the most successful aspects of lean production. It is hard but still possible to find companies that maintain huge inventories of supplies or products. But the vast majority of large firms and many small ones use JIT principles, and as discussed earlier, Amazon and Walmart are pushing the envelope even further on deliveries. However, JIT has a major weakness in that it can be disrupted easily by strikes, worker shortages, the weather, poor quality control (e.g., Takata airbag disaster), newly imposed tariffs, and other disruptions. Even the strongest part of lean production has weaknesses.

And sixth, the implications of technology—especially in robotics, automation, additive production (3-D printing), and artificial intelligence—imply that many factories and service provision will be done much more easily or by non-humans (Locke 2014; Lipson and Kurman 2013). Factories already exist in Japan that use few if any humans (see Yamazaki Mazak factories in Japan). The most extreme position on technology is that of Ray Kurzweil (2005; see also Rifkin 1995), who uses Moore's Law on the exponential increase in the power of computer chips to predict that by 2030 and certainly by 2050 artificial intelligence will be far superior to human intelligence in not just computation but also in pure creativity. At that point in time, which he calls "singularity", human and computer brains will merge, and human labor will largely be unneeded except possibly in historical theme parks which demonstrate how work "used to be done" (perhaps like the Henry Museum and Greenfield Village in Dearborn, MI).

Final Words

Replacing Fordism, lean production, and Toyotism have become the dominant form of the new division of labor, with three models that embrace lean principles. Toyotism is the foremost model, followed by Nikeification and Waltonism. Industrial engineering is the most comprehensive approach to lean production, but it and the other disciplines involved with lean production need to broaden their disciplinary lenses to consider work being done in other fields. We are like blind researchers touching different parts of lean production (i.e., the elephant in the parable of the blind men) and not at all seeing the other important research being done on this important topic. This does not erase disagreements, but it does promote understanding and DQP shows how lean might broaden its base of worker participation in the future.

Note

1. Neo-liberals pursue tax cuts, reducing bureaucracy, cutting welfare benefits, deregulation, and a general policy of getting the government out of the lives of citizens.

References

The dates in parentheses are original publication dates, while the non-parenthesized dates are for the actual publication that we used.

Abegglen, James. 1958. *The Japanese Factory*. New York: Free Press.

———. 1985. *Kaisha, the Japanese Corporation*. New York: Basic Books.

———. 2006. *21st Century Japanese Management: New Systems, Lasting Values*. London: Palgrave Macmillan.

Abernathy, Frederick, John Dunlop, Janice Hammond, and David Weil. 1999. *A Stitch in Time*. New York, NY: Oxford University Press.

Abo, Tetsuo. 2015. "Researching International Transfer of the Japanese-style Management and Production System: Hybrid Factories in Six Continents." *Asian Business and Management*. 14(1);5–35.

Acemogulu, Daron and James A. Robinson. 2012. *Why Nations Fail: The Origins of Power, Prosperity and Poverty*. New York: Crown Publishers of Random House.

Adams, Jonathan. 2010. "The Demise of 'Lifetime Employment' in Japan". *PRI*, May 30, 2010. www.pri.org/stories/2010-05-18/temp-nation-demise-lifetime-employment-japan. Accessed January 10, 2019.

Adler, Paul. 1995. "Democratic Taylorism: The Toyota Production System at NUMMI". pp. 207–19 in Steven Babson (ed.) *Lean Work*. Detroit: Wayne State University Press.

———. 2007. "The Future of Critical Management Studies: A Paleo-Marxist Critique of Labour Process Theory". *Organization Studies*. 28(9):1313–45.

———. 2012. "The Sociological Ambivalence of Bureaucracy: From Weber via Gouldner to Marx" *Organizational Science*. 23(1):244–66.

Adler, Paul S., Mary Benner, David J. Brunner, John Paul MacDuffie, Emi Osono, Bradley Staats, Hirotaka Takeuchi, Michael L. Tushman, and Sidney G. Winter. 2009. "Perspectives on the Productivity Dilemma". *Journal of Operations Management*. 27(2):99–113.

Aglietta, Michel. 1978. "Phases of US Capital Expansion". *New Left Review*. 110:17–28.

———. 1979. *A Theory of Capitalist Regulation: The US Experience*. London: New Left Books.

Ahuvia, Aaron and Elif Izberk-Bilgin. 2011. "Limits of the McDonaldization Thesis: eBayization and Ascendant Trends in Post-Industrial Consumer Culture". *Consumption, Markets & Culture*. 14(4):361–84.

Albritton, Robert. 2001 (1995). "Regulation Theory: A Critique". pp. 416–42 in Bob Jessop (ed.) *The Parisian Regulation School, Volume 1* of *The Regulation School and the Crisis of Capitalism*. Cheltenham: Edward Elgar.

Alefarii, Mudhafar, Konstantinos Salonitis, and Yuchun Xu. 2017. "The Role of Leadership in Implementing Lean Manufacturing". *Procedia CIRP*. 63:756–61.

Algelica, Paul D. and Vlad Tarko. 2013. "Co-Production, Polycentricity, and Value Heterogeneity: The Ostroms" Public Choice Institutionalism Revisited". *American Political Science Review*. 107(4):726–41.

Allford, Harry. 2015. "Apple's Strategy Evolves, But Their Business Model Remains Unchanged". *The Hill*. March 3, 2015. https://thehill.com/blogs/congress-blog/technology/234623-apples-strategy-evolves-but-their-business-model-remains Accessed November 9, 2018.

American Honda Motor Company, Inc. 2019. "Confirmed Rupture of Defective Takata Airbag Inflator in Buckeye, Arizona". *Press Statement*. March 29, 2019. http://hondaairbaginfo.com/statement-from-american-honda-motor-co-inc-re-confirmed-rupture-of-defective-takata-airbag-inflator-in-buckeye-az-march-29–2019/ Accessed April 22, 2019.

American Society for Quality. 2018a. *Why ASQ?* Milwaukee, WI: ASQ. https://asq.org/why-asq Accessed December 12, 2018.

———. 2018b. *Your 8 Step Guide to ASQ Certification*. Milwaukee, WI: ASQ. https://asq.org/cert/resource/pdf/certification/ASQ-Journey-Infographic-WEB-READY.pdf Accessed December 12, 2018.

Anderson, Chris. 2012. *Makers: The New Industrial Revolution*. New York: Crown Business.

Anner, Mark. 2012. "Corporate Social Responsibility and Freedom of Association Rights". *Politics and Society*. 40(4):609–44.

Anttila, Timo, Tomi Oinas, and Armi Mustosmäki. 2018. "Towards Formalisation: The Organisation of Work in the Public and Private Sectors in Nordic Countries". *Acta Sociologica*. March 26, 2018.

———. Forthcoming. "Lean in Europe—A New Dominant Division of Labor?". Chapter 18 in Thomas Janoski and Darina Lepadatu (eds.) *The International Handbook of Lean Production*. New York: Cambridge University Press.

Aoki, Katsuki. 2008. "Transferring Japanese Kaizen Activities to Overseas Plants in China". *International Journal of Operations and Production Management*. 28(6):518–39.

Applebaum, Eileen and Rosemary Batt. 1994. *The New American Workplace*. Ithaca, NY: ILR Press of Cornell University Press.

Appelbaum, Steven. 1997. "Socio-Technical Systems Theory: An Intervention Strategy for Organizational Development". *Management Decision*. 35(6):452–63.

Apple Inc. 2013. *Apple Supplier Responsibility: 2013 Progress Report*. Cupertino, CA: Apple Inc.

Apple Press Release. 2017. "Apple Park Opens to Employees in April". February 22. www.apple.com/newsroom/2017/02/apple-park-opens-to-employees-in-april/ Accessed November 9, 2018.

———. 2018. "Apple Accelerates US Investment and Job Creation". January 17. www.apple.com/newsroom/2018/01/apple-accelerates-us-investment-and-job-creation/ Accessed November 9, 2018.

Apple Supplier Responsibility Progress Report (SRPR). 2018. www.apple.com/supplier-responsibility/pdf/Apple_SR_2018_Progress_Report.pdf Accessed November 30, 2018.

Aramaki, Kenji. 2018. *Japan's Long Stagnation, Deflation, and Abenomics: Mechanisms and Lessons*. London: Palgrave MacMillan.

Arrighi, Giovanni. 2007. *Adam Smith in Beijing*. New York: Verso.

Arrighi, Giovanni and Jason Moore. 2001. "Capitalist Development in World Historical Perspective". pp. 56–75 in R. Albritton, M. Itoh, R. Westra, and A. Zuege (eds.) *Phases of Capitalist Development*. London: Palgrave Macmillan.

Arrighi, Giovanni and Beverly Silver. 1999. *Chaos and Governance in the Modern World System*. Minneapolis: University of Minnesota Press.

Asia Floor Wage Alliance (AFWA). 2016. *Precarious Work in the Walmart Global Value Chain*. Hong Kong: Asia Floor Wage Alliance. file:///C:/Users/tjanos/Downloads/Precarious%20Work%20in%20the%20Walmart%20Global%20Value%20Chain%20(1).pdf Accessed March 11, 2017.

Auletta, Ken. 2010. *Googled: The End of the World as We Know It*. New York: Penguin Books.

Auslin, Michael. 2017. *The End of the Asian Century: War, Stagnation, and the Risks to the World's Most Dynamic Region*. New Haven, CT: Yale University Press.

Austrom, Douglas and Carolyn Ordowich. 2018. "Calvin Pava: Sociotechnical Design for the Digital Coal Mines". In David Szabla, William Pasmore, Mary A. Barnes, and Asha N. Gipson (eds.) *The Palgrave Handbook of Organizational Change Thinkers*. London: Palgrave Macmillan.

Babson, Steve (ed.). 1995. *Lean Work: Empowerment and Exploitation in the Global Auto Industry*. Detroit, MI: Wayne State University Press.

Baccaro, Lucio. 2018. "What Happened to Diversified Quality Production?" *Socio-Economic Review*. 16(3):614–18.

Baccaro, Lucio and Chris Howell. 2017. *Trajectories of Neoliberal Transformation. European Industrial Relations Since the 1970s*. Cambridge: Cambridge University Press.

Bain, Marc. 2018. "To see how Asia's Manufacturing Map is being Redrawn, Look at Nike and Adidas" Quartz. May 10. https://qz.com/1274044/nike-and-adidas-are-steadily-ditching-china-for-vietnam-to-make-their-sneakers/ Accessed June 19, 2019.

Baines, Joseph. 2015. "Encumbered behemoth: Wal-Mart Differential Accumulation and International Retail Restructuring". pp. 149–66 in Kees van der Pijl (ed.) *Handbook of International Political Economy of Production*. Cheltenham: Edward Elgar.

Bair, Jennifer and Sam Bernstein. 2006. "Labor and the Wal-Mart Effect". pp. 99–114 in Stanley Brunn (ed.) *Wal-Mart World*. New York: Routledge.

Baja, Vikas. 2012. "Fatal Fire in Bangladesh Highlights the Dangers Facing Garment Workers". *New York Times*. November 25, 2012. www.nytimes.com/2012/11/26/world/asia/bangladesh-fire-kills-more-than-100-and-injures-many.html Accessed March 21, 2018.

Baker, Peter and Keith Bradshear. 2019. "Trump Insists That He Can Make US Companies Leave China". *New York Times*. August 25, 1019.

Ballé, Michael, Daryl Powell, and Kondo Yokozawa. 2019. "Monozukuri, Hitozukuri, Kotozukuri" *Planet Lean: The Lean Global Network Journal*. January 8, 2019. https://planet-lean.com/monozukuri-hitozukuri-kotozukuri/ Accessed April 22, 2019.

Bamber, G.J., P. Stanton, T. Bartram, and Ballardie, R. 2014. "Human Resource Management, Lean Processes and Outcomes for Employees: Towards a Research Agenda". *International Journal of Human Resource Management*. 25(21):2881–91.

Barnett, William P. 2008. *The Red Queen among Organizations: How Competitiveness Evolves*. Princeton, NJ: Princeton University Press.

Baron, James and David Kreps. 1999. *Strategic HR*. New York: Wiley.

Bartholdi, John J. and Kevin R. Gue. 2004. "The Best Shape for a Cross-Dock". *Transportation Science*. 38(2):235–44.

Bartley, Tim. 2014. "Transnational Governance and the Re-Centered State: Sustainability or Legality?" *Regulation and Governance*. 8(1):93–109.

———. 2018a. *Rules without Rights: Land, Labor, and Private Authority in the Global Economy*. New York: Oxford University Press.

———. 2018b. Transnational Corporations and Global Governance". *Annual Review of Sociology*. 44:145–65.

Bartley, Tim and Curtis Child. 2011. "Movements, Markets and Fields: The Effects of Anti-sweatshop Campaigns on U Firms 1993–2000". *Social Forces*. 90(2):425–51.

———. 2014. "Shaming the Corporation: The Social Production of Targets and the Anti-Sweatshop Movement". *American Sociological Review*. 79(4):653–79.

Bartley, Tim and Niklas Egels-Zandén. 2015. "Responsibility and Neglect in Global Production Networks: The Uneven Significance of Codes of Conduct in Indonesian Factories". *Global Networks*. 15:S21–44.

Bartley, Tim and Doug Kincaid. 2015. "The Mobility of Industries and the Limits of Corporate Social Responsibility: Labor Codes of Conduct in Indonesian Factories". pp. 393–429 in Kiyoteru Tsutsui and Alwyn Lim (eds.) *Corporate Social Responsibility in a Globalizing World*. New York: Cambridge University Press.

Bartley, Tim, Sebastian Koos, Hiram Samel, Gustavo Setrini, and Nikolas Summers. 2015. *Looking Behind the Label: Global Industries and the Conscientious Consumer.* Bloomington: Indiana University Press.

Bartley, Tim and Lu Zhang. 2016. "China and Global Labor Standards". pp. 228–49 in Scott Kennedy (ed.) *China and Global Governance*. London: Routledge.

Basker, Emek. 2007. "The Causes and Consequences of Wal-Mart's Growth". *Journal of Economic Perspectives*. 21(3):177–98.

Basker, Emek, Shawn Klimek, and Pham Hoang Van. 2012. "Supersize It: The Growth of Retail Chains the Rise of the Big-Box Store." *Journal of Economics and Management Strategy*. 21(3):541–82.

Basker, Emek and Michael Noel. 2009. "The Evolving Good Chain: Competitive Effects of Wal-Mart's Entry Into the Supermarket Industry". *Journal of Economics and Management Strategy*. 18(4):977–1009.

Battilana, Julie and Tiziana Caseiaro. 2012. "Change Agents, Networks and Institutions". *Academy of Management Journal*. 55(2):381–98.

———. 2013. "Overcoming Resistance to Organizational Change". *Management Science*. 59(4):819–36.

Baumol, William J., Robert Litan, and Carl J. Schramm. 2007. *Good Capitalism Bad Capitalism*. New Haven, CT: Yale University Press.

Baxter, Gordon and Ian Sommerville. 2011. "Socio-Technical Systems: From Design Methods to Systems Engineering". *Interacting with Computers*. 23:4–17.

BBC. 2013. "China Acknowledges Cancer Villages." *BBC News, Asia*. February 22, 2013. www.bbc.com/news/world-asia-china-21545868 Accessed June 20, 2019.

Beckman, Martin, Borje Johansson, Folke Snickars, and Roland Thord. 1998. *Knowledge and Networks in a Dynamic Economy: Festschrift in Honor of Åke E. Andersson*. Heidelberg: Springer-Verlag.

Bell, Daniel. 1973. *The Coming of Post-Industrial Society*. New York, NY: Basic Books.

Bellah, Robert. 1957. *Tokugowa Religion: The Values of Pre-Industrial Japan*. New York, NY: Free Press.

Berg, Axel van den and Thomas Janoski. 2005. "Conflict Theories in Political Sociology". pp. 72–95 in Thomas Janoski, Robert Alford, Alexander Hicks, and Mildred Schwartz (eds.) *The Handbook of Political Sociology*. New York: Cambridge University Press.

Bergdahl, Michael. 2007. *The 10 Rules of Sam Walton*. New York: Wiley.

Berggren, Christian. 1992. *Alternatives to Lean Production*. Ithaca, NY: ILR/Cornell University Press.

———. 1993. "Lean Production: The End of History?" *Work, Employment and Society*. 7(2):163–88.

Berman, Barry. 2011. *Competing in Tough Times: Business Lessons from LL Bean, Trader Joes, Costco and Other World Class Retailers*. Upper Saddle River, NJ: FT Press/Pearson.

Bernier, Barnard. 2009. "The Expansion of Precarious Labor: A Consequence of the Japanese System of Lean Production". *Anthropologie et Societes*. 33(1):229–44.

Besser, Terry. 1993. "The Commitment of Japanese and US Workers: A Reassessment of the Literature". *American Sociological Review*. 58(6):873–81.

———. 1996. *Team Toyota: Transplanting the Toyota Culture to the Camry Plant in Kentucky*. Albany, NY: State University of New York Press.

Bhamu, Jaiprakash. 2014. "Lean Manufacturing: Literature Review and Research Issues". *International Journal of Operations and Production Management*. 34(7):876–940.

Bicheno, John. 2004. *The New Lean Toolbox: Towards Fast, Flexible Flow*, 3rd Edition. Buckingham: PICSIE Books.

Bicheno, John and Matthias Holweg. 2016. *The Lean Toolbox: Towards Fast, Flexible Flow*, 5th Edition. Buckingham: PICSIE Books.

Bindley, Katherine. 2012. "David Sacks, Yammer CEO, Hosts Extravagant Birthday Party". *Huffington Post*. June 19, 2019. https://www.huffpost.com/entry/david-sacks-yammer-ceo-birthday-party_n_1609252 Accessed July 21, 2019.

Bishop, Todd. 2004. "Union says Microsoft work going Off-shore". *Seattle Post-Intelligencer*. July 28, 2004. www.seattlepi.com/business/articles/Union-says-Microsoft-work-going-off-shore-1150405.php Accessed November 11, 2010.

Blanpain, Roger. 2008. *Comparative Labour Law and Industrial Relations in Industrialized Markets*, 9th Edition. A. A. den Rijn, Netherlands: Kluwer Law.

Bloodworth, James. 2018a. "I Worked in an Amazon Warehouse" *The Guardian*. September 17. www.theguardian.com/commentisfree/2018/sep/17/amazon-warehouse-bernie-sanders Accessed January 20, 2019.

———. 2018b. *Hired: Six Months Undercover in Low-Wage Britain*. London: Atlantic Books.

Bloom, Paul and Vanessa Perry. 2001. "Retailer Power and Supplier Welfare: The Case of Wal-Mart". *Journal of Retailing*. 77(3):379–96.

Boccaro, Lucio, Virginia Doellgast, Tony Edwards, and Josh Whitford. 2018. "Diversified Quality Production 2.0: On Arndt Sorge and Wolfgang Streeck". *Socio-Economic Review*. 16(3):613–35.

Bock, Lazlo. 2015. *Work Rules: Insights from Inside Google That Will Transform How You Live and Lead*. New York: Twelve/Hatchette.

Bonanno, Alessandro and Stephan J. Goetz. 2012. "Walmart and Local Economic Development". *Economic Development Quarterly*. 26(4):285–97.

Bonanno, Alessandro and R. A. Lopez. 2012. "Wal-Mart's Monopsony Power in Metro and Non-Metro Labor Markets". *Regional Science and Urban Economics*. 42(94):569–79.

Bonvillian, Willam B. and Peter L. Singer. 2017. *Advanced Manufacturing: The New American Innovation Policies*. Cambridge, MA: MIT Press.

Boothroyd, Geoffrey and Peter Dewhurst. 1990 (1983). *Product Design for Assembly*, 2nd Edition. Wakefield, RI: Boothroyd Dewhurst, Inc.

Boothroyd, Geoffrey, Peter Dewhurst, and Winston A. Knight. 2010. *Product Design for Manufacture and Assembly*, 3rd Edition. CRC Press of Taylor & Francis.

Boudette, Neal. 2007. "Honda and UAW Clash Over New Factory Jobs: Residency Rules Exclude Most Union Members; Indignant in Indiana". *Wall Street Journal*.

October 10. www.wsj.com/articles/SB119196377029953821 Accessed June 21, 2012.

Boyer, Robert. 1990. *The Regulation School: A Critical Introduction*. New York: Columbia University Press.

———. 1993. *L'après-fordisme*. Paris: Styros.

———. 2004. *The Future of Economic Growth: As New Becomes Old*. London: Edward Elgar.

Boyer, Robert, E. Charron, Ulrich Jürgens, and Steve Tolliday (eds.). 1998. *Between Imitation and Innovation: The Transfer and Hybridization of Productive Models in the International Automobile Industry*. Oxford: Oxford University Press.

Boyer, Robert and Daniel Drache (eds.). 1996. *States Against Markets: The Limits of Globalization*. London: Routledge.

Boyer, Robert and Michel Freyssenet. 2000. "A New Approach of Productive Models: The World That Changed the Machine". *Industrielle Beziehungen*. 7(4):385–413.

———. 2002 [2000 in French]. *The Productive Models: The Conditions of Profitability*. Translated by Alan Sitkin. New York: Palgrave Macmillan.

———. 2016. "Special Session: 15 Years of Productive Models" Plenary Session. GERPISA Conference: The New Frontiers of the World Automotive Industry, Puebla, Mexico. May 4, 2016.

Boyer, Robert and Yves Saillard. 2001. *Regulation Theory: The State of the Art*. London: Routledge.

Boyer, Robert and Toshio Yamada. 2000/2003. "Introduction: A Puzzle for Economic Theories". pp. 3–16 in Robert Boyer and Toshio Yamada (eds.) *Japanese Capitalism in Crisis: a Regulationist Interpretation*. New York: Routledge.

Boyle, Matthew. 2018. "Costco's Squeeze on Suppliers Signifies Pain for P&G, Nestle" *Bloomberg Businessweek*. March 7. https://www.bloomberg.com/news/articles/2018-03-07/costco-outshines-other-discounters-but-broader-concerns-linger Accessed July 31, 2019.

Braverman, Harry. 1974. *Labor and Monopoly Capital: The Degradation of Work in the Twentieth Century*. New York: Monthly Review Press.

Bremmer, Ian. 2009. "State Capitalism Comes of Age: The End of the Free Market". *Foreign Affairs*. 88(3):40–55.

———. 2010. *The End of the Free Market: Who Wins the War Between States and Corporations?* New York: Portfolio/Penguin Group.

———. 2018. *US vs. Them: The Failure of Globalism*. New York: Portfolio/Penguin.

Brenner, Robert and Mark Glick. 2001. "The Regulation Approach: Theory and History". pp. 341–415 in Bob Jessop (ed.) *The Parisian Regulation School, Volume 1 of the Regulation School and the Crisis of Capitalism*. Cheltenham: Edward Elgar.

Brown, Karen and Nancy Lea Hyer. 2009. *Managing Projects: A Team-Based Approach*. New York: McGraw-Hill/Irwin.

Bryman, Alan. 1999. "The Disneyization of Society". *Sociological Review*. 47(1):25–49.

———. 2003. "McDonald's as a Disneyized Institution: Global Implications". *American Behavioral Scientist*. 47(2):154–67.

Burawoy, Michael. 1979. *Manufacturing Consent: Changes in the Labor Process Under Monopoly Capitalism*. Chicago: University of Chicago Press.

Burbidge, John. 1978. *Principles of Production Control*, 4th Edition. London: MacDonald and Evans.

———. 1991. "Production Flow Analysis for Planning Group Technology". *Journal of Operations Management*. Special Issue: Group Technological and Cellular Manufacturing. 10(1):5–27.

Bureau of Labor Statistics (BLS). 2019. "Union Members Summary" January 18, www.bls.gov/news.release/union2.nr0.htm Accessed October 14, 2019.

Busemeyer, Marius and Christine Tampusch (eds.). 2012. *The Political Economy of Collective Skill Formation*. New York: Oxford.

Butler, Peter, Linda Glover and Olga Tregaskis. 2011. "'When the Going Gets Tough . . .' Recession and the Resilience of Workplace Partnership". *British Journal of Industrial Relations*. 49(4):666–87.

Butler, Peter, Olga Tregaskis and Linda Glover. 2013. "Workplace Partnership and Employee Involvement: Contradictions and Synergies". *Economic and Industrial Democracy*. 34(1):5–24.

Butollo, Florian. 2015. *The End of Cheap Labor: Industrial Transformation and "Social Upgrading" in China*. Frankfurt-am-Main: Campus Verlag.

Cable, Josh. 2009. "GM, Ford and Chrysler Strive to Become the Lean Three". *Industry Week*. https://www.industryweek.com/public-policy/gm-ford-and-chrysler-strive-become-lean-three Accessed January 10, 2019.

Cable, Josh. 2012. "Honda Celebrates 30 Years on Manufacturing in America . . . the Honda Way". *Industry Week*. May 15. www.industryweek.com/engagement/honda-celebrates-30-years-manufacturing-america-honda-way Accessed August 14, 2018.

Cain, Áine. 2018. "Amazon Will Raise Its Minimum Wage to $15 an Hour: Here's What It's Really Like to Work There, According to Employees". *Business Insider*. October 2. www.businessinsider.com/what-its-like-to-work-at-amazon-2018-2 Accessed January 10, 2019.

Calfas, Jennifer. 2018. "This Company Has the Best Pay and Benefits, According to Employees". *Money*. February 27, 2018.

Calveley, Moira, David Allsop, Natalia Lawton, and Monika Huesmann. 2017. "Managing the Employment Relationship". pp. 281–323 in Gary Rees and Paul E. Smith (eds.) *Strategic Human Resource Management: An International Perspective*. Thousand Oaks, CA: Sage Press.

Canedy, Dana. 1998. "McDonald's Burger War Salvo: Is "Made for You" the Way Folks Want to Have It?" *New York Times*. June 20, 1998. www.nytimes.com/1998/06/20/business/mcdonald-s-burger-war-salvo-is-made-for-you-the-way-folks-want-to-have-it.html Accessed May 15, 2018.

Carayon, Pascale. 2009. "Human Factors of Complex Sociotechnical Systems". *Applied Ergonomics*. 37(4):525–35.

Carayon, Pascale, Peter Hancock, Nancy Leveson, Ian Nov, Laerte Sznelwar, and Geert van Hootegem. 2015. "Advancing a Sociotechnical Systems Approach to Workplace Safety". *Ergonomics*. 58(4):548–64.

Carlisle, Candace. 2017. "With 1 Billion Dollars Invested in Its North American Campus, Toyota Opens Its Doors". *Dallas News*. July 6, 2017. www.bizjournals.com/dallas/news/2017/07/06/with-1-billion-invested-in-its-new-north-american.html. Accessed September 10, 2018.

Carmel, Erran and Paul Tija. 2005. *Off-shoring Information Technology*. New York: Cambridge University Press.

Carpenter, J. William. 2016. "The World's Top 10 Retailers (WMT, COST)". *Investopedia*. www.investopedia.com/articles/markets/122415/worlds-top-10-retailers-wmt-cost.asp accessed 11/13/18 Accessed March 15, 2018.

Carré, Françoise and Chris Tilly. 2017. *Where Bad Jobs Are Better: Retail Jobs Across Countries and Companies*. New York: Russell Sage.

———. 2018. "A Global Look at What Makes US Retail Jobs So Bad". *Perspectives on Work*. 22:30–34.

Carter, Bob, Andy Danford, Debra Howcroft, Helen Richardson, and Andrew Smith. 2013a. "Taxing Times: Lean Working and the Creating of (in)efficiencies in HMRC". *Public Administration*. 91(1):83–97.

———. 2013b. "'Stressed out of my box': Employee Experience of Lean Working and Occupational Ill-Health in Clerical Work in the UK Public Sector". *Work, Employment and Society*. 27(5):747–67.

Carvajal, Raul. 1983. "Systemic-Netfields: The Systems Paradigm Crisis. Part I". *Human Relations*. March 1, 36(3):227–45.

———. 1985. "Systemic-Netfields: The Emergence of New Frames, Part II". *Human Relations*. 38(9):857–75.

Cascio, Wayne. 2006. "Decency Means More Than 'Always Low Prices': A Comparison of Costco to Wall-Mart's Sam's Club". *Academy of Management Perspectives*, August:26–37.

Chalice, Robert. 2008. *Improving Healthcare using Toyota Lean Production Methods*. Milwaukee, WI: ASQ press.

Chalmers, Norma. 1989. *Industrial Relations in Japan*. London and New York: Routledge.

Chambers, Dennis. 2008. *Toyota*. Westport, CT: Greenwood Press.

Chan, Jenny, Pun Ngai, and Mark Selden. 2015. "Apple's iPad City: Subcontracting Exploitation to China". pp. 76–97 in Kees van de Pijl (ed.) *The Handbook of the Political Economy of Production*. Cheltenham: Edward Elgar.

Chandler, Alfred. 1962. *Strategy and Structure: Chapters in the History of the American Industrial Enterprise*. Cambridge, MA: MIT Press.

Chang, Leslie. 2008. *Factory Girls*. New York: Picador.

Chapelle, Lindsay. 2009. "The Honda Way Is Unpredictable, Contrarian and Successful". *Automotive News*. June 8, 83:63.

Cheng, Andria. 2018. "Walmart's E-Commerce Tactics Against Amazon Look to Be Paying Off". *Forbes*. April August 16. www.forbes.com/sites/andriacheng/2018/08/16/walmarts-ecommerce-tactic-against-amazon-is-paying-off/#4d484c94b74d Accessed June 1, 2018.

Cherns, Albert. 1976. "The Principles of Sociotechnical Design". *Human Relations*. 29(8):783–92.

Chowdhury, Subir. 2001. *The Power of Six Sigma*. Chicago, IL: Dearborn Trade of Kaplan Company.

———. 2002a. *The Power of Design for Six Sigma*; Chicago, IL: Kaplan Publishing.

———. 2002b. *Design for Six Sigma*; Chicago, IL: Kaplan Professional.

———. 2011. *The Power of LEO: The Revolutionary Process for Achieving Extraordinary Results*. New York: McGraw-Hill.

———. 2017. *The Difference: When Good Enough Isn't Enough*. New York: Crown Business Books.

Chowdhury, Subir and Shin Taguchi. 2016. *Robust Optimization: World's Best Practices for Developing Winning Vehicles*, with Shin Taguchi. New York: Wiley.

Chowdhury, Subir and Ken Zimmer. 1996. *QS-9000 Pioneers*. New York: McGraw-Hill.

Chozick, Amy and Motoko Rich. 2018. "The Rise and Fall of Carlos Ghosn". *New York Times*. December 31, 2018. www.nytimes.com/2018/12/30/business/carlos-ghosn-nissan.html Accessed December 31, 2018.

Christopherson, Susan. 2007. "Barriers to US Style Lean Retailing: The Case of Wal-Mart's Failure in Germany". *Journal of Economic Geography*. 7:451–69.

Claeson, Bjorn. 2012. *Deadly Secrets: What Companies Know About Dangerous Workplaces and Why Exposing the Truth Can Save Workers" Lives in Bangladesh and Beyond.* Washington, DC: International Labor Rights Forum. http://laborrights.org/sites/default/files/publications-and-resources/DeadlySecrets.pdf Accessed June 24, 2014.

Clarke, Simon. 1990. "The Crisis of Fordism or the Crisis of Social Democracy?" *Telos* 83:71–98.

Clawson, Dan. 1980. *Bureaucracy and the Labor Process.* New York: Monthly Review Press.

Cleland, Scott and Ira Brodsky. 2011. *Search and Destroy: Why You Can't Trust Google Inc.* St. Louis: Telescope Books.

CNN. 2018. "Apple Fast Facts". www.cnn.com/2014/07/01/business/apple-fast-facts/index.html Accessed November 12, 2018.

Cole, Robert E. 1971. *Japanese Blue Collar: The Changing Tradition.* Berkeley: University of California Press.

———. 1979. *Work, Mobility and Participation: A Comparative Study of American and Japanese Industry.* Berkeley: University of California Press.

———. 1989. *Strategies of Learning: Small Group Activities in American, Japanese and Swedish Industry.* Berkeley: University of California Press.

———. 1999. *Managing Quality Fads: How American Business Learned to Play the Quality Game.* New York: Oxford University Press.

———. 2011a. "Who Was Really at Fault for the Toyota Recalls?" *The Atlantic.* May 1, 2011. www.theatlantic.com/business/archive/2011/05/who-was-really-at-fault-for-the-toyota-recalls/238076/. Accessed October 29, 2018.

———. 2011b. "What Happened to Toyota". *MIT Sloan Management Review.* 62(4):29–35.

Cole, Robert E. and Donald Deskins. 1988. "Racial Factors in Site Location and Employment Patterns of Japanese Auto Firms in America". *California Management Review.* 31(1):9–22.

Cole, Robert E. and Yoshifumi Kakata. 2014. "The Japanese Software Industry: What Went Wrong and What Can We Learn From It?" *California Management Review.* Fall:16–43.

Cole, Robert E., Arne Kalleberg, and James R. Lincoln. 1993. "Assessing Commitment in the United States and Japan: A Comment on Besser". *American Sociological Review.* 58(6):882–5.

Cole, Robert E. and Richard Scott (eds.). 2000. *The Quality Movement in America: Lessons for Theory and Research.* Thousand Oaks, CA: Sage.

Cole, Robert E and D. Hugh Whittaker. 2007. "Introduction". pp. 1–28 in D. Hugh Whittaker and Robert Cole (eds.). *Recovering from Success: Innovation and Technology Management in Japan.* New York: Oxford University Press.

Conney, Richard. 2002. "Is 'Lean' a Universal Production System?" *International Journal of Operations and Production Management.* 22(10):1130–47.

Consumer Reports. 2019. "Takata Airbag Recall: Everything You Need to Know". *Consumer Reports,* March 29, 2019. www.consumerreports.org/car-recalls-defects/takata-airbag-recall-everything-you-need-to-know/ Accessed April 22, 2019.

Contractor, Farok, Vikas Kumar, Sumit Kundu, and Torben Pedersen. 2010. *Global Outsourcing and Off-shoring.* New York: Cambridge University Press.

Cooney, Richard and Graham Sewell. 2008. "From Lean Production to Mass Customization: Recent Developments in the Australian Automotive Industry". pp. 127–49 in V. Pulignano, P. Stewart, A. Danford, and M. Richardson, M. (eds.) *Flexibility at Work.* Basingstoke: Palgrave MacMillan.

Cooper, Robin and Robert S. Kaplan. 2008. "Measure Costs Right: Make the Right Decisions". *Harvard Business Review*, September, 66:96–103.

Courtemanche, Charles and Art Carden. 2014. "Competing with Costco and Sam's Club: Warehouse Club Entry and Grocery Prices". *Southern Economic Journal*. 80(3):565–85.

Coutu, Diane. 2009. "Why Teams Don't Work". *Harvard Business Review*, May, 87(5):98–105, 130.

Crosby, Philip. 1977. *Quality Is Free*. New York: McGraw-Hill.

———. 1984. *Quality Without Tears*. New York: McGraw-Hill.

———. 1988. *The Eternally Successful Organization*. New York: McGraw-Hill.

———. 1989. *Let's Talk Quality*. New York: McGraw-Hill.

———. 1990. *Leading, the Art of Becoming an Executive*. New York: McGraw-Hill.

———. 1996. *Quality Is Still Free: Making Quality Certain in Uncertain Times*. New York: McGraw-Hill.

Cusumano, Michael A. 1987. *The Japanese Automobile Industry: Technology and Management at Nissan and Toyota*. Cambridge, MA: Harvard University Asia Center.

Cutcher-Gershenfeld, Joel, Dan Brooks, and Martin Mulloy. 2015. *Inside the Ford-UAW Transformation*. Cambridge, MA: MIT Books.

Cutcher, Gershenfeld, Joel, Alan Gershenfeld, and Neil Gershenfeld. 2018. "Digital Fabrication and the Future of Work". *Perspectives on Work*. 22:8–13.

Cutcher-Gershenfeld, Joel, Michio Nitta, Betty Baasrrett, Nejib Belhedi, Simon Chow, Takashi Inaba, Iwao Ishino, Weng-Jeng Lin, Michael Moore, William Mothersell, Jennifer Palthe; Shobha Romanand, Mark Strolle, and Arthur Wheaton. 1998. *Knowledge-Driven Work: Unexpected Lessons from Japanese and United States Work Practices*. New York: Oxford University Press.

Cutter, Chip. 2019. "The Best-Managed Companies of 2019 and How they Got that Way". *Wall Street Journal*. November 25, 2019, pp. R1–R6.

Danford, Andy, M. Richardson, P. Stewart, S. Tailby, and M. Upchurch. 2005. *Partnership and the High Performance Workplace. Work and Employment Relations in the Aerospace Industry*. Basingstoke: Palgrave Macmillan.

David, Angela, Jennifer Lynaugh, and Andrea Peterson. 2015. *Membership: The Key to Low Prices*. Create Space Independent publishing platform.

Davis, James, K. Diekmann, and C. Tinsley. 1994. "The Decline and Fall of the Conglomerate Firm in the 1980s". *American Sociological Review*. 49(4):547–70.

Davis, James and Tracy Thompson. 1994. "A Social Movement Perspective on Corporate Control". *Administrative Science Quarterly*. 39(1):141–73.

Davis, Mike. 1978. "Fordism in Crisis: A Review of Michel Aglietta's Regulation et crises: L'experience des Etats-unis". *Review*. 11(2):207–69.

DeBonis, Mike. 2013. "DC Council Approves 'living wage' Bill over Wal-Mart Ultimatum". *Washington Post*. July 10. https://www.washingtonpost.com/local/dc-politics/dc-council-approves-living-wage-bill-over-wal-mart-ultimatum/2013/07/10/724aab6e-e96f-11e2-a301-ea5a8116d211_story.html Accessed June 30, 2018.

Defeo, Joseph and Joseph Juran. 2016. *Juran's Quality Handbook: The Complete Guide to Performance Excellence*, 7th Edition. New York: McGraw-Hill.

Delbridge, Rick. 1998. *Life on the Line in Contemporary Manufacturing*. New York: Oxford University Press.

Delery, John and Harold Doty. 1996. "Modes of Theorizing in Strategic HR Management". *Academy of Management Journal*. 390(4):802–35.

Deming, W. Edwards. 1982. *Out of the Crisis*. Cambridge, MA: MIT Press.
———. 1994. *The New Economics for Industry, Government, Education*. Cambridge, MA: MIT Press.
Dennis, Pascal. 2015 (2002). *Lean Production Simplified*, 3rd Edition. Boca Raton, FL: CRC Press of Taylor and Francis.
Desai, Mihir. 2018. "Why Apple Is the Future of Capitalism". *New York Times*. August 7, 2018.
Distelhorst, Greg, R. Locke, T. Pal, and H. Samel. 2015. "Production Goes Global, Compliance Stays Local". *Regulation and Governance*. 9:24–42.
Dixon, Nicole. 2010. "An Analysis of How McDonald's Delivers its Products and Services". *LinkedIn Slideshare*. www.slideshare.net/nicoledixon/an-analysis-of-how-mc-donalds-delivers-its-products-and-services Accessed November 15, 2018.
Dobbin, Frank. 2009. *Inventing Equal Opportunity*. Princeton, NJ: Princeton University Press.
Dobbin, Frank and Jiwook Jong. 2010. "The Misapplication of Mr. Michael Jensen". *Sociology of Organizations*. 30B:29–64.
Doellgast, Virginia. 2010. "Collective Voice under Decentralized Bargaining: A Comparative Study of Work Reorganization in US and German Call Centres". *British Journal of Industrial Relations*. 48(2):375–99.
———. 2018. "How and Why did DQP Influence Employment Relations Research?" *Socio-Economic Review*. 16(3):619–25.
Doerr, John. 2018. *Measure What Matters: How Google, Bono and the Gates Foundation Rock the World with OKRs*. New York: Portfolio/Penguin/Penguin Random House.
Dombrowski, Uwe, Thomas Richter, and Philipp Krenkel. 2017. "Interdependencies of Industries 4.0 and Lean Production Systems: A Use Cases Analysis". *Procedia Manufacturing*. 11:1060–8. https://ac.els-cdn.com/S2351978917304250/1-s2.0-S2351978917304250-main.pdf?_tid=8aba56d0-ea96-4eb1-8d0f-41595371aa63&acdnat=1544642973_92ed2734a7c5a35436aadc6cb6967097 Accessed December 12, 2018.
Donaghu, Michael and Richard Barff. 1990. "Nike just did it: International Subcontracting and Flexibility in Athletic Footwear Production". *Regional Studies*. 26(6):537–52.
Dore, Ronald. 1958. *City Life in Japan*. Berkeley, CA: University of California Press.
———. 1964. *Education in Tokugawa Japan*. London: Routledge and Kegan Paul.
———. 1973. *British Factory, Japanese Factory*. Berkeley, CA: University of California Press.
———. 1978. *Shinohata: Portrait of a Japanese Village*. London: Allen Lane and Pantheon.
———. 1983. "Goodwill and the Spirit of Market Capitalism". *British Journal of Sociology*. 34(4):459–82.
———. 1986/2003. *Flexible Rigidities: Industrial Policy and Structural Adjustment in the Japanese Economy, 1970–1982*. London: Atholone and Stanford/London: Bloomsbury.
———. 1987. *Taking Japan Seriously: A Confucian Perspective on Leading Economic Issues*. Atholone & Stanford University Press.
———. 2000. *Market Capitalism, Welfare Capitalism: Japan and Germany versus the Anglo-Saxons*. New York: Oxford University Press.
———. 2002. *The Collected Writings of Ronald Dore*. London: Curzon Press.
Dougherty, Carter. 2009. "A Happy Family of 8,000, But for How Long?" *New York Times*. July 11, 2009. www.nytimes.com/2009/07/12/business/global/12german.html.
Drucker, Peter. 2007 (1946). *Concept of the Corporation*. Piscataway, NJ: Transaction Press.
Du, Julan and Chenggang Xu. 2005. "Market Socialism or Capitalism? Evidence from Chinese Financial Market Development". IEA 2005 Round Table on Market and Socialism.

Durand, J-P., P. Stewart, and J. J. Castillo (ed.). 1999. *Teamwork in the Automobile Industry*. New York: Palgrave Macmillan.

Durkheim, Emile. 1997 (1893). *The Division of Labor in Society*. New York: Free Press.

Dyer, William G., W. Gibb Dyer, and Jeffrey Dyer. 2015. *Team Building: Proven Strategies for Improving Team Performance*, 5th Edition. SF, CA: Jossey-Bass.

The Economist. 2013. "Norway: The Rich Cousin—Oil Makes Norway Different from the Rest of the Region, But Only Up to a Point". *The Economist*. February 2, 2013.

———. 2019. "The World Is Flat: Special Report. Global Supply Chains". *The Economist*, July 13, 2019, pp. 1–12.

Economix Editors. 2012. "The iEconomy: How Much do Foxconn Workers Make? *New York Times*. February 24. https://economix.blogs.nytimes.com/2012/02/24/the-ieconomy-how-much-do-foxconn-workers-make Accessed May 13, 2019.

Economy, Elizabeth C. 2018. *The Third Revolution: Xi Jinping and the New Chinese State*. New York: Oxford University Press.

Edgell, Stephen. 2011. *The Sociology of Work: Continuity and Change in Paid and Unpaid Work*, 2nd Edition. New York: Sage.

Edwards, Richard. 1979. *Contested Terrain: The Transformation of the Workplace in the Twentieth Century*. New York: Basic Books.

Eisenstein, Paul. 2013. "Honda Moving US Corporate HQ to Ohio". *The Detroit Bureau*, February 22. www.thedetroitbureau.com/2013/02/honda-moving-us-corp-hq-to-ohio/. Accessed August 10, 2018.

Eisner, Peter. 2009. "Sol Price, Philanthropist and Entrepreneur, Dies at 93". *San Diego Jewish World*. December 14, 2009. www.sdjewishworld.com/2009/12/14/sol-price-philanthropist-and-entrepreneur-dies-at-93/ Accessed January 9, 2019.

Elliot, Jay. 2012. *Leading Apple with Steve Jobs: Management Lessons from a Controversial Genius*. Hoboken, NJ: John Wiley & Sons.

English-Leuck, J. A. 2002. *Cultures at Silicon Valley*. Stanford, CA: Stanford University Press.

———. 2010. *Being and Well-being: Health and the Working Bodies of Silicon Valley*. Stanford, CA: Stanford University Press.

Epstein, Gady. 2010. "The Winners and Losers in Chinese Capitalism". *Forbes*. August 31, 2010.

Etzioni, Amitai. 2018. "Apple: Good Business, Poor Citizen?" *Journal of Business Ethics*. 151:1–11.

Eurofound. 2001. "Agreements Signed on Volkswagen's 5000x5000 Project". September 24. www.eurofound.europa.eu/publications/article/2001/agreements-signed-on-volkswagens-5000-x-5000-project Accessed September 22, 2010.

Falloon, Ian. 2005. *The Honda Story: Road and Racing Motorcycles from 1948 to the Present Day*. Newbury Park, CA: Haynes Publishing.

Farber, David. 2002. *Sloan Rules: Alfred P. Sloan and the Triumph of General Motors*. Chicago: University of Chicago Press.

Feigenbaum, Armand. 1945. *Quality Control: Principles, Practice and Administration*. New York: McGraw-Hill.

———. 1961. *Total Quality Control*. New York: McGraw-Hill.

Feld, William M. 2000. *Lean Manufacturing Tools, Techniques and How to Use Them*. Boca Raton, FL: St. Lucie Press.

Ferguson, Niall. 2012. "We're All State Capitalists Now". *Foreign Policy*. February 9. https://foreignpolicy.com/2012/02/09/were-all-state-capitalists-now/ Accessed January 22, 2014.

Fingas, Robert. 2018. “Apple’s 2018 Environmental Responsibility Report Zeroes in On Renewable Energy, Takes Flak from Greenpeace”. https://appleinsider.com/articles/18/04/19/apples-2018-environmental-responsibility-report-zeroes-in-on-renewable-energy-takes-flak-from-greenpeace Accessed November 30, 2018.

Fishman, Charles. 2006. *The Wal-Mart Effect: How the World’s Most Powerful Company Really Works—and How It’s Transforming the American Economy*. Harmondsworth: Penguin Press.

Fitzgerald, Scott W. 2011. *Corporations and Culture Industries: Time-Warner, Bertelsmann and News Corporation*. Lexington, CT: Lexington Books.

Fleming, Lee, Lyra Colfer, Alexandra Marin, and Jonathan McPhie. 2012. “Why the Valley Went First: Aggregation and Emergence in Regional Inventory Networks”. pp. 520–44 in John Padgett and Walter Powell (eds.) *The Emergence of Organizations and Markets*. Princeton, NJ: Princeton University Press.

Fligstein, Neil. 2001. *The Architecture of Markets*. Princeton, NJ: Princeton University Press.

Ford, Henry with Samuel Crowther. 1922. *My Life and Work*. Garden City, NY: Garden City Publishing.

———. 1926 (1988). *Today and Tomorrow*. New York: Productivity Press.

Ford Motor Manufacturing. 2019. “100 Years of the Moving Assembly Line: Experience a Transformative Innovation’s History.” https://corporate.ford.com/articles/history/100-years-moving-assembly-line.html Accessed September 11, 2019.

Forsyth, Donelson. 2019. *Group Dynamics*, 7th Edition. Boston, MA: Cengage.

Fortune. 2012. “Fortune 500 Companies”. May 21, 2012. http://money.cnn.com/magazines/fortune/fortune500/2012/performers/companies/biggest. Accessed March 22, 2014.

Fossati, Flavia. 2018. “How Regimes Shape Preferences: A Study of Political Actors’ Labour Market Policy Preferences in Flexicurity and Dualizing Countries”. *Socio-Economic Review*. 16(3):523–44.

Franz, Timothy. 2012. *Group Dynamics and Team Interventions*. New York: Wiley-Blackwell.

Freeland, Robert. 1996. “The Myth of the M-Form: Governance, Consent and Organizational Change”. *American Journal of Sociology*. 102(2):483–526.

Freeman, Richard. 1984. *Strategic Management: A Stakeholder Approach*. Boston: Pitman.

Freund, William C. and Eugene Epstein. 1982. *People and Productivity*. Homewood, IL: Dow Jones-Irwin.

Freyssenet, Michel. 2003. *Globalization or Regionalization of the European Car Industry*. New York: Palgrave Macmillan.

———. (ed.). 2009. *The Second Automobile Revolution: Trajectories of the World Carmakers in the 21st Century*. New York: Palgrave Macmillan.

Freyssenet, Michel, Kochi Shimizu, and Guiseppe Volpato (eds.). 2003. *Globalization or Regionalization of the European Car Industry*. New York: Palgrave Macmillan.

Frontline. 2004. “Is Walmart Good for America?” *Frontline*. PBS November 28, 2004. www.pbs.org/wgbh/pages/frontline/shows/walmart/ Accessed June 2, 2010.

Fujimoto, Takahiro. 1999. *The Evolution of a Manufacturing System at Toyota*. Oxford: Oxford University Press.

Gall, Gregor and Jane Holgate. 2018. “Rethinking Industrial Relations: Appraisal, Application and Argumentation”. *Economic and Industrial Democracy*. 34(4):561–76.

Galloway, Scott. 2017. *The Four: The Hidden DNA of Amazon, Apple, Facebook and Google*. New York: Penguin/Random House.

Garrahan, Philip and Paul Stewart. 1992. *The Nissan Enigma: Flexibility at Work in a Local Economy*. Boston: Thomson Learning/Cengage.

Garside, Juliette. 2012. "Apple's Efforts Fail to End Grueling Conditions at Foxconn Factories". *The Guardian*. May, 30, 2012. www.theguardian.com/technology/2012/may/30/foxconn-abuses-despite-apple-reforms?INTCMP=SRCH Accessed June 2, 2015.

Gartman, David. 1998. "Postmodernism; or, the Cultural Logic of Post-Fordism". *Sociological Quarterly*. 39(1):119–37.

Gartman, David. 2004. "Three Ages of the Automobile: The Cultural Logics of the Car". *Theory, Culture and Society*. 21(4–5):169–95.

Geels, Frank. 2004. "From Sectoral Systems of Innovation to Socio-Technical Systems: Insights About Dynamics and Change from Sociology and Institutional Theory". *Research Policy* 33:897–920.

Geisinger, Scott. 2012. *The Development of Nike Corporation*. Senior Thesis. Sociology Department, University of Kentucky.

Geissler, Heike. 2018. *Seasonal Associate*. Translated by Katy Derbyshire, Afterword by Kevin Vennemann. Pasadena, CA: Semiotext(e).

Gereffi, Gary. 1994. "The Organization of Buyer-Driven Global Commodity Chains". pp. 95–122 in Gary Gereffi and Miguel Korzeniewicz (eds.) *Commodity Chains and Global Capitalism*. Westport, CT: Greenwood Press.

———. 2005. *The New Off-shoring of Jobs and Global Development*. Geneva: ILO.

Gereffi, Gary and Michelle Christian. 2009. "The Impacts of WalMart: The Rise and Consequences of the World's Dominant Retailer". *Annual Review of Sociology*. 35:573–91.

Gereffi, Gary, J. Humphrey, and Timothy Sturgeon. 2005. "The Governance of Global Value Chains". *Review of International Political Economy*. 12(1):78–104.

Gerschenfeld, Neil, Allan Gerschenfeld, and Joel Cutcher-Gerschenfeld. 2017. *Designing Reality: How to Survive and Thrive in the Third Digital Revolution*. New York: Basic Books.

Gertner, Jon. 2013. *The Idea Factory: Bell Labs and the Great Age of American Innovation*. Harmondsworth: Penguin Books.

Ghosn, Carlos and Philippe Riès. 2005. *Shift: Inside Nissan's Historic Revival*. New York: Doubleday.

Gilder, George. 2018. *Life After Google: The Fall of Big Data and the Rise of the Blockchain Economy*. Washington, DC: Regnery Gateway.

Gilmore, Lee. 2010. *Theater in a Crowded Fire: Ritual and Spirituality at Burning Man*. Berkeley, CA: University of California Press.

Gilmore, Lee and Mark van Proyen (eds.). 2005. *After Burn: Reflections on Burning Man*. Albuquerque, NM: University of New Mexico Press.

Glassdoor. 2013. "Salaries, McDonald's. Salaries, Walmart". Accessed August 5, 2013. www.glassdoor.com/Salaries/index.htm Accessed March 21, 2015.

———. 2019. "Salaries, Apple" https://www.glassdoor.com/Reviews/lexington-apple-store-reviews-SRCH_IL.0,9_IM494_KE10,21.htm Accessed December 11, 2019.

Global Honda. 2018. https://global.honda Accessed August 8, 2018.

Glover, Linda. 2000. "Neither Poison Nor Panacea: Shop Floor Responses to TQM". *Employee Relations*. 22(2):121–41.

———. 2010. "Can Informal Relations Help Explain Worker Relations to Managerial Interventions? Some Case Evidence from a Study of Quality Management". *Economic and Industrial Democracy*. 32(3):357–378.

Glover, Linda and Mike Noon. 2005. "Shop-Floor Worker's Responses to Quality Management Initiatives" *Work, Employment and Society*. 19(4):727–45.

Glover, Linda, Olga Tregaskis, and Peter Butler. 2014. "Mutual Gains? The Workers Verdict: A Longitudinal Study". *International Journal of HR Management*. 25(6):895–914.

Gould, Anthony. 2010. "Working at McDonald's: Some Redeeming Features of McJobs". *Work, Employment and Society*. 24(4):780–802.

Graban, Mark. 2009. *Lean Hospitals*. Boca Raton, FL: CRC Press.

Graham, Laurie. 1995. *On the Line at Subaru-Isuzu: The Japanese Model and the American Worker*. Ithaca, NY: ILR Press.

Greene, Kenyon B. de. 1973. *Sociotechnical Systems: Factors in Analysis, Design, and Management*. Englewood Cliffs, NJ: Prentice-Hall.

Grint, Keith. 2005. *The Sociology of Work*, 3rd Edition. London: Polity Press.

Grint, Keith and Darren Nixon. 2016. *The Sociology of Work*, 4th Edition. London: Polity Press.

Grønning, Terje. 1997. "The Emergence and Institutionalization of Toyotism: Subdivision and Integration of the Labour Force at the Toyota Motor Corporation from the 1950s to the 1970s". *Economic and Industrial Democracy*. 18:423–55.

Gross, Tandiwe. 2013. "Rana Plaza: Private Governance and Corporate Power in Global Supply Chains". *Global Labour Column. Corporate Strategy and Industrial Development*. http://column.global-labor-university.org Accessed March 3, 2015.

Grove, Andrew. 2010. "How Americans Can Create Jobs". *Businessweek*. February:60–64.

Grunden, Naida. 2008. *The Pittsburgh Way to Efficient Health Care: Improving Patient Care Using Toyota Based Methods*. Boca Raton, FL: CRC Press.

Gulowsen, Jon. 1979. "A Measure of Work-Group-Autonomy". pp. 206–18 in L. E. Davis and J. C. Taylor (eds.) *Design of Jobs*, 2nd Edition. Santa Monica: Goodyear.

———. 2000. "Three Kinds of Autonomy at Work in Norway". pp. 193–9 in Michael Beyerlein (ed.) *Work Teams: Past, Present and Future*. Dordrecht: Kluwer.

Hackman, J. Richard. 2002. *Leading Teams: Setting the Stage for Great Performances*. Boston, MA: Harvard Business School Press.

———. 2003. "Learning More from Crossing Levels: Evidence from Airplanes, Orchestras, and Hospitals". *Journal of Organizational Behavior*. 24:1–18.

Haak, René and Markus Pudelko (eds.). 2005. *Japanese Management: The Search for a New Balance Between Continuity and Change*. London: Palgrave Macmillan.

Hackman, J. Richard and A. C. Edmondson. 2008. "Groups as Agents of Change". pp. 167–86 in T. Cummings (ed.) *Handbook of Organization Development*. Thousand Oaks, CA: Sage.

Hackman, J. Richard and K. Katz. 2010. "Group Behavior and Performance". pp. 1208–51 in Susan T. Fiske, D. T. Gilbert, and Gardner Lindzey (eds.) *Handbook of Social Psychology*, 5th Edition. New York: Wiley.

Hackman, J. Richard and Ruth Wageman. 2005. "When and How Team Leaders Matter". *Research in Organizational Behavior*. 26:37–74.

———. 2007. "Asking the Right Questions About Leadership". *American Psychologist*. 62:43–47.

Halberstam, David. 1986. *The Rekoning*. New York: William Morrow and Co.

Hall, Douglas. 1996. *The Career Is Dead: Long Live the Career*. San Francisco, CA: Jossey-Bass.

Hall, Douglas and Philip Mirvis. 1995. "The New Career Contract". *Journal of Vocational Behavior*. 4(7):269–89.

Hall, Peter and David Soskice. 2001. *Varieties of Capitalism*. New York: Oxford University Press.

Hall, Richard H. 1994. *Sociology of Work*. Thousand Oaks, CA: Sage.

Hall, Robert W. 1981. *Driving the Productivity Machines: Production Planning and Control in Japan*. American Production and Inventory.

———. 1982. *Zero Inventories*. New York: McGraw-Hill.

———. 1993. *Attaining Manufacturing Excellence*. New York: McGraw-Hill.

———. 2014. *Beyond Lean: Towards Compression Thinking*. Boca Raton, FL: CRC Press of Taylor Francis.

Halzack. 2019. "Walmart Is Besting Amazon in a Business Worth $35 Billion. Bloomberg Opinion Analysis". *Washington Post*. April 2. www.washingtonpost.com/business/walmart-is-beating-amazon-in-a-business-worth-35-billion/2019/04/01/6fb8cb1e-548b-11e9-aa83-504f086bf5d6_story.html?noredirect=on&utm_term=.ebb6f86798cd Accessed May 11, 2019.

Hamada, Koichi, Anil Kashyap, and David Weinstein (eds.). 2010. *Japan's Bubble, Deflation, and Long-Term Stagnation*. Cambridge, MA: MIT Press.

Hamilton, Nolan. 2013. "Wal-Mart Is Scared of These True Stories from Its Own Employees". *Gawker*. July 11. https://gawker.com/wal-mart-is-scared-of-these-true-stories-from-its-own-e-743832841. Accessed June 11, 2018.

Hamilton v. Wal-Mart Stores, Inc. 2017. Case no. ED CV 17-001415-AB. *US District Court, C. D. California*. September 29, 2017.

Harney, Alexandra. 2008. *The China Price*. Harmondsworth, UK: Penguin.

Harvey, David. 1991. *The Condition of Post-Modernity*. New York: Wiley-Blackwell.

———. 2011. *The Enigma of Capital and the Crises of Capitalism*. New York: Oxford University Press.

Hatamiya, Lon. 2014. "Communities with Walmart Supercenters Experience Increase in Small Business Growth and Sales Tax Revenue with Compared to Communities Without Walmart Supercenters". *Walmart News*. https://news.walmart.com/news-archive/2014/01/28/study-confirms-that-walmart-supercenters-have-positive-economic-impact-in-california Accessed December 11, 2016.

Hauptmeier, Marco. 2012. "Institutions Are What Actors Make of Them: The Changing Construction of Firm Level Employment Relations in Spain". *British Journal of Industrial Relations*. 50(4):737–59.

Hearnshaw, Edward and Mark Wilson. 2013. "A Complex Network Approach to Supply Chain Network Theory". *International Journal of Operations and Production Management*. 33(4):442–69.

Herrigel, Gary. 1996. *Industrial Constructions: The Sources of German Industrial Power*. New York: Cambridge University Press.

———. 2010. *Manufacturing Possibilities: Creative Actions and Industrial Recomposition in the US, Germany and Japan*. New York: Oxford University Press.

———. 2015. "Globalization and the German Industrial Production Model". *Journal of Labour Market Research*. 48:133–49.

Herstaat, Cornelius, Christoph Stockstrom, Hugo Tschirky, and Akio Nagahira (eds.). 2006. *Management of Technology and Innovation in Japan*. Heidelberg: Springer.

Hertwig, Markus. 2015. "European Works Councils and the Crisis: Change and Resistance in Cross-Border Employee Representation at Honda and Toyota". *British Journal of Industrial Relations*. 53(2):326–49.

Hicks, Michael. 2015. "Does Walmart Cause an Increase in Anti-poverty Expenditures?" *Social Science Quarterly*. 96(4):1136–52.

Hino, Satoshi. 2006. *Inside of the Mind of Toyota: Management Principles for Enduring Growth*. New York: Productivity Press.

Hodson, Randy and Teresa Sullivan. 2012. *The Social Organization of Work*. Belmont, CA: Wadsworth.

Holmes, Stanley. 2007. "Can Nike Do It?" *Businessweek*. February 7.

Holusha, John. 1988. "Japanese Faulted Over Black Hiring". *New York Times*. November 27, 1988. www.nytimes.com/1988/11/27/us/japanese-faulted-over-black-hiring.html Accessed August 8, 2018.

Honda Code of Conduct. 2018. https://global.honda/content/dam/site/global/about/cq_img/codeofconduct/pdf/HondaCodeofConduct_en.pdf Accessed August 7, 2018.

Howard, Phoebe Wall. 2019. "Out of Gear: Ford Plunged Ahead with Focus, Fiesta with Known Defects". *Detroit Free Press*, July 14:1, 15–21A.

Hoxie, Robert Franklin. 1966. *Scientific Management and Labor*. New York: A.M. Kelley.

Huang, Yukon. 2017. *Cracking the China Conundrum: Why Conventional Economic Wisdom Is Wrong*. New York: Oxford University Press.

Huber, Vandra L. and Karen A. Brown. 1991. "Human Resource Issues in Cellular Manufacturing: A Sociotechnical Analysis". *Journal of Operations Management*. 10(1):138–59.

Humble, Jez and David Farley. 2011. *Continuous Delivery: Reliable Software Releases Through Build, Test, and Deployment Automation*. Upper Saddle River, NJ: Addison Wesley.

Humble, Jez, Joanne Molesky, and Barry O'Reilly. 2015. *Lean Enterprise: How High Performance Organizations Innovate at Scale*, 10th Release. Sebastopol, CA: O'Reilly Media.

Huselid, Mart. 1995. "The Impact of HR Management Practices on Turnover, Productivity, and Corporate Financial Performance". *Academy of Management Journal*. 38(3):635–72.

Hutchinson, Sue. 2017. "Flexible Working". pp. 193–235 in Gary Rees and Paul E. Smith (eds.) *Strategic Human Resource Management: An International Perspective*. Thousand Oaks, CA: Sage Press.

Hyer, Nancy Lea and Karen A. Brown. 1999. "The Discipline of Real Cells". *Journal of Operations Management*. 17(5):557–74.

Hyer, Nancy Lea, Karen A. Brown, and Sharon Zimmerman. 1998. "A Socio-Technical Systems Approach to Cell Design: Case Study and Analysis". *Journal of Operations Management*. 17(2):179–203.

Hyman, Louis. 2018. *Temp: How American Work, American Business and the American Dream Became Temporary*. New York: Viking.

Hytrek, Gary. 2008. "From Ford to Gates: How Globalization Is Transforming Patterns of Stratification in the United States". pp. 187–214 in Beverly Crawford and Edward Fogarty (eds.) *The Impact of Globalization on the United States, Vol. 3*. Westport, CT: Praeger.

Ingram, Paul, Lori Qingyuan Yue, and Rao Hayagreeva. 2010. "Trouble in Store: Probes, Protests and Store Openings by Wal-Mart, 1998–2007". *American Journal of Sociology*. 116(1):53–92.

Ishikawa, Kaoru. 1968. *Guide to Quality Control*. [in Japanese] Tokyo: Asian Productivity Organization.

———. 1980 (1970, in Japanese). *General Principles of the QC Circles*. Tokyo: Japanese Union of Scientists and Engineers (JUSE).

———. 1985. *How to Operate QC Circle Activities*. Tokyo: Japanese Union of Scientists and Engineers (JUSE).

———. 1986. *Guide to Quality Control*, Revised Edition. Tokyo: Asian Productivity Association.

Ishikawa, Kaoru and David Lu. 1985. *What Is Total Quality Control? The Japanese Way*. Englewood Cliffs, NJ: Prentice Hall.

ISO. 2011. *ISO 13053–1:2011: Quantitative Methods in Process Improvement: Six Sigma*. Geneva: ISO. www.iso.org/standard/52901.html Accessed July 11, 2019.

Jaffee, David. 2001. *Organization Theory: Tension and Change*. New York: McGraw-Hill.

Jakobson, Cathryn Ramin. 1988. "They Can Get It for Your Wholesale". *New York Times Magazine*. December 4. www.nytimes.com/1988/12/04/magazine/they-can-get-it-for-you-wholesale.html Accessed January 9, 2019.

Jameson, Frederic. 1991. *Postmodernism, or the Cultural Logic of Late Capitalism*. London: Verso.

Jamieson, David. 2012. "Target, Union Fight Goes Back to Square One". *Huffington Post*. May 24, 2012. www.huffingtonpost.com/2012/05/24/target-union-ufcw-nlrb_n_1544030.html Accessed June 22, 2015.

Janoski, Thomas. 1990 (reprinted 2018). *The Political Economy of Unemployment: Active Labor Market Policy in the United States and West Germany*. Berkeley, CA: University of California Press.

———. 2010. *The Ironies of Citizenship*. New York: Cambridge University Press.

———. 2016. "The New Division of Labor as Lean Production". *International Journal of Sociology*, Fall. 45(2):85–94.

Janoski, Thomas and Darina Lepadatu. 2009. "Lean Production in the Auto Industry". *NSF Grant ARRA 0940807*. Alexandria, VA: Sociology Section of NSF.

———. 2014. *Dominant Divisions of Labor: Models of Production That Have Transformed the World of Work*. New York: Palgrave Macmillan.

Janoski, Thomas, David Luke, and Chris Oliver. 2014. *The Causes of Structural Unemployment*. Cambridge: Polity Press.

Jantzen, Robert, Donn Pescatrice, and Andrew Braunstein. 2009. "Wal-Mart and the US Economy". *Eastern Economic Journal*. 35:297–308.

Japan Management Association. 1981. *Study of Toyota Production System: From Industrial Engineering Viewpoint*. Tokyo: Japanese Management Association.

Jargon, Julie. 2010. "McDonald's Tackles Repair of 'Broken' Service". *Wall Street Journal*, April 10, 2013.

Jessop, Bob. 1990. *The Regulation School: A Critical Introduction*. New York: Columbia University Press.

———. 2001a. "Series Preface". pp. ix–xxiii in Bob Jessop (ed.) *The Parisian Regulation School, Volume 1* of the *Regulation School and the Crisis of Capitalism*. Cheltenham: Edward Elgar.

———. 2001b. "Introduction". pp. xxv–xxxvi in Bob Jessop (ed.) *The Parisian Regulation School, Volume 1* of the *Regulation School and the Crisis of Capitalism*. Cheltenham: Edward Elgar.

Jessop, Robert and Ngai-Ling Sum. 2006. *Beyond the Regulation Approach*. London: Edward Elgar.

Johnson, Chalmers. 1982. *MITI and the Japanese Miracle*. Stanford, CA: Stanford University Press.

Johnson, Steven. 2012. *Future Perfect: The Case for Progress in a Networked Age*. New York: Riverhead Publishing.

Juran, Joseph. 1951. *The Quality Control Handbook*. New York: McGraw-Hill.

Juran, Joseph and Joseph A. De Feo. 2010. *Juran's Quality Handbook: The Complete Guide to Performance Excellence*, 6th Edition. New York: McGraw-Hill.

Juran, Joseph and A. Blanton Godfrey. 1999. *Juran's Quality Handbook*, 5th Edition. New York: McGraw-Hill.

Juran, Joseph and Frank Gryna (eds.). 2011. *Joseph Juran's Quality Handbook*, 4th Edition. New York: McGraw-Hill.

Juravich, Tom. 1985. *Chaos on the Shop Floor: A Workers View of Quality, Productivity and Management*. Philadelphia, PA: Temple University Press.

Jürgens, Ulrich. 1997. "Germany: Implementing Lean Production". pp. 109–16 in Thomas Kochan, Russell Lansbury, and John Paul MacDuffie (eds.) *After Lean Production*. Ithaca, NY: Cornell University Press.

———. 2004. "An Elusive Model—Diversified Quality Production and the Transformation of the German Automobile Industry". *Competition and Change*. 8(4):411–23.

Jürgens, Ulrich and Michael Krzywdzinski. 2016. *New Worlds of Work: Varieties of Work in Car Factories in the BRIC Countries*. Oxford: Oxford University Press.

Jürgens, Ulrich, Thomas Malsch, and Kurt Dohse. 1993. *Breaking from Taylorism*. Cambridge: Cambridge University Press.

Kalleberg, Arne L. 1989. "Linking Macro and Micro Levels: Bringing the Workers Back Into the Sociology of Work". *Social Forces*. 67:582–92.

———. 1998. "Commentary: The Institution of Gradual Retirement in Japan". pp. 92–100 in K. Warner Schaie and Carmi Schooler (eds.) *Impact of Work on Older Individuals*. New York: Springer Publishing.

———. 2000. "Nonstandard Employment Relations: Part-Time, Temporary and Contract Work". *Annual Review of Sociology*. 26:341–65.

———. 2007. *The Mismatched Worker*. New York: Norton.

———. 2011. *Good Jobs, Bad Jobs: The Rise of Polarized and Precarious Employment Systems in the United States, 1970s–2000s*. New York: Russell Sage Foundation.

———. 2018. *Precarious Lives: Job Insecurity and Well-Being in Rich Democracies*. Cambridge: Polity Press.

Kalleberg, Arne L. and James R. Lincoln. 1988. "The Structure of Earnings Inequality in the U.S. and Japan". *American Journal of Sociology*. 94 (Supplement on Organizations and Institutions):S121–53.

Kalleberg, Arne L. and Torger Reve. 1993. "Contracts and Commitment: Economic and Sociological Perspectives on Employment Relations". *Human Relations*. 46:1103–32.

Kalleberg, Arne L. and Jeremy Reynolds. 2003. "Work Attitudes and Nonstandard Work Arrangements in the United States, Japan and Europe". pp. 423–76 in Susan Houseman and Machiko Osawa (eds.) *Nonstandard Work in Developed Economies: Causes and Consequences*. Kalamazoo, MI: W.E. Upjohn Institute for Employment Research.

Kamata, Satoshi. 1982. *Japan in the Passing Lane*. New York: Pantheon.

Kaner, Cem, James Back, and Bret Pettichord. 2002. *Lessons Learned in Software Testing: A Context-Driven Approach*. New York: John Wiley & Sons.

Kanigel, Robert. 1997. *The One Best Way: Frederick Winslow Taylor and the Enigma of Efficiency*. New York: Viking.

Karatsu, Jajime. 1986. *Tough Words for American Industry*. Forward by Norman Bodek. Cambridge, MA: Productivity Press.

Katz, Harry, Wonduck Lee, and Joohee Lee. 2004. *The New Structure of Labor Relations: Tripartism and Decentralization*. Ithaca, NY: ILR Press.

Kaufman, Alexander. 2017. "Walmart Touts Itself as an Environmental Leader. Its Workers Say Otherwise". *Huffington Post*, May 1, 2017. www.huffpost.com/entry/walmart-environmentalism_n_5907524ae4b02655f83eba40 Accessed June 15, 2019.

Kaufman, Bruce E. 2004. "Toward an Integrative Theory of HR Management". pp. 368–73 in Bruce Kaufman (ed.) *Theoretical Perspectives on Work and the Employment Relationship*. Madison/Urbana: Industrial Relations Research Association.

Keller, Maryann. 1989. *Rude Awakening: The Rise, Fall and Struggle for Recovery of General Motors*. New York: William Morrow & Co.

———. 1993. *Collision: GM, Toyota, Volkswagen and the Race to Own the 21st Century*. New York: Doubleday.

Kenney, Martin. 2000. *Understanding Silicon Valley*. Stanford, CA: Stanford Business Books.

Kenney, Martin and Richard Florida. 1993. *Beyond Mass Production: The Japanese System and its Transfer to the US*. New York: Oxford University Press.

Keys, J. Bernard, Luther Denton, and Thomas Miller. 1994. "The Japanese Management Theory Jungle—Revisited". *Journal of Management*. 20(2):373–402.

Keys, J. Bernard and Thomas. R. Miller. 1984. "The Japanese Management Theory Jungle". *The Academy of Management Review*. 9(2):342–53.

Kharif, Olga. 2007. "Walmart's Latest Sale: Broadband." *Businessweek*. October 8th.

Kidder, John Tracy. 1981. *The Soul of a New Machine*. Boston, MA: Little Brown.

Kiesiel, Ralph. 2007. "Honda to Ala. Workers: UAW Not Needed". *The Automotive News*, October 8. www.autonews.com/article/20071008/ANA/710080335/honda-to-ala.-workers%3A-uaw-not-needed Accessed August 10, 2018.

Kim, Larry. 2017. "The Results of Google's Tam-Effectiveness Research Will Make You Rethink How You Build Teams". *Inc.com*. https://medium.com/the-mission/the-results-of-googles-team-effectiveness-research-will-make-you-rethink-how-you-build-teams-902aa61b33 Accessed December 10, 2018.

Kirk, Richard. 2015. "Cars of the Future: The Internet of Things in the Automotive Industry". *Network Security*. 9:16–18.

Kline, Daniel B. 2018. "Fast Food Future: Your McDonald's Experience Is About to Change". *USA Today*, May 2, 2018. www.usatoday.com/story/money/restaurants/2018/05/02/your-McDonald's-experience-is-about-to-change/34473093/ Accessed May 1, 2019.

Knights, David and Hugh Willmott (eds.). 1990. *Labour Process Theory*. London: Palgrave Macmillan.

Kochan, Thomas and Lee Dyer. 2017. *Shaping the Future of Work: A Handbook for Action and a New Social Contract*. Cambridge, MA: MIT Press.

Kochan, Thomas, Russell Lansbury, and John Paul MacDuffie (eds.). 1997a. *After Lean Production: Evolving Employment Practices in the World Auto Industry*. Ithaca, NY: Cornell University Press.

———. 1997b. "Introduction". pp. 3–8 in Thomas Kochan, Russell Lansbury, and John Paul MacDuffie (eds.) *After Lean Production: Evolving Employment Practices in the world auto industry*. Ithaca, NY: Cornell University Press.

———. 1997c. "Conclusion: After Lean Production". pp. 303–24 in Thomas Kochan, Russell Lansbury, and John Paul MacDuffie (eds.) *After Lean Production: Evolving Employment Practices in the World Auto Industry*. Ithaca, NY: Cornell University Press.

Koeber, Charles, David Wright, and Elizabeth Dingler. 2012. "Self-Service in the Labor Process: Control and Consent in the Performance of "Consumptive Labor." *Humanity & Society*. 36(1):6–29

Köhler, Holm Detlev. 2019. "The Sociology of Work in Germany". pp. 81–124 in Paul Stewart and Jean-Marie Durand (eds.) *The Palgrave Handbook of the Sociology of Work*. London: Palgrave-MacMillan.

Kondo, Yoshio. 1994. "Kaoru Ishikawa: What He Thought and Achieved, a Basis for Further Research". *Quality Management Journal*. 1(4):86–91.

Kono, Toyohiro and Stewart Clegg. 2001. *Trends in Japanese Management*. New York: Palgrave.

Kopp, Rochelle. 2010. "How Do the Japanese Put in Their Two Cents? *Ringi*". *Japan Intercultural Consulting News*. April 27. www.japanintercultural.com/en/news/default.aspx?newsID=68 Accessed January 10, 2019.

———. 2012. "Defining *Nemawashi*". *Japan Intercultural Consulting News*. December 20. www.japanintercultural.com/en/news/default.aspx?newsID=234 Accessed January 10, 2019.

———. 2014. "*Oidashibeya*: Japanese Purgatory". *Japan Intercultural Consulting News*. August 12, 2014. www.japanintercultural.com/en/news/default.aspx?newsID=299 www.japanintercultural.com/en/news/default.aspx?newsID=299 Accessed June 24, 2019.

Kornelakis, Andreas and Michail Veliziotis. 2019. "Job Quality in Europe: Regulation, Workplace Innovation and Human Resources Practices". pp. 173–89 in Gabriele Suder, Monica Riviere and Johan Lindeque (eds.) *The Routledge Companion to European Business*. London: Routledge.

Kotter, John. 1996. *Leading Change*. Boston, MA: Harvard Business School Press.

Krafcik, John F. 1988. "Triumph of the Lean Production System". *Sloan Management Review*. 30(1):41–52.

Krajewski, Lee; Manoj Malhotra, and Larry Rizman. 2015. *Operations Management: Processes and Supply Chains*, 11th Edition. London: Pearson.

Krzywdzinski, Michael. 2011. "Exporting the German Work Model to Central and Eastern Europe". pp. 99–116 in Sylvie Contrepois, Violaine Delteil, Patrick Dieuaide, and Steve Jefferys (eds.) *Globalizing Employment Relations*. Basingstoke/New York: Palgrave Macmillan.

———. Forthcoming. "Alternative Models of Lean Production: Germany". In Thomas Janoski and Darina Lepadatu (eds.) *The International Handbook of Lean Organization*. New York: Cambridge University Press.

Kurzweil, Ray. 2005. *The Singularity Is Near*. Harmondsworth: Penguin.

Labovitz, G., V. Rosansky, and Y. S. Chang. 1993. *Making Quality Work: A Leadership Guide for the Results-Driven Manager*. New York: Harper Business.

LaCroix, Jean Guy and Gaëtan Tremblay. 1997. "The Information Society and Culture Industries Theory". *Current Sociology*. 45(4):162–83.

Lafasto, Frank and Carl Larson. 2001. *When Teams Work Best*. Thousand Oaks, CA: Sage.

Lane, Bill. 2008. *Jacked Up*. New York: McGraw-Hill Books.

Larson, Aaron, Sonal Patel, and Darrell Proctor. 2019. "Plant of the Year: Egypt Megaproject: An Expedited Power Transformation". *Power: News and Technology for the Global Energy Industry Since 1882*. 163(7):24–28.

Lawler, Edward E. 2006. *Built to Change: How to Achieve Sustained Organizational Change*. SF, CA: Wiley/Jossey-Bass

Lawler, Edward E. III and Susan A. Mohrman. 1985. "Quality Circles After the Fad". *Harvard Business Review*, January.

Lean Six Sigma. 2017. "Costco Uses Six Sigma Methodologies for High Achievement". October 30, 2017. Lean Six Sigma Articles. www.6sigma.us/retail/costco-uses-six-sigma-methodologies-high-achievement/ Accessed June 3, 2018.

Leatherbarrow, Charles and Gary Rees. 2017. "HRM: The Added Value Debate". pp. 106–44 in Gary Rees and Paul E. Smith (eds.) *Strategic Human Resource Management: An International Perspective*. Thousand Oaks, CA: Sage Press.

Lee, Ching Kwan. 2007. *Against the Law: Labor Protests in China's Rustbelt and Sunbelt*. Berkeley: University of California Press.

Lemon, Jason. 2018. "Nike Called Out for Low Wages in Asia amid Colin Kaepernick Ad Promotion. *Newsweek*. September 6. https://www.newsweek.com/nike-factory-workers-still-work-long-days-low-wages-asia-1110129 Accessed June 13, 2019.

Leonhardt, David. 2019. "New York Did Us All a Favor By Standing Up to Amazon". *New York Times*, February 18, 2019:A19.

Lepadatu, Darina and Thomas Janoski. 2011. *Diversity at Kaizen Motors: Gender, Race, Age and Insecurity in a Top Japanese Transplant*. Lanham, MD: University Press of America.

———. 2018. "Just in Time Workforce: Temporary Workers as a Structural Aspect of Lean Production in the United States Auto Industry". *International Journal of Automotive Technology and Management*. 18(2):160–77.

Leswing, Kif. 2018. "Apple Just Put the Finishing Touches on Its New 5 Billion $ Headquarters—And the Results Are Stunning". *Business Insider*. January 16, 2018. www.

businessinsider.com/apple-park-spaceship-campus-2-finished-results-photos-2018-1 Accessed November 9, 2018.

Levenson, Alec. 2000. "Long-Run Trends in Part-Time and Temporary Employment: Toward an Understanding". pp. 335–97 in David Neumark (ed.) *On the Job: Is Long-Term Employment a Thing of the Past?* Thousand Oaks, CA: Sage.

Levi, Daniel. 2017. *Group Dynamics for Teams*, 5th Edition. Thousand Oaks, CA: Sage.

Levinson, William A. 2002. *Henry Ford's Lean Vision: Enduring Principles from the First Ford Motor Plant.* New York: Productivity Press.

Levy, Steven. 2011. *In the Plex: How Google Thinks, Works, and Shapes Our Lives.* New York: Simon and Schuster.

Lewin, Arie and Jisung Kim. 2004. "The Nation-State and Culture as Influences on Organizational Change and Innovation". pp. 324–53 in Marshall S. Poole and Andrew van de Ven (eds.) *Handbook of Organizational Change and Innovation.* New York: Oxford University Press.

Lewin, Arie, Silvia Massini, and Carine Peeters. 2009. "Why Are Companies Off-shoring Innovation?" *Journal of International Business Studies.* 40(8):1406–16.

Lichtenstein, Nelson. 2006. *Wal-Mart: The Face of Twenty-First Century Capitalism.* New York: New Press.

———. 2010. *The Retail Revolution.* New York: Picador.

Liker, Jeffrey. 2004. *The Toyota Way.* New York: McGraw Hill.

———. 2011. "Toyota Recall Crisis: What Have We Learned?" *Harvard Business Review.* February 11, 2011. https://hbr.org/2011/02/toyotas-recall-crisis-full-of Accessed October 30, 2018.

Liker, Jeffrey and Gary Convis. 2011. *The Toyota Way to Lean Leadership* New York: McGraw-Hill.

Liker, Jeffrey and James K. Franz. 2012. *The Toyota Way to Continuous Improvement* New York: McGraw Hill.

Liker, Jeffrey and Michael Hoseus. 2008. *The Toyota Culture: The Heart and Soul of the Toyota Way.* New York: McGraw-Hill.

Liker, Jeffrey and David Meier. 2005. *The Toyota Way Fieldbook* New York: McGraw-Hill.

———. 2007. *Toyota Talent: Developing Your People the Toyota Way.* New York: McGraw-Hill.

Liker, Jeffrey and Timothy Ogden. 2011. *Toyota Under Fire: Lessons for Turning Crisis Into Opportunity.* New York: McGraw Hill.

Liker, Jeffrey and Karyn Ross. 2016. *The Toyota Way to Service Excellence: Lean Transformation in Service Organizations.* New York: McGraw-Hill.

Lincoln, James R. 1987. "Japanese Industrial Organization in Comparative Perspective". *Annual Review of Sociology.* 13:289–312.

Lincoln, James R. and Christina Ahmadjian. 2001. "Keiretsu, Governance, and Learning: Case Studies in Change from the Japanese Automotive Industry". *Organizational Science.* 12:683–701.

Lincoln, James R. and Michael Gerlach. 2004. *Japan's Network Economy: Structure, Persistence and Change.* New York: Cambridge University Press.

Lincoln, James, Didier Guillot, and Matthew Sargent. 2017. "Business Groups, Networks, and Embeddedness: Innovation and Implementation Alliances in Japanese Electronics, 1985–1998". *Industrial and Corporate Change.* 26(3):357–78.

Lincoln, James R. and Arne L. Kalleberg. 1985. "Work Organization and Workforce Commitment: A Study of Plants and Employees in the United States and Japan". *American Sociological Review.* 50:738–60.

———. 1990. *Culture, Control, and Commitment: A Study of Work Organization and Work Attitudes in the United States and Japan*. New York: Cambridge University Press.

———. 1996. "Commitment, Quits and Work Organization in Japanese and U.S. Plants". *Industrial and Labor Relations Review*. 50:39–59.

Lincoln, James R. and Masahiro Shimotani. 2010. "Business Networks in Postwar Japan: Whither the Keiretsu?" pp. 127–56 in Asli Colpan, Takashi Hikino and James Lincoln (eds.) *The Oxford Handbook of Business Groups*. New York: Oxford University Press.

Lipietz, Alain. 1987. "Rebel Sons: The Regulation School". pp. 3–12 in Bob Jessop (ed.) *The Parisian Regulation School, Volume 1* of the *Regulation School and the Crisis of Capitalism*. Cheltenham: Edward Elgar.

———. 1997. "The Post-Fordist World: Labour Relations, International Hierarchy and Global Ecology". *Review of International Political Economy*. 4(1):1–41.

Lipson, Hod and Melba Kurman. 2013. *Fabricated: The New World of 3D Printing*. Indianapolis, ID: Wiley.

Locke, Richard M. (ed.) 2014. *Production in the Innovation Economy*. Cambridge, MA: MIT Press.

Locke, Richard, Fei Qin, and Alberto Brause. 2006. "Does Monitoring Improve Labor Standards? Lessons from Nike" MIT Sloan Research Paper No. 4612-06. Cambridge, MA: MIT.

Locke, Richard, Thomas Kochan, and Michael Piore (eds.). 1995. *Employment Relations in a Changing World Economy*. Cambridge, MA: MIT Press.

Lodgaard, Eirin, Jonas Ingvaldsen, Inger Gamme, and Silje Aschehoug. 2016. "Barriers to Lean Implementation: Perceptions of Top Managers, Middle Managers and Workers". *Procedia CIRP*. 57:595–600.

Lombardo, Jessica. 2017. "Ford Motor Company's Organizational Culture Analysis". *Panmore Institute*. February 5, 2017. http://panmore.com/ford-motor-company-organizational-culture-analysis Accessed May 29, 2019.

Lucio, Miguel M. and Robert MacKenzie. 2017. "The State and the Regulation of Work and Employment: Theoretical Contributions, Forgotten Lessons and New forms of Engagement". *International Journal of Human Resource Management*. 28(21):2983–3002.

Lund, Susan, James Manyika, Jonathan Woetzel, Jacques Bughin, Mekala Krishan, Jeongmin Seong, and Mac Muir. 2019. *Globalization in Transition: The Future of Trade and Value Chains*. McKinsey Report, Featured Insights. January 2019.

MacDuffie, John Paul. 1996. "International Trends in Work Organization in the Auto Industry". pp. 71–113 in Lowell Turner and Kirsten Wever (eds.) *The Comparative Political Economy of Industrial Relations*. Ithaca, NY: Cornell University Press.

———. 2013. "Modularity-as-Property, Modularization-as-Process, and "Modularity-as-Frame." *Global Strategy Journal*. 3:1–40.

MacDuffie, John Paul and Fritz Pil. 1995. "The International Assembly Plant Study: Philosophical and Methodological Issues". In S. Babson (ed.) *Lean Production and Labor*. Detroit: Wayne State University Press.

———. 1996. "From Fixed to Flexible: Automation and Work Organization Trends from The International Assembly Plant Survey". pp. 328–54 in Koichi Shimokawa, Ulrich Jürgens, and Takahiro Fujimoto (eds.) *Transforming Auto Assembly: International Experiences with Automation and Work Organization*. Frankfurt: Springer Verlag.

———. 1997. "Changes in Auto Industry Employment Practices: An International Overview". pp. 9–42 in Thomas Kochan, Russell Lansbury, and John Paul MacDuffie (eds.) *After Lean Production: Evolving Employment Practices in the World Auto Industry*. Ithaca, NY: Cornell University Press.

MacDuffie, John Paul, K. Sethuraman, and M. L. Fisher. 1996. "Product Variety and Manufacturing Performance: Evidence from the International Automotive Assembly Plant Study". *Management Science*. 42(3):350–69.

Macrotrends. 2019. "Honda Revenue 2006–2018/HMC". www.macrotrends.net/stocks/charts/HMC/honda/revenue Accessed April 22, 2019.

Madrid, Manuel. 2018. "Low Wages, Sexual Harassment as McDonald's Profits Soar". *The American Prospect*. May 29. https://prospect.org/economy/fast-food-blues-workers-protest-low-wages-sexual-harassment-mcdonald-s-profits-soar/ Accessed October 14, 2019.

Magee, David. 2003. *Turn Around: How Carlos Ghosn Rescued Nissan*. New York: Harper Collins.

Magee, David. 2004. *Ford Tough: Bill Ford and the Battle to Rebuild America's Automaker*. New York: Wiley.

Maher, Kris. 2007. "Wal-Mart Seeks New Flexibility in Worker Shifts". *Wall Street Journal*. November 5. https://www.wsj.com/articles/SB116779472314165646 Accessed July 13, 2014.

Mair, Andrew. 1998. "Internationalization at Honda: Transfer and Adaptation of Management Systems". *Employee Relations*. 20(3):285–302.

———. 1999. "Learning from Honda". *Journal of Management*. 36(1):26–44.

Manjoo, Farhad. 2018. "How Apple Thrived in a Season of Tech Scandals". *New York Times*. September 28, 2018.

Mann, James. 2007. *The China Fantasy: How Our Leaders Explain Away Chinese Repression*. New York: Viking/Penguin.

Marsden, David. 1999. *A Theory of Employment Systems: Micro-Foundations of Societal Diversity*. Oxford: Oxford University Press.

Martin, James William. 2009. *Lean Six Sigma in the Office*. Boca Raton, FL: CRC Press of Taylor Francis.

Martinez, Michael. 2016. "Ford Unveils 10 Year Plan to Transform Dearborn Campus". *Detroit News*. April 12, 2015. www.detroitnews.com/story/business/autos/ford/2016/04/12/ford-transform-campus/82935594/ Accessed March 3, 2018.

Marx, Karl. 1976. *Capital: A Critique of Political Economy Volume One*. Harmondsworth: Penguin Books.

McCann, Hilton. 2009. *Off-shore Finance*. New York: Cambridge University Press.

McCarthy, Niall. 2017. "Costco Named America's Best Employer 2017". *Forbes*. May 10, 2017. www.forbes.com/sites/niallmccarthy/2017/05/10/costco-named-americas-best-employer-2017-infographic/#158365116022 Accessed July 2, 2019.

McDonald's Corporation. 2018. *McDonald's Corporation—US SEC Form 10-K, 2018*. https://corporate.McDonald's.com/content/dam/gwscorp/investor-relations-content/annual-reports/McDonald's_2018_Annual_Report.pdf#MCD-12312018X10K_HTM_SFA5FE41EA0705915B56664233E831CAB Accessed October 30, 2019.

McGrath, Joseph and Franziska Tschan. 2004. "Dynamics in Groups and Teams: Groups as Complex Action Systems". pp. 50–72 in Marshall S. Poole and Andrew van de Ven (eds.) *Handbook of Organizational Change and Innovation*. New York: Oxford University Press.

McKinsey and Company. 2014. *The Lean Management Enterprise: A System for Daily Progress, Meaningful Purpose and Lasting Value*. www.mckinsey.com/~/media/mckinsey/industries/consumer%20packaged%20goods/our%20insights/the%20consumer%20sector%20in%202030%20trends%20and%20questions%20to%20consider/2014_lean_management_enterprise_compendium.ashx Accessed June 3, 2016.

Mehri, Darius. 2005. *Notes from Toyota-Land: An American Engineer in Japan*. Ithaca, NY: ILR and Cornell University Press.

Mejia, Zameena. 2018. "Meet the Family Whose Business Has Been the No. 1 Fortune 500 Company for Six Straight Years". *Make It*. New York: CNBC, May 23, 2018.

Menegus, Bryan. 2018. "Amazon's Aggressive Anti-Union Tactics Revealed in Leaked 45-minute Video". *Gizmodo*, September 25, 2018. https://gizmodo.com/amazons-aggressive-anti-union-tactics-revealed-in-leake-1829305201 Accessed March 3, 2019.

Merk, Jeroen. 2015. "Global Outsourcing and Socialization of Labour: The Case of Nike". pp. 115–31 in Kees van der Pijl (ed.) *The Handbook of International Political Economy of Production*. Cheltenham: Edward Elgar.

Mickle, Tripp. 2018. "Trump Presses Apple to Shift Production to US". *Wall Street Journal*. September 9, 2018. www.wsj.com/articles/trump-presses-apple-to-shift-production-to-u-s-1536432033 Accessed November 6, 2018.

Miller, Bridget. 2018. "Is there a Conflict of Interest in the HR Role?" *HR Daily Advisor*. https://hrdailyadvisor.blr.com/2018/01/04/conflict-interest-hr-roles/ Accessed June 5, 2019.

Miller, Jon. 2013. *Lean Leadership Lessons from Costco Wholesale*. Morro Bay, CA: Gemba Academy, February 15, 2013. https://blog.gembaacademy.com/2013/02/15/lean_leadership_lessons_from_costco_wholesale/ Accessed March 21, 2019.

Mintzberg, Henry. 1987. "Crafting Strategy". *Harvard Business Review*. July–August:66–75.

——— (ed.). 1989. *Mintzberg on Management*. New York: Free Press.

———. 1996. "Introduction: CMR Forum: The Honda Effect Revisited". *California Management Review*. 38(4):78–79.

Moberg, David. 1978. *Rattling the Golden Chains: Conflict and Consciousness of Auto Workers*. University of Chicago, Anthropology Dissertation (unpublished).

Mohr, Bernard and Pierre van Amelsvoort. 2016. *Co-Creating Humane and Innovative Organizations: Evolutions in the Practice of Socio-technical System Design*. Milton, VT: Global STS-D Network Press.

Monden, Yasuhiro. 2011 (1987). *Toyota Production System: An Integrated Approach to Just-in-Time*, 4th Edition. Boca Raton, FL: CRC Press of Taylor Francis.

Montgomery, David. 1997. *The Fall of the House of Labor*. New York: Cambridge University Press.

Montgomery, Douglas. 2009. *Introduction to Statistical Quality Control*, 6th Edition. New York: Wiley.

Morgan, James and Jeffrey Liker. 2006. *The Toyota Product Development System*. New York: Productivity Press.

Morris, Jonathan and Barry Wilkinson. 1995. "The Transfer of Japanese Management of Alien Institutional Environments". Special Issue, *Journal of Management Studies*. 32(6):719–30.

Mossberg, Walt. 2016. "Mossberg: Five Things I Learned from Jeff Bezos at Code". *The Verge*. June 8, 2016. www.theverge.com/2016/6/8/11879684/walt-mossberg-jeff-bezos-amazon-blue-origin-code-conference-2016 Accessed June 20, 2019.

Moyer, Justin W. 2015. "Alphabet, Now Google's Overlord, Ditches 'Don't Be Evil' for 'Do the Right Thing.'" *Washington Post*, October 5, 2015.

MSNBC. 2007. "CEO Insisted in Bringing the Taurus Back". *MSNBC*. February 2007. www.nbcnews.com/id/17033668/ns/business-autos/t/mulally-insisted-bringing-taurus-back/ Accessed June 2, 2019.

Mulholland, Kate. 2011. "In Search of Teamworking in a Major Supermarket: A Fig-Leaf for Flexibility". pp. 213–31 in Irena Grugulis and Odul Bözkürt (eds.) *Retail Work*. New York: Palgrave Macmillan.

Mulholland, Kate and Paul Stewart. 2013. "Working in Food Distribution: Global Commodity Chains and Lean Logistics". *New Political Economy*. 19(4):534–8.
Murakami, Thomas. 1997. "The Autonomy of Teams in the Car Industry: A Cross National Comparison". *Work, Employment and Society*. 11(4):749–58.
Murphy, Craig and JoAnne Yates. 2009. *The International Organization for Standardization* (ISO). New York: Routledge.
Musacchio, Aldo and Sergio G. Lazzarini. 2014. *Reinventing State Capitalism: Leviathan in Business, Brazil and Beyond*, 1st Edition. Cambridge, MA: Harvard University Press.
———. 2018. "State Ownership Reinvented? Explaining Performance Differences Between State-Owned and Private Firms". *Corporate Governance: An International Review*.
Musacchio, Aldo, Sergio G. Lazzarini, and Ruth Aguilera. 2015. "New Varieties of State Capitalism: Strategic and Governance Implications". *Academy of Management Perspectives*. 29(1):115–31.
Nader, Ralph. 1965. *Unsafe at Any Speed*. New York: Pocket Books.
———. 1970. *Unsafe at Any Speed*, 2nd Edition. New York: Grossman.
Nader, Ralph, and Clarence Ditlow. 2007. *The Lemon Book: Auto Rights*, 4th Edition. Chicago: Moyer Bell Ltd.
Napolitano, Maida. 2000. *Making the Move to Crossdocking: A Practical Guide to Planning, Designing, and Implementing a Cross Dock Operation*. Oakbrook, IL: WERC.
National Retailers Federation (NRF). 2019. "Top 50 Global Retailers". *National Retailers Federation Website*. https://nrf.com/resources/top-retailers/top-50-global-retailers Accessed June 2, 2019.
Natto, Hassan. 2014. "Wal-Mart Supply Chain Management". *International Journal of Scientific and Engineering Research*. 5(1):1023–6.
Neary, Brigitte U. 1992. "Management in the United States and (West) German Machine Tool Industry: Historically Rooted and Socioculturally Contingent". Duke University Dissertation, Sociology Department.
Nelson, Daniel. 1996. *Managers and Workers: Origins of the Twentieth-Century Factory System in the United States, 1880–1920*. Madison: University of Wisconsin Press.
Nelson, Dave, Rick May, and Patricia Moody. 1998/2007. *Powered by Honda: Developing Excellence in the Global Enterprise*. New York: Wiley/Lincoln, NE: Authors Guild Backinprint.com.
Netland, Torbjørn and Daryl Powell (eds.). 2017. *The Routledge Companion to Lean Management*. New York: Routledge.
Neugebauer, John. 2017. "Equality and Diversity in the Workplace". pp. 324–68 in Gary Rees and Paul E. Smith (eds.) *Strategic Human Resource Management: An International Perspective*. Thousand Oaks, CA: Sage.
Neumark, David, Junfu Zhang, and Stephen Ciccarella. 2007. "The Effects of Wal-Mart on Local Labor Markets". *Journal of Urban Economics*. 63:405–30.
Newsom, Gavin with Lisa Dickey. 2013. *Citizenville: How to Take the Town Square Digital and Reinvent Government*. New York: Penguin.
Ng, David, Gord Vail, Sophia Thomas, and Nicki Schmidt. 2015. "Applying the Lean Principles of the Toyota Production System to Reduce Wait Times in the Emergency Department". *Canadian Journal of Emergency Medicine*. 12(1):50–57.
Nguyen, Thi Thu Ha. 2017. "Wal-Mart's successfully Integrated Supply Chain and the Necessity of Establishing the Triple-A Supply Chain in the 21st Century". *Journal of Economics and Management*. 29(3):102–17.
Nicas, Jack. 2018. "Apple, Spurned by Others, Signs Deal with Volkswagen For Driverless Cars". *New York Times*. May 24, 2018. www.cnbc.com/2018/05/23/walmart-is-the-no-1-fortune-500-company-for-the-6th-straight-year.html Accessed December 8, 2018.

Nicas, Jack. 2019. "Apple's Plan to Buy $75 Billion of Its Stock Fuels Spending Debate". *New York Times*, April 30. https://www.nytimes.com/2019/04/30/technology/apple-stock-buyback-quarterly-results.html Retrieved October 1, 2019.

Nike, Inc. 2007. *Corporate Responsibility Report*. Beaverton, OR. http://www.nikebiz.com/crreport/content/strategy/2-1-4-a-new-model-and-shift-to-sustainable-business-and-innovation.php?cat=cr-strategy Accessed March 11, 2014.

Noe, Raymond, John Hollenbeck, Barry Gerhart, and Patrick Wright. 2017. *Fundamentals of Human Resource Management*, 7th Edition. New York: McGraw-Hill.

Nolan, Mary. 1994. *Visions of Modernity: American Business and the Modernization of Germany*. New York: Oxford University Press.

Nollen, Stanley and Helen Axel. 1996. *Managing Contingent Workers: How to Reap the Benefits and Reduce the Risks*. New York: American Management Association.

Nye, David. 2013. *America's Assembly Line*. Cambridge, MA: MIT Press.

O'Boyle, Thomas. 1998. *At Any Cost: Jack Welch, General Electric, and the Pursuit of Profit*. New York: Vintage.

O'Connor, James. 1973. *The Fiscal Crisis of the State*. New York: St. Martins.

OECD. 2009. *State Owned Enterprises and the Principle of Competitive Neutrality*. Paris: OECD.

———. 2018. *Responsible Business Conduct: OECD Guidelines for Multinational Enterprises*. Paris: OECD. http://mneguidelines.oecd.org/themes/human-rights.htm Accessed January 17, 2019.

Offe, Klaus. 1986. *Contradictions of the Welfare State*. London: Hutchinson.

Ohio Honda. 2018. "Honda Builds Its 25 Millionth Vehicle in the US". March 6, 2018. https://ohio.honda.com/article/honda-builds-its-25-millionth-automobile-in-the-u.s Accessed August 8, 2018.

Ohno, Taiichi. 1988a. [1978 in Japanese]. *Toyota Production System: Beyond Large-Scale Production*. Forward by Norman Bodek. Cambridge, MA: Productivity Press.

———. 1988b. *Workplace Management*. Cambridge, MA: Productivity Press.

Onetto, Marc. 2014. "When Toyota Met e-Commerce: Lean at Amazon". *McKinsey Quarterly*. February. www.mckinsey.com/business-functions/operations/our-insights/when-toyota-met-e-commerce-lean-at-amazon Accessed December 12, 2018.

Osono, Emi, Norihiko Shizumi, and Hirotaka Takeuchi. 2008. *Extreme Toyota: Radical Contradictions That Drive Success at the World's Best Manufacturer*. New York: Wiley.

O'Toole, Jack. 1996. *Forming the Future: Lessons from the Saturn Corporation*. London: Blackwell.

Ouchi, William. 1980. *Theory Z: How American Business Can Meet the Japanese Challenge*. New York: Avon Books.

Packer, George. 2013. "Change the World: Silicon Valley Transfers its Slogans—and Its Money—to the Realm of Politics". *The New Yorker*, May 2, 2013:44–55.

———. 2014. "Cheap Words: Amazon Is Good for Customers. But Is It Good for Books?" *The New Yorker*, February 17, 2014.

Padgett, John and Walter Powell (eds.). 2012. *The Emergence of Organizations and Markets*. Princeton, NJ: Princeton University Press.

Page, Alan, Ken Johnston, and B. J. Rollison. 2008. *How We Test Software at Microsoft*. Redmond, WA: Microsoft Press.

Pardi, Tommaso. 2005. "Where Did It Bo Wrong?: Hybridization and Crisis of Toyota Manufacturing UK, 1989–2001". *International Sociology*. 20(1):93–118.

———. 2007. "Refining the Toyota Production System: The European Side of the Story". *New Technology, Work and Employment*. 22(1):2–20.

———. 2010. "Do State and Politics Matter? The Case of Nissan's Direct investment in Great Britain and Its Implications for British Leyland". *Business and Economic History*. 8:1–13.

Pascale, Richard. 1996. "The Honda Effect". *California Management Review*. 38(4):80–91.

Pasmore, William A. 1994. *Creating Strategic Change: Designing the Flexible, High-Performing Organization*. New York: Wiley.

———. 1995. "Social Science Transformed: The Socio-Technical Perspective". *Human Relations*. 48(1):1–21.

———. 2001. "Action Research in the Workplace". pp. 38–46 in Peter Reason and Hilary Bradbury (eds.) *Handbook of Action Research*. Thousand Oaks, CA: Sage.

———. 2015. *Leading Continuous Change: Navigating Churn in the Real World*. Oakland, CA: Berrett Koehler Publishers.

Pasmore, William A. and Susan Minot. 1994. "Developing Self-Managing Work Teams: An Approach to Successful Integration". *Compensation and Benefits Review*. 26(4):1–9. 95 *Human Relations*.

Pasmore, William A., Stu Winby, Susan Albers Mohrman, and Rick Vanasse. 2019. "Reflections: Sociotechnical Systems Design and Organizational Change". *Journal of Change Management*. 19(2):67–85.

Pava, Calvin. 1983. *New Office Technology: An Organizational Strategy*. New York: Free Press.

———. 1986. "New Strategies for Systems Change". *Human Relations*. 39(7):615–33.

Pearce, Jone L. 1993. "Toward and Organizational Behavior of Contract Laborers: Their Psychological Involvement and Effects on Employee Coworkers". *Academy of Management Journal*. 36(5):1082–96.

Peck, Jaime. 1996. *The Work-Place: The Social Regulation of Labor Markets*. New York: Guilford Press.

Perruci, Robert. 1994. *Japanese Auto Transplants in the Heartland: Corporatism and Community*. Hawthorne, NY: Aldine de Gruyter.

Petriglieri, Gianpiero, Susan Ashford and Amy Wrzesniewski. 2018. "Agony and Ecstasy in the Gig Economy: Cultivating Holding Environments for Precarious and Personalized Work Identities". *Administrative Science Quarterly* 64(1): 124–70.

Pettersen, Jostein. 2009. "Defining Lean Production: Some Conceptual and Practical Issues". *The TQM Journal*. 21(2):127–42.

Petterson, Hayley. 2019. "More Than 2,100 Stores Are Set to Open in 2019 as Costco, Dollar General and TJ Maxx Defy the Retail Apocalypse". *Business Insider*. March 13, 2019. www.businessinsider.com/costco-dollar-general-tj-maxx-aldi-stores-opening-2019-3 Accessed June 27, 2019.

Petty Consulting/Productions. 1991. *Deming of America* (Documentary). Cincinnati, OH: Petty Consulting/Productions.

Phillips, Jon. 2015. "Apple Wants Its Cars in Production by 2020, Report Says". *Macworld-Digital Edition*. 32(4).

Phillips, Matt. 2018. "Apple's 1 Trillion Milestone Reflects Rise of Powerful Megacompanies". *New York Times*. August 3, 2018.

Pil, Frits K. and Takahiro Fujimoto. 2007. "Lean and Reflective Production: The Dynamic Nature of Production Models". *International Journal of Production Research*. 45(16):3741–61.

Pil, Fritz K. and John Paul MacDuffie. 1996. "The Adoption of High Involvement Work Practices". *Industrial Relations*. 35(3):423–55.

Piore, Michael and Charles Sabel. 1986. *The Second Industrial Divide*. New York: Basic Books.

Poole, Marshall S. and Andrew van de Ven. 2004. "Theories of Organizational Change and Innovation Processes", pp. 374–98 in Marshall S. Poole and Andrew van de Ven (eds.) *Handbook of Organizational Change and Innovation*. New York: Oxford University Press.

Pope, Devin and Jaren Pope. 2015. "When Walmart Comes to Down: Always Low Housing Prices? Always?" *Journal of Urban Economics*. 87(1-Issue C):1–13.

Poppendieck, Mary and Tom Poppendieck. 2006. *Implementing Lean Software Development*. Boston: Addison-Wesley.

Porter, Michael E. 1980. *Competitive Strategy*. New York: Free Press.

———. 2008. "The Five Competitive Forces That Shape Strategy". *Harvard Business Review*. January 2008.

Price, Robert E. 2012. "Sol Price, Retail Revolutionary and Innovator". San Diego History Center. Sand Diego.

Pruijt, Hans. 2003. "Teams Between Neo-Taylorism and Anti-Taylorism". *Economic and Industrial Democracy*. 24(1):77–101.

Pyzdek, Thomas and Paul Keller. 2012. *The Handbook for Quality Management: A Complete Guide to Operational Excellence*, 2nd Edition. New York: McGraw-Hill.

———. 2014. *The Six Sigma Handbook*, 4th Edition. New York: McGraw Hill.

Quiñones, Renato, Marcelo Fuentes, Rodrigo Montes, Doris Soto and Jorge León-Muñoz. 2019. "Environmental Issues in Chilean Salmon Farming: A Review" *Reviews in Aquaculture*. 11(2):375–402.

Radford, George S. 1922. *The Control of Quality in Manufacturing*. New York: The Ronald Press. https://archive.org/details/controlofquality00radf/page/n5 Accessed June 15, 2017.

Ragin, Charles C. 2008. *Redesigning Social Inquiry: Fuzzy Sets and Beyond*. Chicago: University of Chicago Press.

Rao, Arun and Piero Scaruffi. 2011. *A History of Silicon Valley*. Silicon Valley, CA: Omniware.

Rawlinson, Nik. 2017. "History of Apple: The Story of Steve Jobs and The Company He Founded". www.macworld.co.uk/feature/apple/history-of-apple-steve-jobs-mac-3606104/ Accessed November 12, 2018.

Reich, Adam and Peter Bearman. 2018. *Working for Respect: Community and Conflict at Walmart*. New York: Columbia University Press.

Relihan, Tom. 2018. "How Costco's Obsession with Culture Drove Success". *MIT Management Sloan School: Ideas Made to Matter*. May 11, 2018. https://mitsloan.mit.edu/ideas-made-to-matter/how-costcos-obsession-culture-drove-success Accessed June 27, 2018.

Rice, Murray, Anthony Ostrander, and Chetan Tiwari. 2016. "Decoding the Development Strategy of a Major Retailer: Walmart's Expansion in the United States". *The Professional Geographer*. 68(4):640–9.

Rich, Nick, Nicola Bateman, Ann Esain, Lynn Massey, and Donna Samuel. 2012. *Lean Evolution: Lessons from the Workplace*. New York: Cambridge University Press.

Rifkin, Jeremy. 1995. *The End of Work*. New York: Putnam and Son.

Rinehard, James, Christopher Huxley, and David Robertson. 1997. *Just Another Car Factory? Lean Production and Its Discontents*. Ithaca, NY: ILR Press.

Ritzer, George. 2004 (1993). *The McDonaldization of Society*, Revised. Thousand Oaks, CA: Sage Press.

———. 2011. *Sociological Theory*. New York: McGraw-Hill.

———. 2019. *The McDonaldization of Society: Into the Digital Age*. Thousand Oaks, CA: Sage.

Rogers, Jackie Krasas. 1995. "Just a Temp: Experience and Structure of Alienation in Temporary Clerical Employment". *Work and Occupations*. 22(2):137–66.

———. 2000. *Temps: The Many Faces of the Changing Workplace*. Ithaca, NY: Cornell University Press.

Rohlen, Thomas. 1974. *For Harmony and Strength: Japanese White-Collar Organization in Anthropological Perspective*. Berkeley, CA: University of California Press.

Rossi, Monica, James Morgan, and John Shook. 2017. "Lean Product and Process Development". pp. 55–74 in Torbjørn Netland and Daryl Powell (eds.) *The Routledge Companion to Lean Management*. New York: Routledge.

Roth, Daniel. 2018. "LinkedIn Top Companies 2018: Where the US Wants to Work Now". *LinkedIn*, March 21, 2018. www.linkedin.com/pulse/linkedin-top-companies-2018-where-us-wants-work-now-daniel-roth/ Accessed June 7, 2019.

Rother, Mike. 2010. *Toyota Kata: Managing People for Improvement, Adaptiveness, and Superior Results*. New York: McGraw-Hill.

Rother, Mike, John Shook, Jim Womack, and Dan Jones. 2003 (1999). *Learning to See: Value Stream Mapping to Add Value and Eliminate MUDA, Version 1.3*. Cambridge, MA: Lean Enterprise Institute.

Rothfender, Jeffrey. 2014. *Driving Honda: Inside the World's Most Innovative Car Company*. New York: Penguin.

Rothschild, Emma. 1973. *Paradise Lost: The Decline of the Auto-Industrial Age*. New York: Random House.

Rothstein, Jeffrey. 2016. *When Good Jobs Go Bad: Globalization, De-Unionization and Declining Job Quality in the North American Auto Industry*. New Brunswick, NJ: Rutgers University Press.

Rubinstein, Saul and Thomas Kochan. 2001. *Learning from the Rings of Saturn*. Ithaca, NY: ILR of Cornell University Press.

Rushe, Dominic. 2018. "'It's a Huge Subsidy': The $4.8 bn Gamble to Lure Foxconn to America". *The Guardian*. July 2, 2018. www.theguardian.com/cities/2018/jul/02/its-a-huge-subsidy-the-48bn-gamble-to-lure-foxconn-to-america Accessed January 9, 2019.

Sabel, Charles. 1984. *Worker Politics*. New York: Cambridge University Press.

Sabel, Charles and Jonathan Zeitlin. 2002. *World of Possibilities: Flexibility and Mass Production in Western Industrialization*. New York: Cambridge University Press.

Sachs, Sybille and Edwin Rühli. 2011. *Stakeholders Matter: A New Paradigm for Strategy in Society*. New York: Cambridge University Press.

SACOM. 2018a. "Investigative Report: Apple Watch Series 4—Still Failed to Protect Teenage Student Workers". 22 pp. Hong Kong: SACOM. http://sacom.hk/wp-content/uploads/2018/10/Apple-Watch-Series-4-Still-Failed-to-Protect-Teenage-Student-Workers.pdf Accessed May 5, 2019.

———. 2018b. "Throw Away the Bad Apple: Labour Groups Call for Ethical Production". *Hong Kong, SACOM*. http://sacom.hk/2018/06/25/throw-away-the-bad-apple-labour-groups-call-for-ethical-production/ Accessed May 5, 2019.

Sainato, Michael. 2018. "Exploited Amazon Workers Need a Union: When Will They Get One?" *The Guardian*. July 8. www.theguardian.com/commentisfree/2018/jul/08/amazon-jeff-bezos-unionize-working-conditions Accessed January 10, 2019.

Salary.com. 2019. "For Walmart, Amazon and Costco". https://www1.salary.com/Amazon-Com-Inc-Executive-Salaries.html Accessed May 10, 2019.

Samaniego, Mayra and Ralph Deters. 2016. "Management and Internet of Things". *Procedia: Computer Science*. 94:137–43. https://ac.els-cdn.com/S1877050916317628/1-s2.0-S1877050916317628-main.pdf?_tid=ccff694a-b8b0-4fa1-a813-49fa17a560a1&acdnat=1546006123_5c281be93a766559da566eabc59194b3 Accessed December 10, 2018.

Samuel, Donna, Pauline Found, and Sharon Williams. 2015. "How Did the Publication of the Book *the Machine That Changed the World* Change Management Thinking?

Exploring 25 Years of Lean Literature". *International Journal of Operations and Production Management*. 35(10):1386–407.

Sandoval, Marisol. 2013. "Foxconned Labour as the Dark Side of the Information Age: Working Conditions at Apple's Contract Manufacturers in China". *tripleC*. 11(2):318–47.

Satariano, Adam and Peter Burrows. 2011. "Apple's Supply Chain's Secret? Hoard Lasers". *Bloomberg Businessweek*. November 7. (4253):35–37.

Sato, Masaaki. 2006. *The Honda Myth: The Genius and His Wake*. Translated by Hiroko Yoda with Foreword by Paul Ingrassia. New York: Vertical Inc.

Schmidt, Eric and Jonathan Rosenberg. 2014. *How Google Works*. New York: Grand Central Publishing.

Schneider, Michael. 2017. "Google Spent 2 Years Studying 180 Teams. The Most Successful Ones Shared These 5 Traits". *Inc.com* July 19, 2017. www.inc.com/michael-schneider/google-thought-they-knew-how-to-create-the-perfect.html Accessed December 10, 2018.

Schonberger, Richard J. 1982. *Japanese Manufacturing Techniques: Nine Hidden Lessons in Simplicity*. New York: Free Press.

———. 1986. *World Class Manufacturing: The Lessons of Simplicity Applied*. New York: Free Press.

———. 2007. "Japanese Production Management: An Evolution—With Mixed Success". *Journal of Operations Management*. 25:403–19.

———. 2011. "ASP, The Art and Science of Practice: Taking the Measure of Lean: Efficiency and Effectiveness, Part I". *Interfaces*. 41(2):182–93.

———. 2012. "Measurement of Lean Value Chains: Efficiency and Effectiveness". pp. 65–75 in Herbert Jodlbauer, Jan Olhager, and Richard J. Schonberger (eds.) *Modelling Value*. Heidelberg, FRG: Physica-Verlag/Springer,

———. 2018. "Frustration-Driven Process Improvement". *Business Horizons*. 61(2): 297–307.

Schonberger, Richard J. and Karen A. Brown. 2017. "Missing Link in Competitive Manufacturing Research and Practice: Customer-Responsive Concurrent Production". *Journal of Operations Management*. 49–51:83–87. https://w2.engr.uky.edu/lean/reference/terminology/ Accessed October 22, 2018.

Schwemmer, Carsten and Oliver Wieczorek. 2019. "The Methodological Divide in Sociology: Evidence from Two Decades of Journal Publications". *Sociology*. Online first. https://journals.sagepub.com/doi/pdf/10.1177/0038038519853146 Accessed July 30, 2019.

Scheiber, Noam. 2019. "Inside an Amazon Warehouse: Robots Ways Rub Off on Humans". *New York Times*. July 3, 2019. www.nytimes.com/2019/07/03/business/economy/amazon-warehouse-labor-robots.html Accessed July 8, 2019.

Schuh, Günther, Stefan Rudolf, and Chrisitan Mattern. 2017. "Lean Innovation". pp. 44–54 in Torbjørn Netland and Daryl Powell (eds.) *The Routledge Companion to Lean Management*. New York: Routledge.

Schumann, Michael, Martin Kuhlmann, Frauke Sanders, and Hans Joachim Sperling. 2006. *Auto5000: ein neues Produktionskonzept: Die deutsche Antwort auf den Toyota-Weg?* [Auto5000: A New Production Concept: The German Answer to the Toyota Way] Hamburg: VSA Verlag.

Schwartz, Nelson and Charles Duhigg. 2013. "Apple's Web of Tax Shelters Saved It Billions, Panel Finds". *New York Times*. May 21, 2013.

Sciara, Kristin Lovejoy and Susan Handy. 2018. "The Impacts of Big Box Retail on Downtown: A Case Study of Target in Davis (CA)". *Journal of the American Planning Association.* 84(1):45–60.

Scissors, Derek. 2009. "Deng Undone: The Costs of Halting Market Reform in China". *Foreign Affairs.* 88(3):24–33.

Segall, Ken. 2012. *Insanely Simple: The Obsession That Drives Apple's Success.* New York: Portfolio/Penguin.

Sellers, Patricia. 2009. "Sol Price, in Sam Walton's Memory". *Fortune.* December 17, 2009. http://fortune.com/2009/12/16/sol-price-in-sam-waltons-memory/ Accessed June 20, 2019.

Shaban, Hamz. 2019. "Amazon Workers in Minnesota Plan to Strike Over Working Conditions during Prime Day Sale". *Washington Post.* July 8, 2019. www.washington-post.com/business/2019/07/08/amazon-workers-plan-strike-during-prime-day-sale-over-working-conditions/?utm_term=.2ce25df44f6c Accessed July 10, 2019.

Shah, Rachna and Peter T. Ward. 2003. "Lean Manufacturing: Context, Practice Bundles, and Performance". *Journal of Operations Management.* 21(2):129–49.

———. 2007. "Defining and Developing Measures of Lean Production". *Journal of Operations Management.* 25(4):785–805.

Shatkin Gavin. 2011. "The True Meaning of the Singapore Model: State Led Capitalism and Urban Planning". Association of American Geographers.

Sherif, Muzafer, O. J. Harvey, B. Jack White, William Hood, and Carolyn Sherif. 1988. *The Robbers Cave Experiment: Intergroup Conflict and Cooperation.* Middletown, CT: Wesleyan University Press.

Sherman, Joe. 1994. *In the Rings of Saturn.* New York: Oxford University Press.

Shewhart, Walter A. 1980 (1936). *Economic Control of Quality of Manufactured Product.* Milwaukee, WI: ASQ Quality Press.

———. 2011 (1939). *Statistical Method from the Viewpoint of Quality Control.* Mineola, NY: Dover Publishing.

Shimokawa, Koichi, Ulrich Jürgens, and Takahiro Fujimoto. 1997. *Transforming Automobile assembly: Experience in Automation and Work Organization.* Berlin: Springer Verlag.

Shingo, Shigeo. 1984. *A Study of the Toyota Production System from an Industrial Engineering Viewpoint.* Tokyo: Japan Management Association.

———. 1985. *The Sayings of Shigeo Shingo: Key Strategies for Plant Improvement.* Cambridge, MA: Productivity Press.

———. 1988. *Non-Stock Production: The Shingo System for Continuous Improvement.* Cambridge, MA: Productivity Press.

———. 2017 (1959 in Japanese). *Kaizen and the Art of Creative Thinking.* Bellingham, WA: Enna Products *Corporation.*

Shook, John. 2008. *Managing to Lean: Using the A3 Management Process to Solve Problems, Gain Agreement, Mentor, and Lead.* Cambridge, MA: Lean Enterprise Institute.

Shulzhenko, Elena. 2017. *Reforming the Russian Industrial Workplace.* New York: Routledge.

Signorelli, Andrea. 2019. "Explaining Variation in the Social Performance of Lean Production: A Comparative Case Study of the Role Played by Workplace Unions' Framing of the System and Institutions". *Industrial Relations* 50(2):126–49.

Silver, Hilary. 2008. *Forces of Labor.* New York: Cambridge University Press.

Six Sigma. 2017. "Six Sigma Syllabus: Orange Belt". *Six Sigma Articles*. May 22, 2017. www.6sigma.us/six-sigma-articles/six-sigma-syllabus-orange-belt/ Accessed June 17, 2019.

Slater, Robert. 1998. *Jack Welch and the GE Way*. New York: McGraw-Hill.

Sloan, Alfred. 1986. *My Years with General Motors*. Harmondsworth: Penguin.

Smit, Barbara. 2008. *The Sneaker Wars: The Enemy Brothers Who Founded Adidas and Puma and the Family Feud That Forever Changed the Business of Sports*. New York: Harper Perennial.

Smith, Aaron. 2016. "Report Slams Walmart for 'Exploitative' Conditions in Asia Factories". *CNN Business*, June 1, 2016. https://money.cnn.com/2016/05/31/news/companies/walmart-gap-hm-garment-workers-asia/index.html Accessed March 15, 2019.

Smith, Adam. 1976 (1776). *An Inquiry Into the Nature and Causes of the Wealth of Nations*. Chicago: University of Chicago Press.

Smith, Chris. 2015. "Continuity and Change in Labour Process Analysis Forty Years After *Labor and Monopoly Capital*". *Labor Studies Journal*. 40:222–42.

Smith, Chris and Paul Thompson. 1999. "Re-evaluating the Labor Process Debate". pp. 205–31 in M. Wardell, T. L. Steiger, and P. Meiksins. (eds.) *Rethinking the Labour Process*. Albany, NY: State University of New York Press.

Smith, Chris and Matt Vidal. Forthcoming. The Labor Process View of Lean Production". In Thomas Janoski and Darina Lepadatu (eds.) *The International Handbook of Lean Production*. New York: Cambridge University Press.

Smith, Vicki and Esther Neuwirth. 2008. *The Good Temp*. Ithaca, NY: ILR Press or Cornell University Press.

Sobek, Durward K. II. 1997. *Principles That Shape Product Development: A Toyota-Chrysler Comparison*. Ann Arbor, MI: UMI Dissertation Services.

Soni, Phalguni. 2016. "How Costco Manages Inventory and Supply Chains". *Market Realist*, January 2016. https://marketrealist.com/2016/01/analyzing-costcos-inventory-supply-chain-management-strategies/ Accessed April 15, 2019.

Sorensen, Charles. 1956. *My Forty Years with Ford*. New York: Norton.

Sorge, Arndt and Wolfgang Streeck. 1988. "Industrial Relations and Technical Change: The Case for an Extended Perspective". pp. 19–44 in Richard Hyman and Wolfgang Streeck (eds.) *New Technology and Industrial Relations*. Oxford: Blackwell.

———. 2018. "Diversified Quality Production Revisited: Its Contribution to German Socio-Economic Performance Over Time". *Socio-Economic Review*. 16(3):587–612.

Sports Business Daily. 2012. "Marketing and Sponsorship at Nike". *Sports Business Daily*. October 22, 2012. www.sportsbusinessdaily.com/Daily/Issues/2012/10/22/Marketing-and-Sponsorship/Nike.aspx Accessed August 4, 2013.

St. John, Graham. 2017. "Civilized Tribalism: Burning Man, Event-Tribes and Maker Culture". *Cultural Sociology*. 12(1):3–21.

Stanton, Pauline, Richard Gough, Ruth Ballardie, Timothy Bartram, Greg Bamber and Amrik Sohal. 2014. "Implementing Lean Management/Six Sigma in Hospitals: Beyond Employment or Work Intensification". *International Journal of Human Resource Management*. 25(21):2926–40.

Stalk, Gorge and Thous Hout. 2003. *Competing Against Time: How Time-Based Competition is Reshaping Global markets*. New York: Free Press.

Statista. 2019a. "Statistical Portal for Market Data, Market Research". Statista GmbH, Hamburg, Germany.

———. 2019b. Costco www.statista.com/topics/4399/costco/ Accessed May 15, 2019.

———. 2019c. Ford www.statista.com/topics/1886/ford/ Accessed May 15, 2019.

———. 2019d. Honda www.statista.com/statistics/267275/worldwide-number-of-honda-employees/ Accessed May 15, 2019.

———. 2019e. Nissan. www.statista.com/statistics/232958/revenue-of-the-leading-car-manufacturers-worldwide/ Accessed May 15, 2019.

———. 2019f. Toyota www.statista.com/statistics/262752/total-net-revenues-of-toyota/ Accessed May 15, 2019.

———. 2019g. Walmart www.statista.com/topics/1451/walmart/ Accessed May 15, 2019.

Statt, Nick. 2017. "Amazon Will Start Collecting Sales Tax Nationwide Starting April 1st." *The Verge*. March 24. https://www.theverge.com/2017/3/24/15055662/amazon-us-sales-tax-collection-all-states Accessed May 30, 2017.

Stephan, Ute, Malcolm Patterson, Ciara Kelley, and Johanna Mair. 2016. "Organizations Driving Positive Social Change: A Review and an Integrative Framework of Change Processes". *Journal of Management*. 42(5):1250–81.

———. 2019. "Foxconn Finally admits its Empty Wisconsin 'Innovation Centers' Aren't Being Developed. *The Verge*. October 23, 2019. https://www.theverge.com/2019/10/23/20929453/foxconn-innovation-centers-on-hold-wisconsin-mount-pleasant-trump-deal Accessed December 19, 2019.

Stevenson, Richard W. 1992. "Guilty Plea Arranged in Israel Kickback Case". *New York Times*. July 3, 1992.

Stewart, Paul, Adam Mrozowicki, Andy Danford, and Ken Murphy. 2016. "Lean as Ideology and Practice: A Comparative Study of the Impact of Lean Production on the Working Life in Automotive Manufacturing in the United Kingdom and Poland". *Competition and Change*. 20(3):147–65.

Stewart, Paul, Mike Richardson, Andy Danford, Ken Murphy, Tony Richardson, and Vicki Wass. 2009. *We Sell Out Time No More: Workers' Struggles Against Lean Production in the British Car Industry*. London: Pluto Press.

Stewart, Thomas and Anand Raman. 2007. "Lessons from Toyota's Long Drive: Interview with Toyota President Katsuaki Watanabe". *Harvard Business Review*. July 2007.

Stone, Brad. 2013. *The Everything Store: Jeff Bezos and the Age of Amazon*. Boston: Little, Brown and Company.

———. 2014. *The Everything Store: Jeff Bezos and the Age of Amazon*. New York/Boston: Back Bay Books of Little Brown.

Stone, Kenneth E. 1997. "Impact of the Wal-Mart Phenomenon on Rural Communities". *Proceedings of Increasing Understanding of Public Problems and Policies*. The Farm Foundation, Chicago, IL.

Stout, Lynn. 2012. *The Shareholder Value Myth*. San Francisco, CA: Berrett-Koehler.

Stouten, Jeroen, Denise Rousseau, and David de Cremer. 2018. "Successful Organizational Change: Integrating the Management Practice and Scholarly Literatures". *Academy of Management Annals*. 12(2):752–88.

Strang, David and Michael Macy. 2001. "In Search of Excellence: Fads, Success Stories, and Adaptive Emulation". *American Journal of Sociology*. 107(1):147–82.

Streeck, Wolfgang. 1991. "On the Institutional Conditions of Diversified Quality Production". pp. 21–61 in Egon Matzner and Wolfgang Streeck (eds.) *Beyond Keynesianism: The Socio-Economics of Production and Full Employment*. Aldershot, UK: Edward Elgar.

———. 2008. *Re-Forming Capitalism*. New York: Oxford University Press.

———. 2012. "Skills and Politics: General and Specific". pp. 317–52 in Marius Busmeyer and Christine Tampusch (eds.) *The Political Economy of Collective Skill Formation*. New York: Oxford.

———. 2014. "How will Capitalism End?" *New Left Review*. 87(May–June).

Streitfeld, David. 2013. "Amazon Workers in Germany Strike Again". *New York Times*. December 16, 2013. https://bits.blogs.nytimes.com/2013/12/16/amazon-strikers-take-their-fight-to-seattle/ Accessed May 1, 2019.

———. 2014. "Amazon and Hachette Resolve Dispute". *New York Times*. November 13, 2014. www.nytimes.com/2014/11/14/technology/amazon-hachette-ebook-dispute.html Accessed May 15, 2019.

Stross, Randall. 1997. *The Microsoft Way*. New York: Basic Books.

Sturgeon, Timothy J. 2002. "Modular Production Networks: A New American Model of Industrial Organization". *Industrial and Corporate Change*. 11(3):451–96.

———. 2006. "Reflections on 'Profiting from Innovation.'" *Research Policy*. 35(8):1131–46.

Sugimori, Y. K. Kusunoki, F. Cho, and S. Uchikawa. 1977. "Toyota Production System and Kanban System Materialization of Just-in-Time and Respect-for-Human System". *International Journal of Production Research*. 15(6):553–64.

Sunder, Vijaya and Jiju Antony. 2018. "A Conceptual Lean Six Sigma Framework for Quality Excellence in Higher Education Institutions". *International Journal of Quality and Reliability Management*. 35(4):857–74.

Sussman, Gerald I. 1976. *Autonomy at Work: A Sociotechnical Analysis of Participative Management*. New York: Praeger.

Tabuchi, Hiroko. 2013. "Layoffs Taboo, Japan Workers are Sent to the Boredom Room". *New York Times*. August 16, 2013. www.nytimes.com/2013/08/17/business/global/layoffs-illegal-japan-workers-are-sent-to-the-boredom-room.html Accessed June 24, 2019.

———. 2014. "Airbag Flaw, Long Known to Honda and Takata, Led to Recalls". *New York Times*. September 11, 2014:A1. www.nytimes.com/2014/09/12/business/air-bag-flaw-long-known-led-to-recalls.html Accessed April 22, 2019.

Taguchi, Genichi. 1992. *Taguchi on Robust Technology Development: Bringing Quality Engineering Upstream*. Milwaukee, WI: ASME Press.

Taguchi, Genichi, Subir Chowdhury and Shin Taguchi. 1999. *Robust Engineering: Learn How to Boost Quality While Reducing Costs and Time to Market*. New York: McGraw-Hill.

Taguchi, Genichi, Subir Chowdhury, and Yuin Wu. 2000. *The Mahalanobis-Taguchi System*. New York: McGraw-Hill.

———. 2005. *Taguchi's Quality Engineering Handbook*. New York: Wiley.

Taguchi, Genichi and Rajesh Jugulum. 2002. *The Mahalanobis-Taguchi Strategy: A Pattern Technology System*. New York: Wiley.

Taguchi, Genichi, Rajesh Jugulum, and Shin Taguchi. 2004. *Computer-Based Robust Engineering: Essential for DFSS*. New York: American Society for Quality.

Takeuchi, Hirotaka, Emi Osono, and Norihiko Shimizu. 2008. "The Contradictions That Drive Toyota's Success". *Harvard Business Review*. https://hbr.org/2008/06/the-contradictions-that-drive-toyotas-success Accessed September 11, 2018.

Tang, Thomas Ling-Ping and E. A. Butler. 1997. "Attributions of Quality Circles' Problem-Solving Failure: Differences Among Management, Supporting Staff, and Quality Circle Members". *Public Personnel Management*. 26:203–25.

Tang, Thomas Ling-Ping, Peggy Tollison, and Harold Whiteside. 1987. "The Effect of Quality Circle Initiation on Motivation to Attend Quality Circle Meetings and on Task Performance". *Personnel Psychology*. 40:799–814.

———. 1989. "Quality Circle Productivity as Related to Upper-Management Attendance, Circle Initiation, and Collar Color". *Journal of Management*. 15:101–13.

———. 1991. "Managers Attendance and the Effectiveness of Small Groups: The Case of Quality Circles". *Journal of Social Psychology*. 131(3):335–44.

———. 1993. "Differences Between Active and Inactive Quality Circles in Attendance and Performance". *Public Personnel Management*. 22:579–90.

———. 1996. "The Case of Active and Inactive Quality Circles". *Journal of Social Psychology*. 136:57–67.

Taub, Eric. 1991. *Taurus: The Making of the Car That Saved Ford*. New York: Dutton.

Taylor, Frederick Winslow. 1911. *Principles of Scientific Management*. New York: Norton.

Taylor, James and David F. Felten. 1992. *Performance By Design: Sociotechnical Systems in North America*. Englewood Cliffs, NJ: Prentice Hall.

Taylor, Kate. 2019. "Jeff Bezos Called for Amazon's Competitors to Raise Their Minimum Wage". *Business Insider*. April 11, 2019. www.businessinsider.com/amazon-costco-walmart-target-compare-minimum-wage-2019-4 Accessed May 15, 2019.

Thelen, Kathleen. 2004. *How Institutions Evolve: The Political Economy of Skills in Germany, Britain, the United States and Japan*. New York: Cambridge University Press.

Thomasson, Emma. 2019. "Amazon Workers Strike at Four German Warehouses". *Reuters, Technology News*. April 15, 2019. www.reuters.com/article/us-amazon-com-germany/amazon-workers-strike-at-four-german-warehouses-idUSKCN1RR13D Accessed May 25, 2019.

Thomasson, Emma and Matthias Inverardi. 2018. "Amazon's Treatment of Sellers Comes Under Scrutiny in Germany". *Reuters, Technology News*. November 29, 2018.

Todd, L., T. Sitthichok, K. Mottus, G. Mihlan, and S. Wing. 2008. "Health Survey of Workers Exposed to Mixed Solvent and Ergonomic Hazards in Footwear and Equipment Factory Works in Thailand." *Annals of Occupational Hygiene*. 52(3):195–205.

Ton, Zeynep. 2014. *The Good Jobs Strategy: How the Smartest Companies Invest in Employees to Lower Costs and Boost Profits*. New York: Houghton Mifflin Harcourt.

Toulouse, Stephen. 2012. *A Microsoft Life*. Amazon Digital. Kindle.

Toussaint, John. (forthcoming). "Evidence-based Services and Lean in Hospitals and Service Industries". Chapter 12 in Thomas Janoski and Darina Lepadatu (Eds.) *The International Handbook of Lean Production*. New York: Cambridge University Press.

Toyoda, Aiko. 2010. "Toyota President Akio Toyoda's Statement to Congress". *The Guardian*. February 24, 2010. www.theguardian.com/business/2010/feb/24/akio-toyoda-statement-to-congress Accessed May 15, 2018.

Toyota Corporate News. 2017. "Toyota North America Headquarters Grand Opening: Jim Lentz Remarks". *Toyota Corporate News*, July 6. https://corporatenews.pressroom.toyota.com/releases/2017+toyota+motor+north+america+headquarters+grand+opening+jim+lentz+remarks.htm Accessed September 10, 2018.

Toyota Global. 2018. "Toyota Traditions and Philosophy". www.toyota-global.com/company/toyota_traditions/philosophy/ Accessed October 26, 2018.

Toyota Global Newsroom. n.d.a "The Development of the Prius, parts I and II. *Toyota Motor Corporation Global Newsroom*. https://newsroom.toyota.co.jp/en/prius20th/challenge/birth/01/; https://newsroom.toyota.co.jp/en/prius20th/challenge/birth/02/ Accessed December 27, 2018.

———. n.d.b The Evolution of the Prius. *Toyota Motor Corporation, Global Newsroom* https://newsroom.toyota.co.jp/en/prius20th/evolution/ Accessed December 27, 2018.

———. n.d.c "Toyota Sports 800 Gas Turbine-Hybrid". *Toyota Motor Corporation, Global Newsroom.* https://newsroom.toyota.co.jp/en/prius20th/innovation/gas/ Accessed December 27, 2018.

Toyota Motor Manufacturing Kentucky (TMMK). 2004. *Employees Handbook.* Georgetown, KY: TMMK.

Tremblay, Gaëtan. 1995. "The Information Society: From Fordism to Gatesism: The 1995 Southam Lecture". *Canadian Journal of Communication.* 20(4):461–82.

———. 2008a. "Gatesisme et informationnalisation social". In Philippe Boquillion and Yolanda Combes (eds.) *Les Industries de la Cultural de la Communication an Quebec et Canada.* Paris: L'Harmattan.

———. 2008b. "Industries Culturelles economie crehe et societe de l'Information". *Global Press Journal, Canadian Issue.* 1(1):65–88.

Trist, Eric. 1981. *The Evolution of Socio-Technical Systems: A Conceptual Framework and an Action Research Program.* Toronto, CA: Ontario Quality of Working Life Centre, Occasional Paper No. 2.

———. 1997. *The Social Engagement of Social Science: A Tavistock Anthology.* Philadelphia: University of Pennsylvania Press.

Trist, Eric and Ken W. Bamford. 1951. "Some Social and Psychological Consequences of the Longwall Method of Coal Getting". *Human Relations.* 4(1):3–38.

Trist, Eric, G. W. Higgin, H. Murray, and A. B. Pollock. 2013 (1963). *Organizational Choice: The Loss, Rediscovery, and Transformation of a Work Tradition.* New York: Routledge.

Trist, Eric and Hugh Murray (eds.). 1990. *The Social Engagement of Social Science, A Tavistock Anthology, Volume 1. The Socio-Psychological Perspective.* Philadelphia: University of Pennsylvania Press.

———. (eds.). 1993. *The Social Engagement of Social Science, A Tavistock Anthology, Volume 2. The Socio-Technical Perspective.* Philadelphia: University of Pennsylvania Press.

Trottman, Melanie and Kris Maher. 2013. "Agency Admonishes Wal-Mart Over Protests". *Wall Street Journal.* November 18. https://www.wsj.com/articles/walmart-violated-workers8217-rights-nlrb-says-1384814939 Accessed June 30, 2019.

Trumpf, A. G. 2019. *Company Profile: Facts and Figures.* www.trumpf.com/en_US/company/trumpf-group/company-profile/ Accessed July 10, 2019.

Turner, Fred. 2009. "Burning Man at Google: A Cultural Infrastructure for New Media Production". *New Media and Society.* 11(1&2):73–94.

———. 2017. "Don't be Evil: Interview with Fred Turner". *Logic: A Magazine about Technology.* Issue 3: Justice.

Turner, Lowell. 1993. *Democracy at Work.* Ithaca, NY: Cornell University Press.

Turner, Tim and Colleagues. 2012. *One Team on All Levels: Stories from Toyota Team Members.* Boca Raton, FL: CRC Press/Taylor and Francis.

Twentieth Century Fox. 2013. *The Internship.* Movie, 119 Minutes. Los Angeles: Twentieth Century Fox.

Uchitelle, Louis. 2011a. Working for Less: Factory Jobs Gain, but Wages Retreat". *New York Times.* December 29, 2011.

———. 2011b. "Once Made in the USA". *The American Prospect.* June 9, 2011.

Ulrich, David and Wayne Brockbank. 2005. *The HR Value Proposition.* Cambridge, MA: Harvard Business Review Press.

US Census Bureau. 2013. "Income Statistics". Accessed August 2, 2013. www.census.gove/hhes/www/income/data/historical/measures.

USA Today. 2006. "How Ford Starved Its Taurus". *USA Today*. October 25, 2006. https://usatoday30.usatoday.com/news/opinion/editorials/2006-10-24-ford-taurus_x.htm Accessed May 15, 2019.

Useem, Jerry. 2019. "The End of Expertise: The Workplace Report". *The Atlantic*. July:56–65.

University of California, Berkeley. n.d. "Robert E. Cole, Faculty Web Page". https://haas.berkeley.edu/faculty/cole-robert/ Accessed May 15, 2019.

University of Kentucky Lean Systems Program. 2018. "Lean Reference Terminology". https://w2.engr.uky.edu/lean/reference/terminology/ Accessed October 22, 2018.

Vallas, Steven. 1999. "Rethinking Post-Fordism: The Meaning of Workplace Flexibility". *Sociological Theory*. 17(1):68–101.

———. 2001. "Symbolic Boundaries and the New Division of Labor: Engineers, Workers and Restructuring of Factory of Life". *Research in Social Stratification and Mobility*. 18:3–37.

———. 2006a. "Empowerment Redux: Structure, Agency and the Remaking of Managerial Authority". *American Journal of Sociology*. 111(6):1677–717.

———. 2006b. "Theorizing Teamwork under Contemporary Capitalism". In Vicki Smith (ed.) *Worker Participation*. Special issue of *Research in the Sociology of Work*. Volume 16. Oxford: JAI/Elsevier Press.

Vallas, Steven, William Finlay, and Amy Wharton. 2009. *The Sociology of Work*. New York: Oxford.

Vandergrift, Donald and John Loyer. 2014. "The Effect of Walmart and Target on the Tax Base: Evidence from New Jersey". *Journal of Regional Science*. 55(2):159–87.

Veblen, Thorstein. 1958 (1904). *Theory of the Business Enterprise*. New York: Mentor Books.

Vidal, Matt. 2006. "Theorizing Teamwork Under Contemporary Capitalism". In *Research in the Sociology of Work*. Vol. 16. San Diego, CA: JAI Press.

———. 2007. "Manufacturing Empowerment? "Employee Involvement" in the Labour Process After Fordism". *Socio-Economic Review*. 5:197–232.

———. 2010. "On the Persistence of Labor Market Insecurity and Slow Growth in the US: Reckoning with the Waltonist Growth Regime". *Institute for Research on Labor and Employment, Working Paper 2010–16. UCLA*. www.irele.ucla.edu accessed June 1, 2018.

———. 2011. "Reworking Post-Fordism". *Sociological Compass*. 5(4):273–86.

———. 2020. "Work Exploitation in Capitalism: The Labor Process and the Valorization Process". pp. xx–178 in Matt Vidal, Tony Smith, Tomás Rotta, and Paul Prew (eds.) *The Oxford Handbook of Karl Marx*. Oxford: Oxford University Press.

Vidal, Matt and Leann Tigges. 2007. "Lean Production, Worker Empowerment and Job Satisfaction". *Critical Sociology*. 33(1–2):247–78.

Vogel, Ezra. 1963. *Japan's New Middle Class*. Berkeley, CA: University of California Press.

———. 1971. *Japan as Number One: Lessons for America*. New York: Harper/Colophon.

———. 2001. *Is Japan Still Number One?* Subang Jaya: Pelanduk Publications.

Volti, Rudi. 2008. *An Introduction to the Sociology of Work and Occupations*. Thousand Oaks, CA: Pine Forge Press.

Wageman, Ruth. 2001. "How Leaders foster Self-Managing Teams: Design Choice vs. Hands on Coaching". *Organizational Science*. 12(5):559–77.

Wageman, Ruth, J. Richard Hackman, and Erin Lehman. 2005. "Team-Diagnostic Survey: Development of an Instrument". *Journal of Applied Behavioral Science*. 41:373–98.

Wageman, Ruth, D. A. Nunes, J. A. Burruss, and J. Richard Hackman. 2008. *Senior Leadership Teams: What It Takes to Make Them Great*. Boston, MA: Harvard Business Review Press.

Wailgum, Thomas. 2007. "45 Years of Wal-Mart History: A Technology Time Line". *CIO Magazine* (Chief Information Officer Magazine). www.cio.com/article/2437873/45-years-of-wal-mart-history—a-technology-time-line.html Accessed May 15, 2019.

Wakatabe, Masazumi. 2015. *Japan's Great Stagnation and Abenomics: Lessons for the World*. London: Palgrave Macmillan.

Walker, Guy. 2015. "Come Back Sociotechnical Systems Theory, All is Forgiven . . .". *Civil Engineering and Environmental Systems*. 32(1–2):170–9.

Walker, Guy, Neville Stanton, Paul Salmon, and David Jenkins. 2010. "A Review of Sociotechnical Systems Theory: A Classic Concept for New Command and Control Paradigms". *Theoretical Issues in Ergonomics Science*. 9(6):479–99.

Wall Street Journal (WSJ). 2013. "A Revolution in the Making. Journal Report: Unleashing Innovation−Manufacturing". Articles by John Koten (3), Bob Tita, Shelly Banjo, James Hagerty (2), Deborah Gage, Geoffrey Fowler. *The Wall Street Journal*. June 11, 2013:R1–R7.

Wallerstein, Immanuel. 1974. *The Modern World System*. New York: Academic Press.

Walmart. 2019. Website on Corporate Responsibility. https://corporate.walmart.com/sourcing/promotingresponsibility Accessed May 15, 2019.

Walters, Helen. 2007. "Nike's New Down-Market Strategy". *Businessweek*. February 27.

Walton, Mary. 1986. *The Deming Management Method*. Harmondsworth: Penguin Group.

Walton, Sam. 1992. *Made in America: My Story*. New York: Bantam/Doubleday.

Watson, James. 1997. "McDonald's in Hong Kong: Consumerism, Dietary Change, and the Rise of a Children's Culture". In James Watson (ed.) *Golden Arches East*. Stanford, CA: Stanford University Press.

Weber, Austin. 2018a. "Lean Culture Drives Continuous Improvement". *Assembly Magazine*. December 10, 2018. www.assemblymag.com/articles/94587-lean-culture-drives-continuous-improvement Accessed July 10, 2019.

———. 2018b. "2018 Assembly Plant of the Year: Ford Shifts Flexible Assembly Into High Gear". *Assembly Magazine*. October 2, 2018. www.assemblymag.com/articles/94505-assembly-plant-of-the-year-ford-shifts-flexible-assembly-into-high-gear Accessed on January 30, 2019.

Weirgin, Niels-Erik. 2003. "Teamwork in the Automobile Industry: An Anglo-German Comparison". *European Political Economy Review*. 1(2):152–90.

Weisbord, Marvin. 2012. *Productive Workplaces: Organizing and Managing for Dignity, Meaning and Community*, 3rd Edition. San Francisco, CA: Jossey-Bass.

West, Michael A. 2012. *Effective Teamwork: Practical Lessons from Organizational Research*. New York: Wiley.

Whittaker, D. Hugh. 2009. *Comparative Entrepreneurship: The UK, Japan and the Shadow of Silicon Valley*. New York: Oxford University Press.

———. n.d. "Why I Left Google Redux". www.docjamesw.com/why-i-left-google-redux/ Accessed December 10, 2018.

Whittaker, James. 2009. *Exploratory Software Testing*. Upper Saddle River, NJ: Addison-Wesley

———. n.d. a. "Turning Quality on Its Head: How Google Tests Software I". www.youtube.com/watch?v=KXGnXq5uXR4 Accessed January 2, 2019.

———. n.d. b. "Turning Quality on Its Head: How Google Tests Software II". www.youtube.com/watch?v=UzGhW8Tbs0E Accessed January 2, 2019.

Whittaker, James, Jason Arbon, and Jeff Carollo. 2012. *How Google Tests Software*. Upper Saddle River, NJ: Addison-Wesley. docjamesw@gmail.com Accessed January 2, 2019.

Whitworth, Brian and Aldo de Moor (eds.). 2009. *Handbook of Research on Socio-Technical Design and Social Network Systems*. Hershey, NY: Information Science Reference.

Wickens, Peter and Alan Robert Lopez. 1988. *The Road to Nissan: Flexibility, Quality, Teamwork*. London: Palgrave Macmillan.

Winkler, Ingo. 2016. *ALDI: A Case of Rigorous Employee Control?* Thousand Oaks, CA: Sage Business Cases.

Wolfe, S. E. and D. C. Pyrooz. 2014. "Rolling Back Prices and Raising Crime Rates? The Walmart Effect on Crime in the United States". *British Journal of Criminology*. 54(2):199–221.

Womack, James and Daniel Jones. 2004. "From Lean Production to the Lean Enterprise". *Harvard Business Review*. March/April.

———. 2003. *Lean Solutions: How Companies and Customers Can Create Value and Wealth Together*. New York: Free Press.

Womack, James, Daniel Jones, and Daniel Roos. 1990. *The Machine That Changed the World*. New York: Harper Collins.

Wood, Alex, Mark Graham and Vili Lehdonvirta. 2019. "Networked but Commodified: The (Dis)Embeddedness of Digital Labour in the Gig Economy" *Sociology*. 53(5): 931–50.

Woollard, Frank. 2009. *Principles of Mass and Flow Production*, introduction by Bob Emiliani. 55th Anniversary Special Reprint Edition. Wethersfield, CN: The CLBM, LLC.

Woolridge, Adrian. 2012. "State Capitalism: The Visible Hand". *Special Report from the Economist*. Penguin. January 21, 2012.

Worldwide Responsible Apparel Production (WRAP). n.d. a "History". www.wrapcompliance.org/en/history. Accessed January 10, 2019.

———. n.d.b "WRAP Certified". www.servicewearapparel.com/about/wrap.php Accessed January 10, 2019.

Worstall, Tim. 2013. "If the Welfare System Subsidizes WalMart Then Should We Ban Welfare to Stop Subsidizing WalMart?" *Forbes Magazine*. June 10, 2013 Accessed August 7, 2013. www.forbes.com/sites/timworstall/2013/06/10/if-the-welfare-system-subsidises-walmart-then-should-we-ban-welfare-to-stop-subsidising-walmart/ Accessed January 2, 2019.

Wrege, Charles and Amedeo Perroni. 1974. "Taylor's Pig-Tale: A Historical Analysis of Frederick W. Taylor's Pig-Iron Experiments". *Academy of Management Journal*. 17(1):6–27.

Wright, Christopher. 2011. "Historical Interpretations of the Labour Process: Retrospect and Future Research Directions". *Labour History*. 100:19–32.

Wulfraat, Marc. 2014. "The Secret to Costco's Success Lies in Supply Chain Efficiency". *Canadian Grocer*. May 13, 2014. www.canadiangrocer.com/blog/the-secret-to-costcos-success-lies-in-supply-chain-efficiency-40691 Accessed January 2, 2019.

Yardley, Jim. 2012. "Horrific Fire Revealed a Gap in Safety for Global Brands". *New York Times*. December 8, 2012.

———. 2013. "Justice Still Elusive in Factory Disasters in Bangladesh". *New York Times*. June 29, 2013.

Zeitlin, Jonathan. 1998. "Shop Floor Bargaining and the State: A Contradictory Relationship". pp. 1–45 in Steven Tolliday and Jonathan Zeitlin (eds.) *Shop Floor Bargaining and the State*. Cambridge: Cambridge University Press.

Zeitlin, Jonathan and Steven Tolliday. 1992. *Between Fordism and Flexibility*. London: Berg.

Zhang, Lu. 2015a. *Inside China's Automobile Factories*. New York: Cambridge University Press.

———. 2015b. "Lean Production with Chinese Characteristics: A Case Study of China's Automobile Industry". *International Journal of Comparative Sociology*. 45(2):152–70.

Zola, Emil. 1995 (1864). *The Ladies Paradise (Au Bonheur des Dames)*. New York: Oxford University Press.

Zwolinski, Matt. 2007. "Sweatshops, Choice and Exploitation". *Business Ethics Quarterly*. 17(4):689–727.

Subject Index

Page numbers in *italics* indicate a figure and page numbers in **bold** indicate a table

Name Index